T0275446

A Singular Remedy

Stefanie Gänger explores how medical knowledge was shared across societies tied to the Atlantic World between 1751 and 1820. Centred on Peruvian bark or cinchona, *A Singular Remedy* shows how that remedy and knowledge about its consumption – formulae for bittersweet, 'aromatic' wines, narratives about its discovery or beliefs in its ability to prevent fevers – were understood by men and women in varied contexts: Peruvian academies and Scottish households, Louisiana plantations and Moroccan court pharmacies alike. This study in plant trade, therapeutic exchange and epistemic brokerage exposes how knowledge weaves itself into the fabric of everyday medical practice in different places.

STEFANIE GÄNGER is Professor of Modern History at Heidelberg University in Germany.

SCIENCE IN HISTORY

Series Editors

Simon J. Schaffer, University of Cambridge
James A. Secord, University of Cambridge

Science in History is a major series of ambitious books on the history of the sciences from the mid-eighteenth century through the mid-twentieth century, highlighting work that interprets the sciences from perspectives drawn from across the discipline of history. The focus on the major epoch of global economic, industrial and social transformations is intended to encourage the use of sophisticated historical models to make sense of the ways in which the sciences have developed and changed. The series encourages the exploration of a wide range of scientific traditions and the interrelations between them. It particularly welcomes work that takes seriously the material practices of the sciences and is broad in geographical scope.

A Singular Remedy

Cinchona Across the Atlantic World, 1751–1820

Stefanie Gänger

Heidelberg University

CAMBRIDGE
UNIVERSITY PRESS

University Printing House, Cambridge CB2 8BS, United Kingdom

One Liberty Plaza, 20th Floor, New York, NY 10006, USA

477 Williamstown Road, Port Melbourne, VIC 3207, Australia

314–321, 3rd Floor, Plot 3, Splendor Forum, Jasola District Centre, New Delhi – 110025, India

79 Anson Road, #06–04/06, Singapore 079906

Cambridge University Press is part of the University of Cambridge.

It furthers the University's mission by disseminating knowledge in the pursuit of education, learning, and research at the highest international levels of excellence.

www.cambridge.org
Information on this title: www.cambridge.org/9781108842167
DOI: 10.1017/9781108896269

© Stefanie Gänger 2021

First published 2021

A catalogue record for this publication is available from the British Library.

Library of Congress Cataloging-in-Publication Data
Names: Gänger, Stefanie, 1983– author.
Title: A singular remedy : cinchona across the Atlantic World, 1751-1820 / Stefanie Gänger.
Description: New York : Cambridge University Press, 2021. | Series: Science in history | Includes bibliographical references and index.
Identifiers: LCCN 2020022779 (print) | LCCN 2020022780 (ebook) | ISBN 9781108842167 (hardback) | ISBN 9781108816335 (paperback) | ISBN 9781108896269 (epub)
Subjects: LCSH: Cinchona bark. | Cinchona bark–Therapeutic use–Early works to 1800. | Medicine–History. | Drugs–History.
Classification: LCC RM666.C53 G36 2021 (print) | LCC RM666.C53 (ebook) | DDC 615.3/23956–dc23
LC record available at https://lccn.loc.gov/2020022779
LC ebook record available at https://lccn.loc.gov/2020022780

ISBN 978-1-108-84216-7 Hardback

Contents

Illustrations

Acknowledgements

I have researched, written and rewritten this book over a period of seven years, and, in the process, have incurred many debts. I laid the foundation for the project at the University of Konstanz, in the context of the research group 'Global Processes', funded by the Gottfried Wilhelm Leibniz Prize, in 2012. The book would not have come to fruition in its present shape without the knowledgeable scholars assembled there – especially Jürgen Osterhammel, Jan C. Jansen, Martin Rempe and Franz L. Fillafer – and our stimulating debates on world and global history. I would also like to thank Gabriela Ramos, my former supervisor at the University of Cambridge, for encouraging me to choose this particular research topic, and for her unstinting kindness and support for the last decade and a half. I wrote a first version of the manuscript at the University of Cologne, where affable colleagues, a diverse environment, and the generous support of the Global South Studies Center – special thanks are due to its speakers, Barbara Potthast and Michael Bollig – made the process of researching and writing up a great pleasure. I finished the manuscript at the University of Heidelberg, where I found a permanent, happy home in April 2019.

I presented initial outlines of the project and earlier versions of chapters at various seminars and symposia held at the universities of Cambridge, Yale, Tucumán, Munich, Heidelberg, Basel, Dresden, Edinburgh, Birmingham, Zurich, and Brown, and in Paris, Boston, New York, La Plata, Berlin, Vienna, Santiago de Chile and Chicago and I am grateful for the comments and feedback I received on these occasions. I am particularly indebted to Lea Beiermann, Dagmar Wujastyk, José Pardo-Tomás, Rachel E. Black, Jonathan Reinarz, Dagmar Schäfer, Anne-Isabelle Richard, Emile Chabal, Barbara E. Mundy, Michael Zeheter, Nuria Valverde Pérez, Paola Bertucci, Miruna Achim, Nina Verheyen, Irina Podgorny, Iris Montero, Sarah Albiez-Wieck, Neil Safier, Timothy D. Walker, Sebastian Haumann, Gesine Müller, Clare Griffin, Arndt Brendecke, Tim Le Cain and Elaine Leong, for guidance and support at various stages of the project. Matthew Crawford and

Samir Boumediene, who had researched and written about other aspects of the history of cinchona long before I came to the topic, were so kind as to share their work, and to offer their advice. I am especially grateful to the colleagues who have taken the time to read chapters and to suggest changes: Heiner Fangerau, Raquel Gil Montero, Jürgen Osterhammel, Sebastian Haumann, Jan C. Jansen, Martin Rempe, Franz L. Fillafer and Elaine Leong. Barbara Orland, Agnes Gehbald, Caterina Mantilla and Melina Teubner have kindly taken the trouble to read and comment upon nearly the entire manuscript. I particularly owe a great debt to Cambridge University Press. I cannot thank the two anonymous readers and the Series Editors – Jim Secord and Simon Schaffer – enough for their valuable, perceptive comments and suggestions, which have made this a much better book than I could have written on my own. I am also deeply grateful to Lucy Rhymer and her editorial team for their steady support and kind guidance throughout the publication process.

Furthermore, I would like to thank the staff at the archives I visited in Spain, Portugal, Peru, Ecuador, England, France and Switzerland in the course of my research. I owe particular gratitude to Gudrun Kling at the Zurich Archive of the History of Medicine, Esther García Guillén at the archives of the Royal Botanical Garden in Madrid, Ada Arrieta Álvarez at the archive pertaining to the Riva Agüero Institute in Lima, Damien Blanchard at the Academy of Medicine Library in Paris and Isabel Amado at the Portuguese Overseas Archive in Lisbon. I would not have been able to navigate the Ecuadorian national archive in Quito so well without the generous help and hospitality of a colleague, Christiana Borchert de Moreno, and her husband, Segundo Moreno Yáñez. I am also grateful to Katalin Pataki and Reina María Pacheco Olivera, who were kind enough to share archival materials from Mexico and Hungary with me, or to point me towards them. I particularly would like to thank Meike Knittel for making me aware of the manuscript recipe collections kept at the Zurich Archive of the History of Medicine. Ruby Ellis prepared the index and, together with Albert Loran, helped format the manuscript and secure copyright permissions.

I am grateful to the Royal Botanical Garden in Madrid, the Wellcome Library and the Science Museum in London, the Academy of Medicine Library in Paris, Lisbon's Pharmacy Museum (*Museu da Farmacia*) and Spain's National Heritage body (*Patrimonio Nacional*) for granting copyright permissions for the illustrations in this book. I also thank Miguel Jaramillo Baanante for allowing me to adapt his statistics for Figure 2.1 and Monika Feinen for crafting Figures 2.2 and 5.1. I also thank Oxford University Press for permission to include parts of an article I wrote for the journal *Environmental History* ('Cinchona Harvest, Deforestation, and Extinction in the Viceroyalty of New Granada, 1752–1811', Vol. 24/4, 2919) in Chapter 5.

Finally, and once again, I thank my friends and family, particularly my husband, Arne, for his companionship, quiet faith and untiring support. I finished writing the first draft of the manuscript for this book when I was pregnant with our son, Elias, and revised it – again, and again – in the months and years after he was born. It is dedicated to them both, with all my love.

A Note on Source Material

This book brings together evidence from a wide array of sources, variable with respect to genre, language and origin – ranging from imperial bureaucratic reports to domestic recipe collections – and scattered, much like its subject matter, across the Atlantic World. In order to reassemble the history of the bark's harvest, and of Spanish trade and Portuguese, British, Dutch and French contraband in it, the book relies primarily on official correspondence and reports, legal files and royal orders from the General Archive of the Indies (*Archivo General de Indias*) in Seville, the Spanish state archives in Simancas (*Archivo General de Simancas*), the archives pertaining to Spain's Royal Palace (*Archivo del Palacio Real*) in Madrid, the Portuguese Overseas Archive (*Arquivo Histórico Ultramarino*) in Lisbon and the Ecuadorian National Historical Archive (*Archivo Nacional de la Historia*) in Quito. The records contained in these archives revealed the bark's passage into a vast array of territories far beyond the Spanish and Portuguese empires – primarily, though not exclusively, across Europe and the societies within or adjoining its colonial possessions and commercial and evangelizing entrepôts rimming the Atlantic basin. Moving beyond traditional archival research, the book follows the pathways alluded to in these Iberian records, alongside those mentioned in the extant historiography, to archives, libraries and repositories, both physical and digital, elsewhere, for print and manuscript sources on bark use in these other societies. I primarily searched pharmacopoeias, medical treatises and various genres of popular print, especially published self-help manuals, recipe collections and almanacs, for bark recipes, stories and medical understandings. Specifically, I selected and probed into medical and popular print sources from Spain and Portugal, the Viceroyalties of Peru, Brazil, New Granada and New Spain, the Dutch, British and French West Indian possessions, the Kingdom of France, England and Scotland, British India, the Habsburg territories and the Swiss Confederacy, the Italian Peninsula, the Portuguese and British enclaves along the African coast and the French and British North American colonies – or, after 1776, the United States. I also drew on

Russian pharmacopeias, wherever they were available in Latin. To verify, at least selectively, whether readers heeded the medical advice these formats dispensed I consulted a selection of manuscript notebooks of medical recipes kept by women and men for domestic use – from Britain, France, various German- and Italian-speaking territories, the Portuguese Empire and the Viceroyalties of Mexico and Peru – kept in the archives and manuscript collections of London's Wellcome Library, Portugal's National Library (*Biblioteca Nacional*), Peru's National Library (*Biblioteca Nacional de Lima*) and the University of Zurich (*Archiv für Medizingeschichte*), all of which hold national as well as international collections of late eighteenth- and early nineteenth-century recipe books. Studies that operate on a geographical and socio-political canvas more extensive than one empire, or nation-state require a measure of reliance on the work of specialists in other world areas. On account of the impossibility of mastering all the languages and visiting all the archives of the places where the bark's pathways lead, I relied primarily on secondary literature for evidence on bark use in Sweden, the Sultanate of Morocco, the Dutch colonies, the Ottoman Empire, the Chinese and Mughal empires and Tokugawa Japan.

A Note on Language and Translation

The research for this project was conducted in many languages. I have worked with primary sources in English, Spanish, Portuguese, French, Latin, German and Italian. All translations of these sources into English are my own, unless otherwise noted. I also benefitted from the translations, and linguistic advice, of other scholars with sources and literature in languages I have not mastered: Maike Lehmann and Ingrid Schierle with Russian, Meike Knittel with Dutch, Jean-Baptiste Pettier with Chinese and Katalin Pataki with Hungarian. I am particularly grateful to Sırma Hasgül and Sibylla Wolfgarten for their translation of Bursalı Ali Münşî's eighteenth-century essay on cinchona – 'Tuhfe-i Aliyye', or 'Kına Kına Risâlesi' – from Ottoman Turkish. To avoid ambiguity, I have usually given the English translation or transliteration of quotations (or proper names, for that matter) and provided the original in parentheses.

A Note on Weights, Currencies and Measures

The situation with regard to weights, currencies and measures in the late 1700s and early 1800s is generally one of great complexity, to say the least, and the more so in a book that adopts a transatlantic perspective. Weights, currencies and measures varied not only from area to area during that period, but to some extent also over time. I have converted the various historical units of weight and mass into kilograms, the base unit of mass in the current metric system, and units of length into metres. To avoid ambiguity, and ensure transparency, I have indicated both the conversion and the original units of mass and length throughout the text. I have chosen not to convert historical to other, or present-day, currencies, but found it preferable to give the reader an idea of their value by referencing their purchasing power in the period and society under consideration.

Introduction
A Singular Remedy

> What commerce [...] for the people that are the sole proprietors of the
> most powerful remedy that medicine possesses to restore the health of
> mankind in the four corners of the Earth.
> – Francisco José de Caldas, *Memoria sobre el estado de las quinas*, 1809.

By the late 1700s and early 1800s, cinchona bark was, to many, 'the
most important, and the most usual remedy that medicine possessed'.[1]
Though of limited repertoire – cinchona trees prospered only on the
precipitous eastern slopes of the Andes at the time, in the Spanish
American Viceroyalties of Peru and New Granada – and comparatively
recent acceptance into Old World materia medica, the bark had, by the
turn of the eighteenth century, woven itself into the texture of everyday
medical practice in a wide range of societies within, or tied to, the
Atlantic World. It was everywhere attributed 'wonderful',[2] 'singular',[3]
even 'divine'[4] medicinal virtues, the knowledge of which, so it was said,
had come to mankind from its simplest, and humblest, specimens,
'wild Indians'[5] close to nature and privy to its most coveted secrets.
Bittersweet 'febrifugal lemonades' and bottled wines of the bark sat on
the shelves of Lima apothecaries, the counters of Cantonese market

---

[1] Luis de Rieux, 'Carta a Miguel Cayetano de Soler,' *Archivo General de Indias*, Indiferente 1557, Aranjuez, 1800-05-14, 346 v.

[2] Antonio Caballero y Góngora, Archbishop and Viceroy of New Granada, referred to the bark's 'wonderful effects (*sus maravillosos efectos*)' in a 1788 letter. Antonio Caballero y Góngora, 'Copia de Carta Reservada,' *Archivo del Palacio Real*, Papeles del Almacén de la Quina, Caja 22283 / Expediente 2, Turbaco, 1788-05-28.

[3] Baltasar de Villalobos, *Método de curar tabardillos, y descripción de la fiebre epidemica, que por los años de 1796 y 97 afligio varias poblaciones del partido de Chancay* (Lima: Imprenta Real del Telégrafo Peruano, 1800), 117; Edward Rigby, *An Essay on the Use of the Red Peruvian Bark in the Cure of Intermittents* (London: J. Johnson, 1783), 6.

[4] Simon André Tissot, *Aviso al pueblo acerca de su salud ó Tratado de las enfermedades mas frequentes de las gentes del campo*, trans. Juan Galisteo y Xiorro (Madrid: Imprenta de Pedro Marin, 1790), 161.

[5] William Cockburn, *The Present Uncertainty in the Knowledge of Medicines in a Letter to the Physicians in the Commission for Sick and Wounded Seamen* (London: Benj[amin] Barker, 1703), Preface I. A1.

stands and in the medicine chests of Luanda hospital orderlies. They were routinely concocted, and administered at the bedside, by Moroccan court physicians, French housewives and slave healers alike and they accompanied, tucked into their pouches, Dutch sailors to febrile environs, Peruvian soldiers to the battlefield and North American settlers westward. Scottish physicians, creole botanists and French writers alike were unanimous not only in according the bark 'singularity',[6] and 'the first place among the most effective remedies' (*die erste Stelle unter den würksamsten Arzneimitteln*),[7] but also in holding it to be 'more generally useful to mankind than any in the materia medica'.[8] It was commonly agreed upon that there was 'no febrifuge of such well-known virtue in all of medicine' (*por que no se halla en la Medicina febrífugo de virtud tan conocida*),[9] and that not a single remedy 'more estimable and precious [than the bark] had been discovered unto this day'.[10]

For decades now, historians of science, medicine and technology have insisted on the epistemological lesson that science and knowledge are the result of specific circumstances and close, local settings, situated and bound 'ineluctably to the conditions of their production' – historically contingent, idiosyncratic 'form[s] of practice', rooted in a particular time and place.[11] The field is at present said to be in the midst of a fundamental turn toward global approaches that straddle traditional spatial boundaries but, as some of its most prominent advocates have cautioned, practitioners have hardly begun to understand the consequences of that shift for the field's most basic values and principles, especially its

[6] Aylmer Bourke Lambert, *A description of the genus Cinchona, comprehending the various species of vegetables from which the Peruvian and other barks of a similar quality are taken* (London: B. and J. White, 1797), 1.

[7] Samuel Auguste André Tissot, *Anleitung für das Landvolk in Absicht auf seine Gesundheit* (Zürich: Heidegger und Compagnie, 1763), 288–89.

[8] Rigby, *An Essay on the Use of the Red Peruvian Bark*, 6.

[9] Manuel Hernandez de Gregorio, 'Dn. Manuel Hernandez de Gregorio, Boticario de Camara presenta una memoria compuesta de 37 artículos, queriendo persuadir las grandes conveniencias de la estancación general, y parcial de la Quina en beneficio de la salud publica, y del interés del Real Erario, detallando las reglas gubernativas para su administración,' *Archivo General de Indias*, Indiferente 1556, Madrid, 1804.

[10] Hipólito Ruiz López, *Quinología O Tratado del Arbol de la Quina o Cascarilla, con su descripción y la de otras especies de quinos nuevamente descubiertas en el Perú, del modo de beneficiarla, de su elección, comercio, virtudes, y extracto elaborado von cortezas recientes* (Madrid: La viuda é hijo de Marin, 1792), 38.

[11] For that diagnosis, see James A. Secord, 'Knowledge in Transit,' *Isis* 95, no. 4 (2004), 657. See also Lorraine Daston, 'Science Studies and the History of Science,' *Critical Inquiry* 35, no. 4 (2009). The term 'situated knowledge' is commonly associated with the work of Donna Haraway; see her 'Situated Knowledges: The Science Question in Feminism and the Privilege of Partial Perspective,' *Feminist Studies* 14, no. 3 (1988).

emphasis on locality.[12] This book is an attempt at writing a history of how medical knowledge – in the shape of matter, words and practices – was shared between and across a wide range of geographically disperse and socially diverse societies within the Atlantic World and its Asian entrepôts between 1751 and 1820. Centred on the Peruvian bark, or cinchona, it exposes and examines how that medicine and the imaginaries, therapeutic practices and medical understandings attendant to its consumption, were 'part of the taken-for-granted understanding'[13] of people in many different social and cultural contexts: at Peruvian academies and in Scottish households, on Louisiana plantations and in Moroccan court pharmacies alike. Much of the book is concerned with the conditions, contingency and idiosyncrasy of the prevalence and movement of bark knowledge – through contingent 'act[s] of communication',[14] 'brokerage'[15] and sociality,[16] 'between [...] settings' tied together by Atlantic trade, proselytizing, and imperialism[17] – as well as with the variability of the knowledge in motion. Indeed, the book suggests that cinchona's wide spread owed less to its utter immutability and consistency than, as historians have argued for other tools and substances, to a measure of malleability, and multivalence: its ability to 'subtly adapt', be refashioned, or tinkered with.[18] Scholarship on modern and early modern

[12] Kapil Raj, 'Beyond Postcolonialism ... and Postpositivism. Circulation and the Global History of Science,' *Isis* 104 (2013), 341; Secord, 'Knowledge in Transit,' 660. See also Fa-ti Fan, 'The Global Turn in the History of Science,' *East Asian Science, Technology and Society: An International Journal* 6 (2012).

[13] Secord, 'Knowledge in Transit,' 655.   [14] Ibid., 661.

[15] On the 'historically situated work of mediation', and brokerage, in the history of science, see Simon Schaffer et al., introduction to *The Brokered World. Go-Betweens and Global Intelligence, 1770–1820*, ed. Simon Schaffer et al. (Sagamore Beach: Watson Publishing International, 2009), xx.

[16] Marcy Norton has stressed the role that sustained, and persistent, exposure to substances, especially through social relationships and practices, played for their spread. Marcy Norton, 'Tasting Empire: Chocolate and the European Internalization of Mesoamerican Aesthetics,' *The American Historical Review* 111, no. 3 (2006).

[17] On debates about 'Atlantic interdependence' around 1800, see Richard J. Blakemore, 'The Changing Fortunes of Atlantic History,' *English Historical Review* CXXXI, no. 551 (2016), 855. See also D'Maris Coffman and Adrian Leonard, 'The Atlantic World: Definition, Theory, and Boundaries,' in *The Atlantic World: 1400–1850*, ed. D'Maris Coffman, Adrian Leonard and William O'Reilly, The Routledge Worlds (London: Routledge, 2015), 3. On knowledge not as 'abstract doctrine but as communicative practice in a range of well-integrated and closely understood settings', see Secord, 'Knowledge in Transit,' 671.

[18] David Kaiser, *Drawing Theories Apart: The Dispersion of Feynman Diagrams in Postwar Physics* (Chicago: University of Chicago Press, 2005), 7. This alludes to the work of Bruno Latour, who argued that practices of 'inscription' produced 'immutable mobiles'. The idea was originally formulated in Bruno Latour, *Science in Action: How to Follow Scientists and Engineers Through Society* (Cambridge, Mass.: Harvard University Press, 1987).

globalization, with its liquid language of elusive flows and unconstrained circulation, still tends to evoke an idea of movement as erosive and antithetical to place, and of 'the very idea of locality [...] as a form of opposition or resistance to the [...] global', a gesture towards the discrete, and authentic.[19] It was in large measure the bark's ability to tie itself to locales, however, to settle and become situated,[20] again and again, that accounted for its prevalence and mobility. Science and knowledge are not bound to one time and place, this book holds. They may be unmoored and moved – become well known and generally useful elsewhere – but they will invariably do so in ways that are just as contingent, situated and local as those traditionally associated with their production.

The Outlines of Cinchona

It may appear redundant for the historical account of a plant component to further define the outlines of its object of study. The seeming definitional sharpness of cinchona is deceptive, however.[21] Because the bark was, by the late 1700s and early 1800s, spoken of, sought after and studied in countless tongues across the Atlantic World and beyond, there were considerable shifts in its epistemic, chemical and medical contours, its nomenclature and, not least, its therapeutic indications. This is not to say that cinchona was not a distinct, identifiable object by the late 1700s and early 1800s.[22] Indeed, though its passage into the wider Galenic medical repertoire during the late 1600s had been attended by

---

[19] For a critique of how mobility serves as an antithesis to 'space' in scholarship on globalization, see Stuart Alexander Rockefeller, 'Flow,' *Current Anthropology* 52, no. 4 (2011). On place and the 'liquid' language of global history, see Stefanie Gänger, 'Circulation: Reflections on Circularity, Entity and Liquidity in the Language of Global history,' *Journal of Global History* 12, no. 3 (2017), 316. On 'the very idea of locality [...] as a form of opposition or resistance to the [...] global', see Roland Robertson, 'Glocalization: Time-Space and Homogeneity-Heterogeneity,' in *Global Modernities*, ed. Mike Featherstone, Scott Lash and Roland Robertson (London: Sage, 1995), 30.

[20] This responds in part to Kapil Raj's question of how to tackle to the 'concomitant situatedness and movement of science'. Raj, 'Beyond Postcolonialism ... and Postpositivism,' 337–41.

[21] On the often 'labile' and unstable qualities of substances in movement, see Guy Attewell, 'Interweaving Substance Trajectories: *Tiryaq,* Circulation and Therapeutic Transformation in the Nineteenth Century,' in *Crossing Colonial Historiographies: Histories of Colonial and Indigenous Medicines in Transnational Perspective*, ed. Anne Digby and Waltraud Ernst (Cambridge Scholars Publishing, 2010), 2; Carla Nappi, 'Winter Worm, Summer Grass: *Cordyceps,* Colonial Chinese Medicine, and the Formation of Historical Objects,' in *Crossing Colonial Historiographies : Histories of Colonial and Indigenous Medicines in Transnational Perspective*, ed. Anne Digby, Projit B. Muhkarji and Waltraud Ernst (Newcastle upon Tyne: Cambridge Scholars, 2010).

[22] Nappi, 'Winter Worm, Summer Grass'.

controversy over its nature, virtues and properties,[23] by the late 1700s and early 1800s, medical practitioners, both lay and professional, across the Atlantic World generally agreed on the bark's utility as a remedy and its coherence as a category.[24] Rather, the very latitude and cosmopolitanism of the bark's pathways entailed acts of adaptation, customizing and calibration, and, with them, a measure of variability and volatility that compels us to handle both the subject and the term, cinchona, advisedly, and with a measure of care.[25] As much recent scholarship reminds us, objects exist both in space and in time. They have a diachronic quality; are possessed of lives and biographies;[26] and accrete new meanings, names and properties, as they are identified, translated or 'adjust [...] to context' in the process.[27] They ought thus to be understood as malleable to a point: as multiple yet coherent, as liminal yet recognizable.[28]

---

[23] See in particular Saul Jarcho's 1993 study on the plant's 'discovery', its transmission to and within western Europe and its incipient establishment as a canonical part of medical practice through the lens of Francesco Torti's *Therapeutice specialis* (1712). Saul Jarcho, *Quinine's Predecessor. Francesco Torti and the Early History of Cinchona* (Baltimore: Johns Hopkins University Press, 1993). On the bark's gradual acceptance, see also Andreas-Holger Maehle, *Drugs on Trial: Experimental Pharmacology and Therapeutic Innovation in the Eighteenth Century* (Amsterdam: Editions Rodopi, 1999), 1. See also Harold J. Cook, 'Markets and Cultures. Medical Specifics and the Reconfiguration of the Body in Early Modern Europe,' *Transactions of the Royal Historical Society* 21 (2011), 208–09; Samir Boumediene, *La colonisation du savoir. Une histoire des plantes médicinales du 'Nouveau Monde' (1492–1750)* (Vaulx-en-Velin: Les Éditions des Mondes à Faire, 2016).

[24] Lorraine Daston has written about how phenomena 'amalgamate into a coherent category'. Lorraine Daston, 'Introduction. The Coming into Being of Scientific Objects,' in *Biographies of Scientific Objects*, ed. Lorraine Daston (Chicago: University of Chicago Press, 2000), 6.

[25] Guy Attewell, 'Interweaving Substance Trajectories', 2; Nappi, 'Winter Worm, Summer Grass'.

[26] This is an allusion to studies devoted to the 'lives' and 'biographies' of objects and things. See Igor Kopytoff, 'The Cultural Biography of Things: Commoditization as Process,' in *The Social Life of Things: Commodities in Cultural Perspective*, ed. Arjun Appadurai (Cambridge: Cambridge University Press, 1986).

[27] Lorraine Daston, 'Introduction. Speechless,' in *Things That Talk. Object Lessons from Art and Science* ed. Lorraine Daston (New York: Zone Books, 2004), 18. On substances in motion, see also Carla Nappi, 'Surface Tension. Objectifying Ginseng in Chinese Early Modernity,' in *Early Modern Things. Objects and Their Histories, 1500–1800*, ed. Paula Findlen (London: Routledge, 2012), 34; Barbara Orland and Kijan Espahangizi, 'Pseudo-Smaragde, Flussmittel und bewegte Stoffe. Überlegungen zu einer Wissensgeschichte der materiellen Welt,' in *Stoffe in Bewegung. Beiträge zu einer Wissensgeschichte der materiellen Welt*, ed. Barbara Orland and Kijan Espahangizi (Zürich: diaphanes, 2014).

[28] Historians have in recent years suggested replacing the 'notion of an object as always singular with that of an object as always multiple', and malleable. Nappi, 'Surface Tension,' 46. See also Orland and Espahangizi, 'Pseudo-Smaragde, Flussmittel und bewegte Stoffe.' On the difficulties of 'locating' substances, see also Erika Monahan, 'Locating Rhubarb. Early Modernity's Relevant Obscurity,' in *Early Modern Things. Objects and Their Histories, 1500–1800*, ed. Paula Findlen (London: Routledge, 2013), 239. See also Daston, 'Introduction. Speechless,' 18.

As with other introduced exotic commodities – coffee, rhubarb or pineapple[29] – by the late 1700s and early 1800s appellations for the bark across languages varied, if seldom beyond recognition. *Cinchona* was the standard botanical name for the bark after Carl Linnaeus (1707–1778) first defined the genus in the second, 1742 edition of his *Genera Plantarum*, naming it after the Countess of Chinchón, Francisca Fernández de Ribera, for her legendary and, by all accounts, imaginary role in drawing attention to the bark's virtues sometime between 1632 and 1638.[30] The bark also continued to be referred to by the older name of *quinquina* – from *Quina-Quina*, a Quechua word that actually referred to the balsam tree, and had been misapplied to cinchona by the Genoese physician Sebastianus Badus (fl. 1643–1676) in his 1663 *Anastasis Corticis Peruviae*.[31] *Quinquina* persisted in various guises, coterminous with and alongside cinchona, particularly in French[32] and Italian,[33] into the early nineteenth century, while Spanish[34] and Portuguese[35] sources employed the shorter *quina*. German and Dutch texts, presumably onomatopoetically with the Iberian term, likewise referred in common parlance to *China*[36] – or *Chinarinde*[37] – and *kina*,[38] respectively, and to

---

[29] Monahan, 'Locating Rhubarb,' 232.

[30] Jaime Jaramillo-Arango, 'A Critical Review of the Basic Facts in the History of Cinchona,' *Journal of the Linnaean Society* 53 (1949); Alex Haggis, 'Fundamental Errors in the Early History of Cinchona,' *Bulletin for the History of Medicine* 10 (1941). Linnaeus relied on the description and drawing by Charles-Marie de La Condamine to classify *Cinchona officinalis*, which erroneously merged two distinct cinchona varieties. Spanish botanists would later seek to revise Linnaeus's misapprehension. Matthew Crawford, 'Empire's Experts: The Politics of Knowledge in Spain's Royal Monopoly of Quina (1751–1808)' (unpublished PhD dissertation, University of California, San Diego, 2009), 18–19.

[31] Various historians have examined this early misapprehension: Jaramillo-Arango, 'A critical review'; Haggis, 'Fundamental Errors,' 421–29.

[32] For French uses of the term 'quinquina', see, for instance, M. Mallet, *Sur le Quinquina de la Martinique, connu sous le nom de Quinquina-Piton* (Paris: 1779).

[33] Italian sources frequently referred to 'kinakina'. See, for instance, Enrico Tegut, *Le mirabili virtú della Kinakina, con la maniera di servirsene in qualunque sorte di Febbre, e complessione* (Venice: Presso Antonio Zatta, e Figli, 1785).

[34] See, for instance, Ruiz López, *Quinología*; Pedro Crespo Nolasco, 'Carta apologética de la quina o cascarilla,' *Mercurio Peruano (Lima)* 8 (1795 [1861]).

[35] See, for instance, Jose Mariano Velloso, *Quinografia Portugueza ou Colleccao de varias memorias sobre vinte e duas especies de quinas, tendentes ao seu descobrimiento nos vastos dominios do Brasil, copiada de varios authores modernos, enriquecida com cinco estampas de Quinas verdadeiras, quatro de falsas, e cinco de Balsameiras* (Lisboa: Impressor da Santa Igreja Patriarcal, 1799).

[36] See, for instance, Heinrich von Bergen, *Versuch einer Monographie der China* (Hamburg: Hartwig & Müller, 1826); Tissot, *Anleitung für das Landvolk*, 288.

[37] See, for instance, E. G. Baldinger, 'Geschichte der Chinarinde und ihrer Wirkungen,' *Magazin vor Aerzte* 7 (1778).

[38] For references to 'kina' in Dutch sources, see, for instance, C. Terne, *Verhandelingen over de Vraage, in hoe verre zou men, by gebrek van de Apotheek, uit kelder en keuken de vereischte*

cinchona in jargon. Some European languages possessed other alternate terms for cinchona, revolving around its provenance, medicinal properties or materiality. In English, for instance, its popularity allowed it to be known simply as the 'bark' or, owing to its supposed provenance, as the 'Peruvian bark'. On account of its close association with the Jesuit order, particularly in earlier sources, it was also referred to as the 'Jesuit's bark' or, since it was often available in the pulverized form, the 'Jesuit's powder'.[39] Spanish sources, too, often spoke rather than of *quina* of *cascarilla*, a diminutive of the Spanish word for 'tree bark' (*cascara*), while German sources occasionally referred to it as *Fieberrinde*, that is, 'fever bark'.[40] Nomenclature maintained a measure of coherence and kinship even beyond these earlier consumer societies by virtue of linguistic relationships – translation equivalence, or onomatopoeia – references to geographical provenance, or therapeutic indications. Slavic, Turkic or Asian-language renderings in particular appear to have had onomatopoetic qualities. Eighteenth-century Chinese sources referred to '金鸡勒' ('chin-chi-lei' in Wade-Giles, 'jin ji lei' in pīnyīn),[41] for instance, Russian sources to 'хина' (khina), or 'перуанская хина' (peruanskaya khina),[42] while in the Ottoman Empire the bark was referred to as 'kına' (kina), or 'kûşûru'l-Peruviyane', a literal translation of 'Peruvian bark'.[43] Equations are, to be sure, fraught with difficulty, and these various terms were idiosyncratic and part of widely divergent epistemic systems. They were also, however, cognate appellations, fragments of discourse that reveal networks of production,[44] threaded together by men and women from

---

Geneesmiddelen, ook tegen de zwaarste ziekten en kwaalen, zo uit- als inwendig, kunnen bekomen, mits uitzondere de volgende middelen, Kina, Kwik, Opium, Staal, Delfzuuren, Rhabarber en Ipecacoanna (Amsterdam: Petrus Conradi, 1788).

[39] See, for instance, John Gray, William Arrot and Phil Miller, 'An Account of the Peruvian or Jesuits Bark,' *Philosophical Transactions* 40 (1737/38).

[40] Georg Leonhart Huth, *Sammlung verschiedener die Fieberrinde betreffender Abhandlungen und Nachrichten* (Nürnberg: Seeligmann, 1760); Tissot, *Anleitung für das Landvolk*, 288; Alexander von Humboldt, *Ideen zu einer Geographie der Pflanzen: Nebst einem Naturgemälde der Tropenländer* (Tübingen / Paris: F. G. Cotta / F. Schoell, 1807), 63–67.

[41] The term is mentioned in the *Pen-ts'ao kang mu shih-I*, compiled in 1765 by Chao Hsüeh-min (1719–1805). Cited in Paul Unschuld, *Medicine in China. A History of Pharmaceutics* (Berkeley: University of California Press, 1986), 166.

[42] See, for instance, John T. Alexander, *Bubonic Plague in Early Modern Russia: Public Health and Urban Disaster* (New York: Oxford University Press, 2003), 183.

[43] Feza Günergun and Şeref Etker, 'From Quinaquina to "Quinine Law": A Bitter Chapter in the Westernization of Turkish Medicine,' *Osmanlı Bilimi Araştırmaları* XIV, no. 2 (2013), 47; Salim Aydüz and Esma Yildirim, 'Bursalı Ali Münşî ve Tuhfe-i Aliyye. Kına Kına Risâlesi Adlı Eserinin Çevirisi,' *Yeni Tıp Tarihi Araştırmaları* 8 (2002), 93.

[44] On practices of equation in the history of medicine, see Nappi, 'Winter Worm, Summer Grass,' 29–30.

various world regions who had evidently long engaged with and relied upon one another – not only in apprehending that substance's 'admirable effects'[45] but also in crafting a name for it.

Significant, and growing, world market demand for the bark in the late 1700s and early 1800s – from buyers in Portuguese Luanda, at the Ottoman Porte and in the Archduchy of Austria alike – rendered cinchona's botanical classification and demarcation both imperative and difficult. As with other plant-based medicinal substances of the period,[46] there was considerable controversy not only over the boundary of cinchona via-à-vis other plants but also over the varieties cinchona was to encompass – the kinds and number of species that were to be contained in the genus *Cinchona*, to resort to the period's botanical lexis.[47] It was in particular the repeated removal to novel bark-growing regions in the Spanish American Viceroyalties of New Granada and Peru – on account of the bark's worldwide appeal, and resultant overexploitation – and with it, the encounter with divergent varieties of cinchona, that distressed consumers, medical practitioners and naturalists alike.[48] The Spanish, British and French commercial quest for substitutes also yielded several South Asian, Filipino, and Caribbean cinchonas – from St Lucia, Saint Domingue, Guadeloupe and Martinique – that were subject to clinical trials and chemical analyses, but eventually, for the most part, discarded.[49] In 1805, as the result of a two-decades-long quest, two tree species supposed to be cinchona varieties – *Cinchona macrocarpa* and

---

[45] Note dated as of February 12, 1773, in 'Varios Papeles pertenecientes á la Quina del Péru,' *Archivo del Palacio Real*, Papeles del Almacén de la Quina, Caja 22282 / Expediente Número 6, Madrid, 1773-02-12.

[46] On the difficulties of identifying species of rhubarb, and determining which varieties were the 'true rhubarb', see Monahan, 'Locating Rhubarb,' 229.

[47] In common parlance – the lexis of Spanish colonial officials, harvesters and Creole merchants – the term 'species' was also often applied to cinchona at large – 'the said species cinchona (*la d[ic]ha especie de cascarilla*)'. See, for instance, 'Sobre el acopio de la Quina de los Montes de Loxa Callysalla y otros que la produzcan de buena calidad, y su envio a Espana de cuenta de la Rl. Hazienda,' *Archivo Nacional de la Historia*, Quito, Fondo General, Serie Cascarilla, Caja 3, Expediente 13, Cuenca, 1790-08-26, ff. 34–36; 'Expediente sobre el corte de cascarilla en los Montes de Loxa,' *Archivo Nacional de la Historia*, Quito, Fondo General, Serie Cascarilla, Caja 2, Expediente 5, Loja, 1779-08-19, f. 1.

[48] For a detailed account of the removal from one harvest area to another, see Chapter 5.

[49] On botanical descriptions of 'supposed cinchonas' in the late 1700s, see Luis Alfredo Baratas Díaz and Joaquín Fernández Pérez, 'Conocimiento botánico de las especies de cinchona entre 1750 y 1850: Relevancia de la obra botánica española en América,' *Estudios de historia de las tecnicas, la arqueología industrial y las ciencias* 2 (1998), 648–50. On the French quest, see James E. McClellan and François Regourd, *The Colonial Machine: French Science and Overseas Expansion in the Old Regime* (Turnhout: Brepols Publishers, 2012), 260–62. For an example, see 'Séance du Mardi 30 Juin. La Société m'a chargé de porter sur ses plumitifs le résumé suivt. concernant les différentes especes

Cinchona pubescens – were discovered on Portuguese territory in Rio de Janeiro.[50] Other than to the general limitations of Linnaean taxonomy and the difficulty of examining live plant specimens,[51] it was owing to the variation in properties[52] (bark colour, taste and texture), presented by the proliferation of newly found cinchonas by the beginning of the nineteenth century, that caused contemporaries to continue to differ – in some measure, increasingly so – on how to delineate and group that plant. Opinions on the sheer quantity of extant cinchona species varied from author to author, from two to twenty-two.[53] While the inner and outer botanical outlines of cinchona remained elusive, fragile and tenuous in the eyes of botanists from Uppsala to Santa Fé de Bogotá into the

de quinquina qui ont été soumises á son examen,' *Bibliothèque de l'Académie de médecine*, Procès-verbaux des séances de la Société Royale de la Médicine, Ms 11/11, Paris, 1789-06-30. On the British quest for substitutes, see Maehle, *Drugs on Trial*, 277; Pratik Chakrabarti, 'Empire and Alternatives: Swietenia febrifuga and the Cinchona Substitutes,' *Medical History* 54, no. 1 (2010).

50 Vera Regina Beltrão Marques, *Natureza em Boiões: medicinas e boticários no Brasil setecentista* (Campinas: Editora da Unicamp / Centro de Memória–Unicamp, 1999), 134.

51 Baratas Díaz and Fernández Pérez, 'Conocimiento botánico de las especies de cinchona,' 649.

52 On the 'perceptible dimension' of materials in eighteenth-century chemistry, see Ursula Klein and Wolfgang Lefèvre, *Materials in Eighteenth-Century Science. A Historical Ontology* (Cambridge, Mass.: MIT Press, 2007), 58–59.

53 According to Padréll et Vidal, by 1802, there were between four and seven varieties; see Joseph Padréll et Vidal, 'Dissertation sur l'usage et l'abus du quinquina dans le traitement des fièvres intermittentes; présentée et soutenue á l'École de Médicine de Montpellier le 23 Prairial an 10 (de la République),' in *Collection des thèses soutenues a l'École de Médicine de Montpellier*, ed. L'École de Médicine de Montpellier (Montpellier: Imprimerie de G. Izar e A. Ricard, 1802), 7–14. José Celestino Mutis defined seven species, but found only four of them to be 'medicinal': *Cinchona lancifolia*, *Cinchona oblongifolia*, *Cinchona cordifolia*, and *Cinchona ovalifolia*. Josè Celestino Mutis, *Instrucción formada por un facultativo existente por muchos años en el Perú, relativa de las especies y virtudes de la quina* (Cádiz: Don Manuel Ximenez Careño, 1792); Manuel Hernández de Gregorio, ed., *El arcano de la quina. Discurso que contiene la parte médica de las cuatro especies de quinas oficinales, sus virtudes eminentes y su legítima preparación. Obra póstuma del doctor D. José Celestino Mutis* (Madrid: Ibarra, Impresor de Cámara de S. M., 1828). Hipólito Ruiz López organized his findings into seven types of cinchona in 1792, and revised them in 1801 to include nine. Ruiz López, *Quinología*, vol. 2, 50–54; Hipólito Ruiz López and José Antonio Pavón Jiménez, *Suplemento a la quinologia, en el qual se aumentan las Especies de Quina nuevamente descubiertas en el Perú por Don Juan Tafalla, y la Quina naranjada de Santa Fé con su estampa* (Madrid: Imprenta de la viuda e hijo de Marín, 1801). By 1797, Aylmer B. Lambert had written of eleven species of cinchona; by 1821 he had come to think there were as many as twenty-two kinds. Aylmer B. Lambert, *An illustration of the genus Cinchona: Comprising Descriptions of all the Officinal Peruvian Barks, incl. Several New Species* (London: Searle, 1821). For discussions of the debates about cinchona classification in the Iberian world around 1800, see Baratas Díaz and Fernández Pérez, 'Conocimiento botánico de las especies de cinchona'; Mauricio Nieto Olarte, *Remedios para el imperio. Historia natural y la apropiación del Nuevo Mundo* (Bogotá: Universidad de los Andes – FLACSO-CESO, 2006), 83, 173–95.

early nineteenth century, however, constant debate about its varieties also reified the idea of cinchona as a single object. As historians have argued for other plants, the very discussion of its instantiations – in continuously referencing the category they instantiate – also contributed to stabilizing and objectifying the bark as a recognizable thing.[54]

London physicians,[55] creole bark merchants in the Viceroyalty of New Granada,[56] and Chinese medical authors[57] alike commonly circum-scribed the bark's identity in the late 1700s and early 1800s, like botanists, by virtue of its geographical provenance as well as its material properties – texture, taste, consistency and colour. Genuine cinchona was supposed to have the same shape as cinnamon; a rough, splintery and mealy texture; and to be of either white, pale-yellow, reddish or orange colour, according to species (FIGURE O.1).[58] When chewed, it was to be of a bitter, aromatic and astringent taste.[59] In conjunction with the rise of clinical pharmacology, experimenters also began to define the bark chemically, through experiments and the testing of properties – its acidity, solubility in various solvents or reaction with other substances, particularly bodily fluids.[60] At a time when simple clinical observations, experiences and statistics to evaluate treatments were gradually being introduced, doctors, botanists and surgeons in Madrid, Cartagena de Indias, London, Saint Domingue, New York, Rio de Janeiro or Lyon also increasingly conducted clinical trials – 'exact, and repeated observations', 'by means of a general, extensive administration' of the bark – among the populations of hospitals, slave plantations, or the military to

---

[54] Nappi, 'Surface Tension,' 41.
[55] Robert John Thornton, *New Family Herbal: Or Popular Account of the Natures and Properties of the Various Plants Used in Medicine, Diet and the Arts* (London: Richard Phillips, 1810), 117.
[56] Matthew Crawford, *The Andean Wonder Drug. Cinchona Bark and Imperial Science in the Spanish Atlantic, 1630–1800* (Pittsburgh, Pa.: University of Pittsburgh Press, 2016), 103.
[57] Chao Hsüeh-min described cinchona as 'consist[ing] of thin, hollow twigs' that 'resembled the drug *yüan-chih*, after one ha[d] removed from it the marrow' and affirmed that 'the taste [was] slightly acrid'. Cited in Unschuld, *Medicine in China*, 166.
[58] William Buchan advised his readership to learn to 'distinguish' 'genuine' barks from 'false' ones. William Buchan, *Domestic Medicine: Or, a treatise on the prevention and cure of diseases* (London: W. Strahan, 1774), 169.
[59] See, for instance, Johan Andreas Murray, *Johan Andreas Murray's Vorrath an einfachen, zubereiteten und gemischten Heilmitteln, zum Gebrauche praktischer Aerzte bearbeitet*, ed. Ludwig Christoph Althof, 2 vols., vol. 1 (Göttingen: Johann Christian Dieterich, 1793), 1118; Padréll et Vidal, 'Dissertation sur l'usage et l'abus du quinquina,' 7–14. Aydüz and Yildirim, 'Bursalı Ali Münşî ve Tuhfe-i Aliyye,' 94; Crawford, *The Andean Wonder Drug*, 101–02.
[60] Chakrabarti, 'Empire and Alternatives,' 89; Maehle, *Drugs on Trial*, 8, 27; Klein and Lefèvre, *Materials in Eighteenth-Century Science*.

Figure 0.1 *Cinchona rosea Flor. Peruviana*. Sample collected under the aegis of the Botanical Expedition to the Viceroyalty of Peru (1778–1816), under the command of Hipólito Ruiz López and José Antonio Pavón. MA-780943. *Herbario del Real Jardín Botánico, CSIC.* © *RJB-CSIC*

put different or newly discovered varieties of cinchona on trial and gain 'a proper understanding of their virtues' (*o devido conceito das virtudes*).[61] None of these criteria of demarcation was absolute or definite, however.

---

[61] On cinchona testing in English, German and French language contexts, see Maehle, *Drugs on Trial*, 268–75. On cinchona testing in hospitals of the Spanish Empire, in what signified a shift away from the mere observation of the bark's physical characteristics, see Crawford, *The Andean Wonder Drug*, 117–18; Rosario Terreros Gomez Maria and Andrés Turrión Maria Luisa, 'First Hospital Experiences with Cinchona Ordered by Spanish Court (ca. 1770),' *Revue d'histoire de la pharmacie* 84, no. 312 (1996). The Portuguese Crown also tested the bark on sufferers. See, for instance, 'Decretos do príncipe regente,' *Arquivo Histórico Ultramarino*, 076 – REINO RESGATE 20121023 / Cx. 30-A, Pasta 18, Queluz, 1804-09-22. For an instance of cinchona testing on Saint Domingue, see Joseph Gauché, 'Description d'un Quinquina indigène á St. Domingue, par Joseph Gauché, habitante, concessionnaire et administrateur des eaux thermales de Boynes, membre du Cercle des Philadelphes du Cap Français. Mémoire lu à l'Académie des Sciences, le 24 juillet 1787,' *La Bibliothèque centrale du Muséum national d'histoire naturelle*, Ms 1275, n.p., c. 1787.

Plant materials belonged in the world of commodities and trade, and human indiscretion, as well as natural variation in their materiality – or 'perceptible qualities', to use the period's lexis – rendered them as resistant to epistemic and medical stabilization as they did to botanical classification. Other than the removal to novel bark-growing regions and the commercial quest for substitutes, by the late eighteenth century, instances of wilful fraud – the addition of poor-quality cinchona or other, non-medicinal barks – by Caribbean pirates,[62] Habsburg customs officials[63] and London apothecaries[64] alike, as well as of deterioration in transport, further induced the authors of health advice manuals and popular recipe collections to advise caution in selecting cinchona bark.[65] Cinchona was 'now for the most part adulterated', as the author of an Italian manuscript recipe collection phrased it.[66] Readers were well advised to take care that the bark they purchased not be 'spoiled by moisture,'[67] that its taste be neither 'nauseous, or [...] mucilaginous', nor its surface too tough or too 'spongy, [...] woody, or powdery'[68] – that the bark be, in short, neither false nor deteriorated. Cinchona, as it was conveyed across landmasses and bodies of water, and taken into hospitals, laboratories and apothecary shops the Atlantic World over, thus exhibited a material tendency to decay and a natural and circumstantial bent for variation that hinged on the very breadth of its acceptance and the steadfastness of its appeal. Discourses and practices attendant to the bark's propensity to decay, and its bent for variation,

[62] See Chapter 2.

[63] Luis Martínez de Beltrán, 'Oficio de D. Luis Martínez de Beltrán a Manuel Muzquiz, comunicándole que cuando lleguen los dos cajones de quina regalada a la Emperatriz Reina de Hungría, los hará seguir a su destino,' *Archivo General de Simancas*, Legajo 907, Genova, 1771-04-27.

[64] See the extract from a circular letter, dated as of November 15, 1799, by the Royal College of Physicians, on the yellow bark's liability to 'adulteration', in 'Receipts copied from Miss Myddleton's Book, August 15th, 1785. With many added receipts for remedies by various later hands, extracts, and pasted-in cuttings from newspapers, etc.', *Wellcome Library*, Archives and manuscripts, Closed stores WMS 4, MS.3656, n.p., c. 1785–1818.

[65] Matthew Crawford has studied the problem of cinchona fraud in the Spanish Empire in detail. Matthew Crawford, '"Para desterrar las dudas y adulteraciones": Scientific Expertise and the Attempts to Make a Better Bark for the Royal Monopoly of *Quina*,' *Journal of Spanish Cultural Studies* 8, no. 2 (2007); Crawford, *Empire's Experts*. On the problem of counterfeit drugs in the period more generally, see Roy Porter and Dorothy Porter, *Patient's Progress. Doctors and Doctoring in Eighteenth-Century England* (Cambridge: Polity Press, 1989), 167.

[66] 'Collection of medical receipts and prescriptions: in Italian, by various hands.' Wellcome Library, Archives and manuscripts, Closed stores WMS 4, MS.4105, n.p., n.d.

[67] Thornton, *New Family Herbal*, 117.

[68] Ibid. On counterfeit bark, see also Murray, *Vorrath von einfachen, zubereiteten und gemischten Heilmitteln*, 1, 1118, 21.

certainly encumbered and delayed its epistemic and medical delineation, and stabilization in ways that render any account of it a 'history of likenesses rather than [...] of an object',[69] of a historical category rather than of a specific kind of matter. They also indicate, however, the extent to which cinchona had, by the decades around 1800, become an object that trained observation could discern, and the integrity of which it was considered necessary, and ultimately possible, to maintain, police and regulate.[70]

Cinchona was extensive not only in its geographic reach by the late 1700s and early 1800s – enjoying popularity in societies the Atlantic World over – but also in its therapeutic indications. As historians of pharmacology have shown, while in the seventeenth century physicians had still taken cinchona to be a 'specific' – a remedy that targeted and extinguished one particular kind of disease, 'intermittent fevers'[71] – by the eighteenth, medical practitioners from Britain to Muscovy, and from the sultanate of Morocco to the Viceroyalty of New Spain, would have agreed that the bark was effective for various types of fevers – intermittent, but also remittent,[72] bilious,[73] nervous[74] or yellow[75] (FIGURE O.2). Some practitioners suggested that cinchona could also

---

[69] Nappi, 'Winter Worm, Summer Grass,' 29.   [70] Nappi, 'Surface Tension,' 41.

[71] The very concept of 'specific' changed around 1800: while since Thomas Sydenham it had denoted a remedy that extinguished the *species morbi*, regardless of the patient's individual constitution, it came to mean any medicine that was uniquely powerful and that united its known pharmacological properties – astringency, antiseptic power, etc. – in such 'an inimitable way that it was superior to all other drugs sharing those properties'. Maehle, *Drugs on Trial*, 287. On medical specifics, see also Cook, 'Markets and Cultures.'

[72] See, for instance, Thomas Dancer, *The Medical Assistant; or Jamaica Practice of Physic: Designed chiefly for the Use of Families and Plantations* (Kingston, Jamaica: Alexander Aikman, 1801), 87; Jose Pinto de Azeredo, *Ensaios sobre algumas enfermidades d'Angola* (Lisboa: Na Regia Officina Typografica, 1799), 64. See also the 1808 Russian pharmacopoeia: James Wylie, *Pharmacopoeia castrensis Ruthena* (St Petersburg: Typographia Medica, 1808), 41.

[73] See, for instance, James Clark, *A Treatise on the Yellow Fever, as it appeared in the Island of Dominica, in the Years 1793-4-5-6; to which are added, Observations on the Bilious Remittent Fever, on Intermittents, Dysentery, and Some Other West Indian Diseases* (London: J. Murray & S. Highley, 1797), 81; Gilbert Blane, *A Short Account of the Most Effectual Means of Preserving the Health of Seamen, particularly in the Royal Navy, to the Flag-Officers and Captains of his Majesty's Ships of War on the West-India Station* (Sandwich, off Antigua: – 1780), 33.

[74] Dancer, *The Medical Assistant; or Jamaica Practice of Physic*, 72.

[75] See, for instance, Padréll et Vidal, 'Dissertation sur l'usage et l'abus du quinquina,' 14; José Celestino Mutis, 'Borrador del oficio de José Celestino Mutis al virrey Pedro Mendinueta y Muzquis,' *Archivo del Real Jardín Botánico*, Real Expedición Botánica del Nuevo Reino de Granada (1783–1816), José Celestino Mutis, Correspondencia, RJB03/0002/0002/0172, Santa Fé de Bogotá (Colombia), 1801-10-24.

Figure 0.2 The 'Fever Tree (*Lignum Febrium*)' by Francisco Torti, which supplemented the author's taxonomy of fevers. Branches covered with bark, occupying the left part of the picture, represent fevers curable by Peruvian bark, whereas denuded, leafless branches represent continued fevers not curable by cinchona. At the centre are trunks and branches partly covered by bark, corresponding to the 'proportionate fever', in which susceptibility varied. Branches that anastomose represent fevers that change from one category to another, 1712. Francisco Torti *Therapeutice Specialis Ad Febres Periodicas Perniciosas*. Credit: Wellcome Collection. CC BY

be useful in other diseases: in gangrene,[76] haemorrhages,[77] dysentery,[78] epilepsy,[79] smallpox,[80] rheumatism,[81] consumption,[82] scurvy,[83] jaundice,[84] the gout[85] or in obstructions of the menstrual flux, that is, to induce the menses.[86] Novel indications were brought on both inadvertently, by 'chance observations' and therapeutic experience,[87] and on

[76] See, for instance, Thornton, *New Family Herbal*, 123. The Edinburgh new dispensatory likewise advised the bark 'in gangrenous sore throats, as [...] in every species of gangrene'. William Lewis and John Rotheram, *The Edinburgh new dispensatory: with the additions of the most approved formulae, from the best foreign pharmacopoeias; the whole interspersed with practical cautions and observations; and enriched with the latest discoveries in natural history, chemistry, and medicine; with new tables of elective attractions of antimonial and mercurial preparations, &c.* (Walpole, Newhampshire: Thomas & Andrews, 1796), 144. Maehle discusses the administration of the bark in 'gangrene' at length. Maehle, *Drugs on Trial*, 247–58.

[77] Murray, *Vorrath von einfachen, zubereiteten und gemischten Heilmitteln*, 1, 1193. See also Ralph Irving, *Experiments on the Red and Quill Peruvian Bark: with Observations on its History, Mode of Operation, and Uses* (Edinburgh: C. Elliot, 1785), 174–75; Wylie, *Pharmacopoeia castrensis Ruthena*, 41.

[78] See, for instance, Wylie, *Pharmacopoeia castrensis Ruthena*, 41; Lewis and Rotheram, *The Edinburgh new dispensatory*, 144.

[79] Murray, *Vorrath von einfachen, zubereiteten und gemischten Heilmitteln*, 1, 1191.

[80] According to Thornton, in 'confluent small-pox it promotes languid eruption and suppuration, diminishes the fever, and prevents or corrects putrescence and gangrene.' Thornton, *New Family Herbal*, 123; Irving, *Experiments on the Red and Quill Peruvian Bark*.

[81] Thornton quoted a Dr Haygarth, who had had 'lately extolled its use in acute rheumatism, from the very commencement, even without premising venesection'. Thornton, *New Family Herbal*, 123.

[82] On cinchona in consumption, see, for instance, William Buchan, *Domestic Medicine, or, the Family Physician: Being an Attempt To Render the MEDICAL ART more generally useful, by shewing people what is in their own power both with respect to the PREVENTION and CURE of Diseases. CHIEFLY Calculated to recommend a proper attention to REGIMEN and SIMPLE MEDICINES* (Edinburgh: Balfour, Auld, and Smellie, 1769), 206; Murray, *Vorrath von einfachen, zubereiteten und gemischten Heilmitteln*, 1, 1186.

[83] Murray cited cases where practitioners had employed the bark with varying degrees of success in scurvy. Murray, *Vorrath von einfachen, zubereiteten und gemischten Heilmitteln*, 1. According to Lewis, some practitioners had 'great confidence in it joined with the acid of vitriol, in cases of phthisis, serophula, ill conditioned ulcers, rickets, scurvy, and in states of convalescence'. Lewis and Rotheram, *The Edinburgh new dispensatory*, 144.

[84] Murray, *Vorrath von einfachen, zubereiteten und gemischten Heilmitteln*, 1, 1202; Thornton, *New Family Herbal*, 123.

[85] See, for instance, Murray, *Vorrath von einfachen, zubereiteten und gemischten Heilmitteln*, 1, 1204. For an example from the Portuguese context, see Francisco Tavares's treatise on the 'profitable, and wholesome use of cinchona in the gout'. Francisco Tavares, *Observações, e reflexões sobre o uso proveitoso, e saudavel da quina na gota* (Lisbon: Regia Oficina Typografica, 1802).

[86] Buchan, *Domestic Medicine*, 361. Cinchona was used as an emmenagogue, or menstrual regulator, in eighteenth-century Europe, where menstruation was considered a necessary cleansing process that, if missed, would cause a multitude of afflictions. Londa Schiebinger, *Plants and Empire* (Cambridge, Mass: Harvard University Press, 2004), 182.

[87] Maehle, *Drugs on Trial*, 247–58. See also Boumediene, *La colonisation du savoir*, 235.

account of alterations in interpretations of the bark's mode of operation –
it was increasingly thought to act through general, not specific, tonic, or
stimulant, antiseptic, astringent or corroborant properties[88] – as well as
in the understanding of the causes of fevers. In the late eighteenth
century, fevers came to be seen as the effect of conditions such as debility
of the fibres, recurrent 'atony' or putridity, the same disorders that were
thought to produce ailments like gangrene, smallpox or dysentery.[89] This
is not to say that cinchona ceased to be the remedy of choice in
intermittent fevers. As a matter of fact, while it was 'pretty generally
agreed' among medical practitioners, both lay and professional, from
the West Indies to the Ottoman Porte, that cinchona was the remedy
they could 'most certainly rely on for the cure of intermittent fevers',[90] its
propriety and effects in other disorders, particularly those that were not
fevers, were considered uncertain, 'various and often opposite in different patients, and in different states of the same patients'.[91] It is to say,
however, that the bark's curative indications expanded considerably,
rendering it, by all accounts, a broad-spectrum febrifuge by the turn of
the eighteenth century, and, at least temporarily[92] and to some practitioners, also a panacea and universal remedy.

We tend to think of substances as durable kinds of matter with uni-
form, definite properties: as foundational, fundamental entities, and as
ontologically basic[93] – everything that cinchona, in its evident variability
and ambiguity, its shifting epistemic, chemical and medical contours,
was not. The Peruvian bark was not so much a specific kind of matter by
the late 1700s and early 1800s. It was, rather, a specific historical
category that encompassed various kinds of matter: a number of dried,

---

[88] Maehle, *Drugs on Trial*, 258; 63–66. Antiseptics derived their name from 'septic', which
meant putrefactive. Pratik Chakrabarti, *Medicine and Empire 1600–1960* (Basingstoke:
Palgrave Macmillan, 2014), 45.

[89] Maehle, *Drugs on Trial*, 264.

[90] Irving, *Experiments on the Red and Quill Peruvian Bark*, 176. For similar remarks, see
Robert Jackson, *An Outline of the History & Cure of Fever, Endemic and Contagious; More
Expressly the Contagious Fevers of Jails, Ships, & Hospitals, the Concentrated Endemic,
Vulgarly Called the Yellow Fever of the West Indies* (Edinburgh: John Meir, 1808), 276;
Padréll et Vidal, 'Dissertation sur l'usage et l'abus du quinquina,' 1–2; Tissot, *Anleitung
für das Landvolk*, 288–89; Johann Jacob Rambach, *Versuch einer physisch-medizinischen
Beschreibung von Hamburg* (Hamburg: Carl Ernst Bohn, 1801), 310–12; Aydüz and
Yildirim, 'Bursalı Ali Münşî ve Tuhfe-i Aliyye,' 96. See also Maehle, *Drugs on Trial*,
246; 85.

[91] Irving, *Experiments on the Red and Quill Peruvian Bark*, 176.

[92] 'Collection of medical receipts and prescriptions: in Italian, by various hands,' *Wellcome
Library*, Archives and manuscripts, Closed stores WMS 4, MS.4105, n.p., n.d.

[93] Theodore Schatzki, 'Nature and Technology in History,' *History and Theory*, no. 42
(2003), 82–93; Howard Robinson, 'Substance,' in *Stanford Encyclopedia of Philosophy*,
ed. Edward N. Zalta (2014).

bitter-tasting shreds of tree bark marketed, dispensed and classified under the name of cinchona – or, indeed, one of that name's alternate and foreign equivalents – inclusive and aware of their shifting therapeutic attributes and of their material tendencies toward decay or variation over time and space.

An Appraisal of the Historiography

The singular medicinal virtues ascribed to cinchona have, over the centuries, attracted a considerable number of historians to the subject. The historiography has suffered, however, from a tendency toward presentism on the one hand – a subservience to quinine, one of cinchona's active compounds, and malaria – and a close association with particular empires and states on the other, the British and Spanish especially. Particularly with an Anglo-American reading public, cinchona is still closely associated with the British Empire and the salvation of the lives and minds of Englishmen in the malaria-stricken Raj of the late nineteenth century.[94] *A Singular Remedy* breaks with these two historiographical traditions, in that it centres on the knowledge movement that limitation to particular empires has largely obliterated and on the contingency, variability and idiosyncrasy of bark knowledge that presentism has so often obscured. At the very heart of *A Singular Remedy* is the richness and latitude of the life of a substance that habitually crossed imperial and medical boundaries: the diversity of therapeutic practices and routines of medication pertaining to cinchona, the variety of ailments and conditions in which it was employed, and the wide range of creole, French, Cantonese, Portuguese or Levantine experts, sufferers and vendors given to its consumption, sale or advocacy.

The book breaks, first, from an important sector of the historiography that has reduced the bark to its part as the source of, and precursor to, quinine, and proceeded on the assumption that it would have been, like that active compound it contained, effective against malaria. Indeed, cinchona has long occupied a prominent place in presentist histories of medicine chronicling 'the ideas and events which brought medicine ever closer to the secrets of disease and health'.[95] Along those same lines, a popular, laudatory, genre of historiography has celebrated the bark as

---

[94] 'Lives and minds', attributed to Winston Churchill (1874–1965), reads: 'The gin and tonic has saved more Englishmen's lives, and minds, than all the doctors in the Empire.'

[95] Morris J. Vogel, introduction to *The Therapeutic Revolution. Essays in the Social History of Medicine*, ed. Morris J. Vogel and Charles Rosenberg (Philadelphia: University of Pennsylvania Press, 1979), viii.

'the remedy that has spared, or at least ameliorated, the greatest number of lives in human history'.[96] That historiography has also celebrated its discoverers, advocates and pioneers: the friars of the Jesuit order who first appreciated its true value, the visionary physicians and apothecaries – Robert Talbor (1642–1681) and Thomas Sydenham (1624–1689) – who overcame widespread resistance to it, and French and Prussian naturalists – Charles-Marie de La Condamine (1701–1774), Joseph de Jussieu (1704–1779) and Alexander von Humboldt (1769–1859) – who 'braved swamps, [...] dangerous animals, and wild river rapids' to bring back specimens, and observations, of cinchona plants in their natural habitat.[97] Much of the academic historiography, too, though far from embracing the same triumphalist rhetoric, has proceeded on the assumption that the bark was a natural remedy against malaria.[98] Even where historians have doubted the bark's efficacy, they have reduced it largely to its administration in ailments retrospectively diagnosed as malaria. Many of the earliest historical studies of the bark,[99] as well as some of the most conspicuous recent publications that make reference to it by environmental and global historians of disease,[100] have come out of the historiography pertaining to malaria. It is, to be sure, perfectly plausible to assume that the various barks contemporaries consumed under the designation of cinchona effectively contained, like their modern-day

---

[96] Mark Honigsbaum and Merlin Willcox, 'Cinchona,' in *Traditional Medicinal Plants and Malaria*, ed. Merlin Willcox, Gerard Bodeker and Philippe Rasoanaivo (Boca Raton, Fla.: CRC Press, 2004), 22. For a Spanish-language example of the laudatory genre, see Plutarco Naranjo, 'Pedro Leiva y el secreto de la quina,' *Revista Ecuatoriana de Medicina* XV, no. 6 (1979).

[97] Steven Lehrer, *Explorers of the Body. Dramatic Breakthroughs in Medicine from Ancient Times to Modern Science* (New York: iUniverse, 2006), 236–43. Cinchona has inspired many similar, often popular accounts by historians or doctors. See, for instance, Honigsbaum and Willcox, 'Cinchona,' 25–26; Leonard Jan Bruce-Chwatt, 'Cinchona and Quinine: A Remarkable Anniversary,' *Interdisciplinary Science Review* 15, no. 1 (1990); T. W. Keeble, 'A Cure for the Ague: The Contribution of Robert Talbor (1642–81),' *Journal of the Royal Society of Medicine* 90, no. 5 (1997).

[98] Honigsbaum and Willcox, 'Cinchona,' 21. For a similar discourse, see also the chapter on quinine in Lucille H. Brockway, *Science and Colonial Expansion* (New Haven: Yale University Press, 2002 (1979)), 108.

[99] The earliest relevant publications on cinchona by Alex W. Haggis and Jaime Jaramillo-Arango were partly stimulated by the military importance of malaria control in the Second World War. Haggis, 'Fundamental Errors'; Jaramillo-Arango, 'A Critical Review.' See also L. W. Hackett, *Malaria in Europe. An Ecological Study* (London: Oxford University Press, 1937).

[100] For general histories of malaria that include chapters on or references to cinchona, see Randall M. Packard, *The Making of a Tropical Disease: A Short History of Malaria* (Baltimore: Johns Hopkins University Press, 2007); Leonard Jan Bruce-Chwatt and Julian de Zulueta, *The Rise and Fall of Malaria in Europe* (Oxford: Oxford University Press, 1980); James L. A. Webb, *Humanity's Burden. A Global History of Malaria* (Cambridge: Cambridge University Press, 2009), 92–196.

equivalents and in varying proportions according to species, natural alkaloids (among them, quinine and quinidine, cinchonine and cinchonidine), which, in an isolated and crystallized state, are at present thought to interfere with the growth and reproduction of malarial parasites.[101] It is also reasonable to assume a relationship between intermittent fevers, the ailments most commonly treated with the bark, and forms of malaria, or rather, the disease consequences of the proto-zoan parasite species *Plasmodium vivax*, which occurs with 48-hour periodicity, *Plasmodium malariae*, which causes paroxysms every 72 hours, and *Plasmodium falciparum*, respectively.[102] For intermittent fevers, as distinguished from continual or remitting fevers, had 'intervals or remissions of the symptoms'.[103] There were tertian fevers, so-called because febrile accessions recurred on the third day, and quartan fevers, so-called because they came with attacks on the first and fourth days, as well as several less clearly synchronous, malignant forms of intermittent fevers.[104] It is not pertinent, however, to reduce the bark to its adminis-tration in intermittent fevers, when it was by all accounts a broad-spectrum febrifuge and panacea by the turn of the eighteenth century. Nor is it commensurate to assume that the bark cured men and women in the past, nor that it even afforded them relief. Not only is there in fact only limited clinical evidence to support ideas about the efficacy of whole cinchona bark extracts, even in the treatment of uncomplicated *P. falci-parum* and *vivax* malaria, especially since the last extensive clinical trials with whole cinchona bark extracts were conducted in the 1930s.[105]

---

[101] Jane Achan et al., 'Quinine, an Old Anti-malarial Drug in a Modern World: Role in the Treatment of Malaria,' *Malaria Journal* 10, no. 144 (2011).

[102] Frederick L. Dunn, 'Malaria,' in *The Cambridge World History of Human Disease*, ed. Kenneth F. Kiple (Cambridge: Cambridge University Press, 1993), 859.

[103] Whereas a 'continual fever' never left 'the patient during the whole course of the disease', or at least showed 'no remarkable increase or abatement in the symptoms', those suffering from 'remittent fever' experienced variations in the intensity of the fever, yet without any periods of relief. On the eighteenth-century category of 'fevers', see, for instance, William F. Bynum, 'Cullen and the Study of Fevers in Britain, 1760–1820,' in *Theories of Fever from Antiquity to the Enlightenment*, ed. William F. Bynum and Vivian Nutton (London: Wellcome Institute for the History of Medicine, 1981); Johanna Geyer-Kordesch, 'Fevers and Other Fundamentals: Dutch and German Medical Explanations c. 1680 to 1730,' in *Theories of Fever from Antiquity to the Enlightenment*, ed. William F. Bynum and Vivian Nutton (London: Wellcome Institute for the History of Medicine, 1981). On early modern conceptions of 'intermittent fevers' and how settlement and exploration carried that framework into the Atlantic, see Hugh Cagle, *Assembling the Tropics. Science and Medicine in Portugal's Empire, 1450–1700* (Cambridge: Cambridge University Press 2018), 227; 81–82.

[104] Buchan, *Domestic Medicine*, 166–76.

[105] The relevant studies are cited in Philippe Rasoanaivo et al., 'Whole Plant Extracts versus Single Compounds for the Treatment of Malaria: Synergy and Positive Interactions,' *Malaria Journal* 10, no. 1 (2011).

There is also great uncertainty about the concentration of alkaloids in the barks commercially available in the eighteenth century. Even if we were to assume that barks sold under the name of cinchona uniformly contained active compounds and that these were effective in the treatment of malaria, there would still be no way of knowing whether the intermittent fevers for which the bark was ordered were identical with modern-day malaria – retrospective diagnosis based on observation and description of symptoms naturally leads to a wide margin of error[106] – whether contemporaries administered curative doses of the bark, and whether the by all accounts common admixture of purgatives would not have mitigated sufferers' response.[107] Also, the historical record is too incomplete to allow for any kind of quantitative assessment. Comprehensive, systematic military and civilian health records that would allow for conclusions on the impact of medications are essentially creatures of the later nineteenth century.[108] The material point, however, is whether it is the historian's province to pose the essentially ahistorical question of efficacy, and to wrench early modern medicine, and pharmacology, into a twenty-first-century biomedical lexis, and explanatory repertoire, at all. There is overwhelming evidence that 'efficacy and rapid cures were not part of the cultural expectation of the suffering'[109] in the eighteenth century, that our historical subjects' medical horizon of expectation and therapeutic experience differed radically from ours.[110] What is more, historians have long argued that body knowledge is 'in and of itself

[106] On the pitfalls of retrospective diagnosing in the history of malaria, see Guenter B. Risse, *New Medical Challenges during the Scottish enlightenment*, vol. 78, Clio medica: The Wellcome Institute Series in the History of Medicine (Amsterdam: Rodopi, 2005), 173; Mary J. Dobson, *Contours of Death and Disease in Early Modern England* (Cambridge: Cambridge University Press, 1997), 309–27.

[107] Some historians have been more cautious than others about cinchona's 'effectiveness'. Mary Dobson, while she does not doubt that the bark was an 'effective' and 'powerful drug in controlling ague', questions its impact, since it would not have been used sufficiently widely, and would often have been 'adulterated and or used indiscriminately'. Dobson, *Contours of Death and Disease*, 316. Philip Curtin, in a similar vein, points to cinchona barks that contained 'little or none of the effective antimalarial alkaloids'. Philip D. Curtin, *Death by Migration. Europe's Encounter with the Tropical World in the Nineteenth Century* (New York: Cambridge University Press, 1989), 63.

[108] The earliest quantifiable records for the history of disease are military health records kept by the British and other European armies from 1816 onwards. Curtin, *Death by Migration*, xvi.

[109] Martha Baldwin, 'Expanding the Therapeutic Canon: Learned Medicine Listens to Folk Medicine,' in *Cultures of Communication from Reformation to Enlightenment*, ed. James Van Horn Melton (Ashgate: Aldershot, 2002), 255; Maehle, *Drugs on Trial*, 268–72.

[110] Many eighteenth-century physicians would have insisted that 'the timing of remedies rather than the factor of their composition was essential for healing', or that 'only a

constituting', productive rather than merely reflective of versions of the diseased body.[111] Just as any assumption of the constancy of human nature and the human condition is untenable in the face of historians' heightened awareness of historical singularity and discontinuity,[112] the act of collapsing past medical experiences into present categories will invariably distort and obscure our understanding of the corporeal experience of the suffering, their bodily anxieties, knowledge and imaginaries. Cinchona's complexity – the fact that it yields natural alkaloids that are today believed to profoundly affect humans and other living organisms – would have unfolded not only at a scale invisible to the experience of men, women and children in the past, but also at a level that was likely irrelevant to them. This book is greatly indebted to, and draws significantly on, global histories of disease in general, and of malaria in particular. It distances itself, however, from a history written in terms that are not those of its historical subjects and structured in terms of concepts and categories of sickness and therapy not available to past societies.[113] *A Singular Remedy* is concerned precisely with the contingency and peculiarity of medical knowledge and the knowledge movement in the past that a presentist approach will obscure. It studies the variety of illnesses and fevers in which the bark was employed, and the profuse medical vocabulary, rich curative repertoire and influential cultural and topographical imaginary that grounded practitioners' and sufferers' experience of them.

A Singular Remedy breaks, second, from a historiographical tradition confined to imperial boundaries and frameworks, in its attempt at writing a history of how medical knowledge was shared between and across the Atlantic empires. The tendency among historians of the bark to settle for explanations that can be drawn from events and processes within particular national, or imperial, territories is partly symptomatic of the wider field. Indeed, there are few global histories of health, or medicine in

certain application of cinchona in a very particular pattern of febrile illness would lead to health'. Geyer-Kordesch, 'Fevers and Other Fundamentals,' 112.

[111] On the historicity of disease entities and 'versions' of the body, see, for instance, Roger Cooter, 'The Turn of the Body. History and the Politics of the Corporeal,' *ARBOR. Ciencia, Pensamiento y Cultura* CLXXXVI, no. 743 (2010), 396–97. On the fluid, unbounded version of the body that had grown out of humoral pathology, see Barbara Duden, *The Woman beneath the Skin. A Doctor's Patients in Eighteenth-Century Germany* (Cambridge, Mass.: Harvard University Press, 1991).

[112] Michael Pickering, 'Experience as Horizon: Koselleck, Expectation and Historical Time,' *Cultural Studies* 18, no. 2–3 (2004).

[113] See Charles E. Rosenberg, *Explaining Epidemics and Other Studies in the History of Medicine* (Cambridge: Cambridge University Press, 1992), 1.

general,[114] the one exception being the thriving field of historical scholarship on disease, epidemics and contagion.[115] Even the buoyant literature on medicine trade and therapeutic exchange across the Atlantic basin that had taken shape already from the fifteenth century[116] has commonly been written along imperial lines, with studies focusing on the Dutch,[117] Spanish,[118] British[119] or Portuguese[120] contexts.

[114] In the history of health, disease and medicine, studies framed by familiar entities – the colony, or the nation-state, or a particular medical 'tradition' – still 'consistently and predictably' outweigh 'comparative', connected or 'global' studies. Jonathan Andrews, 'History of Medicine: Health, Medicine and Disease in the Eighteenth Century,' *Journal for Eighteenth-Century Studies* 34, no. 4 (2011), 505. For a more recent critique of the fact that 'surprisingly few works in the history of health, disease, and medicine can accurately be described as global histories or claim to be such', see Mark Harrison, 'A Global Perspective: Reframing the History of Health, Medicine, and Disease,' *Bulletin of the History of Medicine* 89, no. 4 (2015), 640.

[115] See Sanjoy Bhattacharya, 'Global and Local Histories of Medicine: Interpretative Challenges and Future Possibilities,' in *The Oxford Handbook of the History of Medicine*, ed. Mark Jackson (Oxford: Oxford University Press, 2011). The earliest, 'global' histories of disease, epidemics and contagion were Alfred W. Crosby, *The Columbian Exchange. Biological and Cultural Consequences of 1492* (Westport, Conn.: Greenwood, 1972); William H. McNeill, *Plagues and Peoples* (Garden City, N.Y.: Anchor Press, 1976). Global histories of particular diseases are a thriving field, too. See, for instance, Christopher Hamlin, *Cholera. The Biography* (New York: Oxford University Press, 2009); Webb, *Humanity's Burden*.

[116] For an overview of the literature on early modern Atlantic drug trade, see Harold J. Cook and Timothy Walker, 'Circulation of Medicine in the Early Modern Atlantic World,' *Social History of Medicine* 26, no. 3 (2013).

[117] See, for instance, A. M. G. Rutten, *Dutch Transatlantic Medicine Trade in the Eighteenth Century under the Cover of the West India Company* (Rotterdam: Erasmus Publishing, 2000); Harold J. Cook, *Matters of Exchange. Commerce, Medicine and Science in the Age of Empire* (Hyderabad: Orient Longman, 2008).

[118] There is a considerable breadth of scholarship on drug trade in the Spanish Atlantic from around 1500. See, for instance, J. Worth Estes, 'The Reception of American Drugs in Europe, 1500–1651,' in *Searching for the Secrets of Nature. The Life and Works of Dr. Francisco Hernández*, ed. Simon Varey, Rafael Chabrán and Dora B. Weiner (Stanford, Calif.: Stanford University Press, 2000); María Luz López Terrada and José Pardo Tomás, 'Las primeras noticias y descripciones de las plantas americanas (1492–1553),' in *Medicina, drogas y alimentos vegetales del Nuevo Mundo. Textos e imágenes españolas que los introdujeron en Europa*, ed. José María López Piñero et al. (Madrid: Ministerio de Sanidad y Consumo, 1998); José María López Piñero, 'Los primeros estudios científicos: Nicolás Monardes y Francisco Hernández,' in *Medicina, drogas y alimentos vegetales del Nuevo Mundo. Textos e imágenes españolas que los introdujeron en Europa*, ed. José María López Piñero et al. (Madrid: Ministerio de Sanidad y Consumo, 1998).

[119] There are several valuable studies of foreign drug imports, and consumption, in England. See, in particular, Patrick Wallis, 'Exotic Drugs and English Medicine: England's Drug Trade, c. 1550 – c. 1800,' *Social History of Medicine* 25, no. 1 (2011).

[120] See, for instance, Timothy Walker, 'The Early Modern Globalization of Indian Medicine: Portuguese Dissemination of Drugs and Healing Techniques from South Asia on Four Continents, 1670–1830,' *Portuguese Literary & Cultural Studies* no. 17/18 (2010); Timothy Walker, 'The Medicines Trade in the Portuguese Atlantic World: Acquisition and Dissemination of Healing Knowledge from Brazil (c. 1580–1800),' *Social History of Medicine* 26 (2013).

A tendency to limit the purview to one imperial context among historians of cinchona is also, however, intrinsic to the subject matter, and an effect of the bark's longstanding association with particular empires and states. Much of the post-1970s English-language historiography, for instance, refers to cinchona almost exclusively as the source of, and precursor to, quinine, a drug that has captured historians' imaginations owing to its alleged role in British imperial expansion in particular and in the high tide of European imperialism after 1878 more broadly. Malaria – or rather, tropical fevers retrospectively diagnosed as *P. falciparum* malaria – historians argued, had long represented perhaps 'the most powerful barrier to the projection of European influence in the tropics'.[121] Increasingly, systematic quinine therapy and prophylaxis after 1820, in reducing Europeans' mortality from the disease, enabled French and British colonizers to finally 'break into the African interior successfully'.[122] In a tendency perhaps symptomatic of a historiographical tradition enamoured with human scientific and technological ingenuity,[123] quinine was often considered, alongside submarine cables, breech-loaders and railroads, as yet 'another technological advance, a triumph over disease', as Daniel R. Headrick put it,[124] or, as Richard Drayton re-phrased it, the 'parable of the relations of mutual benefit struck between science and British Empire'.[125] Along those same lines, historians have devoted considerable attention to imperial rivalries, particularly the British, Dutch, French and Portuguese attempts at breaking the Spanish Empire's natural monopoly on cinchona. The British, Dutch and French quest for cinchona surrogates – plant components that promised similar effects, such as tulip tree bark, quassia, gentian root or Winter's bark – in

---

[121] Richard Drayton, *Nature's Government: Science, Imperial Britain and the 'Improvement' of the World* (New Haven: Yale University Press, 2000), 207.

[122] Lucile Brockway argued in 1997 that the 'availability of increased stores of quinine under British control had a [...] facilitating effect on the British colonial expansion into Africa in the late nineteenth and early twentieth centuries'. Brockway, *Science and Colonial Expansion*, 127–33. Brockway based her argument on the work of Philip Curtin, who pointed to the high share of 'malaria' in soldiers' 'relocation costs', and the supposed role of quinine consumption after 1840 in reducing these. Philip D. Curtin, *The Image of Africa. British Ideas and Action, 1780–1850*, vol. 1 (Madison: University of Wisconsin Press, 1964); Curtin, *Death by Migration*. Historians like Daniel R. Headrick later elaborated on these arguments. Daniel R. Headrick, *The Tools of Empire: Technology and European Imperialism in the Nineteenth Century* (New York: Oxford University Press, 1981), 58–79.

[123] See Tim Ingold, 'Toward an Ecology of Materials,' *Annual Review of Anthropology* 41 (2012), 432.

[124] Headrick, *The Tools of Empire*, 4.

[125] Drayton, *Nature's Government*, 230. See also Chakrabarti, *Medicine and Empire*, 126–28.

the Greater Caribbean and South Asia,[126] the Portuguese pursuit of cinchona varieties on Brazilian territory[127] and the long-standing Dutch and British attempts at smuggling and transplanting cinchona seedlings have long received comparatively abundant consideration.[128] So have, of course, the Spanish Empire's efforts at managing, regulating and preserving the harvest of and trade in cinchona. There are a series of valuable studies on its administrative organization,[129] the five – failed – projects aiming to establish a royal monopoly over the bark[130] and the politics of knowledge, science and expertise attendant to it.[131] Particularly in the latter field of study, historians of Spain's imperial project of economic botany – a Bourbon reform effort centred on the exploitation of profitable natural commodities, of which cinchona exports were one cornerstone – have studied the quest for experiences with and classification of new cinchona varieties in botanical expeditions and studies, especially those in the service of the Spanish Crown. In particular, the cinchona research carried out by José Celestino Mutis (1732–1808) in and beyond the framework of the Royal Botanical Expedition to the

---

[126] On 'Atlantic competitions' over natural commodities, including cinchona, see, Bleichmar, 'Atlantic Competitions.' See also Schiebinger, *Plants and Empire*, 146. On the British quest for substitutes, see Chakrabarti, 'Empire and Alternatives.' On the French quest for substitutes in Canada and Saint-Domingue, see McClellan and Regourd, *The Colonial Machine*, 260–62. Johann Reinhold Forster (1729–1898) and other naturalists for some time set their hopes on 'winter's bark' (*drimys winteri*). Anne Mariss, *A World of New Things. Praktiken der Naturgeschichte bei Johann Reinhold Forster* (Frankfurt: Campus Verlag, 2015), 140. On the quest for cinchona substitutes, see also Maehle, *Drugs on Trial*, 277; 80–81.

[127] Beltrão Marques, *Natureza em Boiões*, 132–35.

[128] See, for instance, Kavita Philip, 'Imperial Science Rescues a Tree: Global Botanic Networks, Local Knowledge and the Transcontinental Transplantation of Cinchona,' *Environment and History* 1, no. 2 (1995), 207–09; Brockway, *Science and Colonial Expansion*, 112–26. There are several somewhat dated, often 'triumphalist' accounts of British and Dutch cunning in smuggling the seeds out of South America. See, for instance, Donovan Williams, 'Clements Robert Markham and the Introduction of the Cinchona Tree into British India, 1861,' *The Geographical Journal* 128, no. 4 (1962); Hilda Knobloch, *Der Wunderbaum im Urwald. Wie die Chinarinde zum Allgemeingut der Menschheit wurde* (Wien: Eduard Wancura Verlag, 1954); Norman Taylor, *Cinchona in Java: The Story of Quinine* (New York: Greenberg, 1945).

[129] On the administrative structures governing cinchona production and trade within the Spanish Empire, see María Luisa Andrés Turrión and María Rosario Terreros Gómez, 'Organización administrativa del ramo de la quina para la Real Hacienda Española en el Virreinato de Nueva Granada,' in *Medicina y Quina en la España del siglo XVIII*, ed. Juan Riera Palmero (Valladolid: Universidad de Valladolid, 1997).

[130] On the five projects after 1752, see Martine Petitjean and Yves Saint-Geours, 'La ecomomía de la cascarilla en el Corregimiento de Loja (Segunda mitad del siglo XVIII-Principios del siglo XIX),' *Revista Cultural del Banco Central del Ecuador* 5, no. 15 (1983), 46.

[131] Crawford, *The Andean Wonder Drug*; Nieto Olarte, *Remedios para el imperio*, 163–206.

Kingdom of Granada (1783–1816),[132] by Jorge Juan y Santacilia (1713–1773) and Antonio de Ulloa (1716–1795), members of the Charles-Marie de La Condamine expedition (1735–1745),[133] and by the 1778–1816 botanical expedition to the Viceroyalty of Peru headed by Hipólito Ruiz López (1754–1816), José Antonio Pavón (1754–1840) and Joseph Dombey (1742–1794) has attracted considerable attention,[134] even if mostly on the margins of studies of these men's wider botanical interests. Scholars from other linguistic or national backgrounds have, likewise, been concerned primarily with the bark's reception in their various domestic contexts: Finland,[135] the Ottoman Empire[136] or the Habsburg territories, with the German-language literature converging on Samuel Hahnemann's (1755–1843) self-experimentation with cinchona and its part in the formation of homeopathy.[137] In virtually all of these

[132] Gonzalo Hernández de Alba, *Quinas Amargas. El sabio Mutis y la discusión naturalista del siglo XVIII* (Bogotá: Academia de Historia de Bogotá / Tercer Mundo Editores, 1991); José Antonio Amaya, ed., *Mutis, Apóstol de Linneo. Historia de la botánica en el virreinato de Nueva Granada (1760–1783)*, 2 vols. (Bogotá: Instituto Colombiano de Antropología e Historia, 2005); Manuel Salvador Vázquez, 'Mutis y las quinas del norte de Nueva Granada,' in *Medicina y Quina en la España del siglo XVIII*, ed. Juan Riera Palmero (Valladolid: Universidad de Valladolid, 1997); Marcelo Frias Núñez, *José Celestino Mutis y la real expedición botánica del nuevo reino de Granada, 1783–1808* (Sevilla: Diputación Provincial de Sevilla, 1994); Daniela Bleichmar, *Visible Empire. Botanical Expeditions and Visual Culture in the Hispanic Enlightenment* (Chicago: University of Chicago Press, 2012).

[133] Luis Javier Ramos Gómez, *El viaje a América (1735–1745), de los tenientes de navío Jorge Juan y Antonio de Ulloa, y sus consecuencias literarias* (Madrid: Consejo Superior de Investigaciones Científicas, Instituto Gonzalo Fernández de Oviedo, 1985), 265–76; Dora Leon Borja, 'Algunos datos acerca de la cascarilla ecuatoriana en el siglo XVIII,' in *Medicina y Quina en la España del siglo XVIII*, ed. Juan Riera Palmero (Valladolid: Universidad de Valladolid, 1997); Boumediene, *La colonisation du savoir*; Eduardo Estrella, 'Introducción de la quina a la terapeutica: misión geodesica y tradición popular,' *Revista de la Facultad de Ciencias Médicas – Quito* 14, no. 1–4 (1989); Paloma Ruiz Vega, 'La quina en la expedición geodésica al Virreinato de Perú (1734–1743),' in *Las Cortes de Cádiz, la Constitución de 1812 y las independencias nacionales en América*, ed. Antonio Colomer Viadel (Valencia: Universidad Politécnica de Valencia, 2011).

[134] Félix Muñoz Garmendia, ed. *La botánica al servicio de la corona. La expedición de Ruiz, Pavón y Dombey al Virreinato del Perú (1777–1831)* (Madrid: CSIC / Real Jardín Botánico, 2003); Arthur Robert Steele, *Flowers for the King: the Expedition of Ruiz and Pavón and the Flora of Peru* (Durham: Duke University Press, 1964); Cesar Gonzalez Gomez, *Aspectos de la labor quinológica de los botánicos Ruiz y Pavón* (Madrid: Imprenta Góngora, 1954); Bleichmar, 'Atlantic Competitions.'

[135] Lena Huldén, 'The First Finnish Malariologist, Johan Haartman, and the Discussion about Malaria in 18th century Turku, Finland,' *Malaria Journal* 10, no. 43 (2011).

[136] Günergun and Etker, 'From Quinaquina to "Quinine Law."'

[137] Birgit Lochbrunner, *Der Chinarindenversuch. Schlüsselexperiment für die Homöopathie* (Essen: KVC Verlag, 2007). See also Georg Bayr, *Hahnemanns Selbstversuch mit der Chinarinde im Jahre 1790. Die Konzipierung der Homöopathie* (Heidelberg: K.F. Haug, 1989).

studies, developments elsewhere serve principally as a distant backdrop against which a given intellectual, political or economic history may unfold – be that elsewhere in the Spanish American natural habitat, as in the case of British historians of empire, or the consumer societies, for Spanish historians of cinchona production and commerce. While it does not deny the relevance of imperial frameworks, or, indeed, the bark's implication in them, the purpose of this book is to bring together, synergistically, cinchona's many elsewheres, in pursuing its object 'across time, space, and specialism'.[138] Inspired by a growing body of research on plant trade, epistemic brokerage and therapeutic exchange across imperial boundaries,[139] *A Singular Remedy* does not settle for explanations drawn from one territory, denomination or linguistic framework. As various historians have argued, though medicine trade across the Atlantic basin was lively and extensive around 1800, few medicinal substances travelled as widely and were traded as massively as the Peruvian bark.[140] With its range and reach fully understood and explored – unfettered by a reductionist emphasis on its collusion, and complicity, with either the British or the Spanish Empire – the bark provides a rare, and valuable, window into drug trade, epistemic brokerage and sustained interaction[141] in the realm of medicine during the late 1700s and early 1800s.

Book Structure

This book covers the period in which the bark was at the height of its popularity: the decades running from 1751, the year when the Spanish Crown issued the royal order that declared cinchona to be 'an object worthy of interest, curiosity and attention',[142] to 1820, when the

---

[138] James Belich, John Darwin and Chris Wickham, 'Introduction. The Prospect of Global History,' in *The Prospect of Global History*, ed. James Belich et al. (New York: Oxford University Press, 2016), 3; Matthew W. Klingle, 'Spaces of Consumption in Environmental History,' *History and Theory. Studies in the Philosophy of History*, no. 42 (2003), 94.

[139] See, for instance, Sarah Easterby-Smith, *Cultivating Commerce. Cultures of Botany in Britain and France, 1760–1815* (Cambridge: Cambridge University Press 2017); Miruna Achim, *Lagartijas medicinales. Remedios americanos y debates científicos en la Ilustración* (Mexico: Conaculta/UAM-C, 2008); Monahan, 'Locating Rhubarb.'; Irina Podgorny, 'The Elk, the Ass, the Tapir, Their Hooves, and the Falling Sickness: A Story of Substitution and Animal Medical Substances,' *Journal of Global History* 13, no. 1 (2018).

[140] For an extensive discussion of this claim, see Chapter 2.

[141] For this paraphrase of 'globalization', see Dennis O. Flynn and Arturo Giráldez, 'Born Again: Globalization's Sixteenth Century Origins (Asian/Global versus European Dynamics),' *Pacific Economic Review* 13, no. 4 (2008), 360.

[142] See the 'Royal Order' issued in Madrid, dated as of August 27, 1751, and addressed to the Viceroys of Peru and New Granada, which proposed the creation of a Royal

dislocations of the struggle for independence in the harvest areas and the isolation of quinine, which unfolded through a sequence of experiments conducted in Lisbon, Paris and Jena,[143] changed the grounds of its production, commerce and consumption. Geographically, it covers an Atlantic World constituted by relations among the continents rimming the Atlantic basin and with other regions, especially the Asian reaches of the Atlantic empires.[144] Its focus rests in particular on the Viceroyalties of Peru, Brazil, New Granada and New Spain, the Dutch, British and French West Indian possessions, and the French and British North American colonies – or, after 1776, the United States; on the Portuguese, Spanish and British enclaves along the African coast, the Sultanate of Morocco and the Ottoman Empire; on France, England and Scotland, the Habsburg territories, Scandinavia, the Swiss Confederacy, the Italian peninsula and Muscovy; on the Spanish, Portuguese, French, British and Dutch colonial possessions and commercial and evangelizing entrepôts in Qing China, Mughal India and Tokugawa Japan, on Java and the Philippines.[145]

At its core, *A Singular Remedy* is concerned with how the Peruvian bark and stories, practices and understandings attendant to its consumption were shared between and across Atlantic societies. The five chapters that follow expose and examine the prevalence and movement of narratives about the discovery of the bark's medicinal properties, of the Peruvian bark as a form of matter, of medical formulae for preparations of the bark, and of understandings of the environs and ailments in which the

---

Monopoly on the bark. 'Real Cedula,' *Archivo General de Indias*, Indiferente 1552, Madrid, 1751-08-27. Cited in Crawford, *The Andean Wonder Drug*, 67. Crawford has studied the order in depth. Crawford, 'Para Desterrar las Dudas y Adulteraciones,' 196.

[143] Quinine was 'discovered' by Pierre-Joseph Pelletier (1788–1842) and Joseph Caventou (1795–1877) through repetition of an experiment devised by the Portuguese naval surgeon Bernardino António Gomes. Walter Sneader, *Drug Discovery: The Evolution of Modern Medicines* (Chichester: John Wiley & Sons, 1985). Gomes published his finds in his 1811/1812 essay 'Ensaio sobre o cinchonino, e sobre sua influencia na virtude da quina e de outras cascas', in the journal *Memórias de ciencias*, edited by the Lisbon Academy of Science. Beltrão Marques, *Natureza em Boiões*, 135.

[144] Historians of the Atlantic World have argued that several, self-sufficient settings were gradually absorbed into a single, interdependent Atlantic World by the eighteenth century. Nicholas Canny and Philip Morgan, 'Introduction. The Making and Unmaking of an Atlantic World,' in *The Oxford Handbook of the Atlantic World: 1450–1850*, ed. Nicholas Canny and Philip Morgan (Oxford: Oxford University Press, 2011), 2.

[145] The 'Atlantic powers', or 'Euro-Atlantic states', were all involved in commerce with regions beyond the Atlantic basin. Bernard Bailyn, 'Introduction. Reflections on Some Major Themes,' in *Soundings in Atlantic History. Latent Structures and Intellectual Currents, 1500–1830*, ed. Bernard Bailyn and Patricia L. Denault (Cambridge, Mass.: Harvard University Press, 2009), 9–10.

use of the Peruvian bark would be most beneficial. *A Singular Remedy* contends not only that bark knowledge – in the shape of matter, words and practices – was movable but that it moved in ways that were contingent upon place and locality – a peculiar culinary lore, cultural imaginary or medical topography. Chapter 1 not only exposes the various story elements present in narratives about the bark's discovery – the natives' alleged secrecy, their closeness to nature and unlettered simplicity – as long-lived *topoi* that served to make sense of, and propagate, the bark's wonderful properties. It also points to these stories' divergent reception across the Atlantic World and the part of cultural, religious or political idiosyncrasies in it. Chapter 2 exposes and examines how bottled compound wines and powdered bark moved along the veins of Atlantic trade, proselytizing and imperialism, with their course defined by the situation of these societies' trade entrepôts, military outposts and diaspora communities. Chapter 3 contends that even though methods for arranging and administering the bark had coalesced into identifiable formulae by the late 1700s and early 1800s – bittersweet febrifugal lemonades, extracts of cinchona and aromatic compound wines of the bark, most notably – these preparations also accommodated a measure of variability. Indeed, medical practitioners tinkered with the particulars of these formulae, adapting them to the religious beliefs, peculiar culinary lore or commercial possibilities of their place of abode. Chapter 4 exposes how bark knowledge was shared in the form of expertise in indications for cinchona, a topographic literacy of sorts that associated certain environments with febrile threat. The chapter holds not only that the period's medical topography, with its distinct contours of insalubrious, febrile environments, directed the bark to particular situations – the world's low-lying marshlands, the sickly air of close, crowded spaces, and the hot and humid climates of the tropics. It also exposes how sufferers adapted modes of administration, depending on the season, place or climate they sought to shield themselves from. There is an unspoken premise in much current scholarship that the local and the global are opposites – the dichotomy of a discrete locale that resists change and a placeless global that imposes it.[146] Not only is that polarity a figment of our scholarly imagination; it is a detrimental one, since it diverts attention from the contingency of both knowledge and its movement. By the late 1700s and early 1800s, knowledge and use of the Peruvian bark was common across a wide range of geographically disperse and socially diverse societies

---

[146] This may well be an inheritance from the globalization discourse of the 1990s, which popularized the 'idea of locality [...] as a form of opposition or resistance to the [...] global'. For a critical discussion of that polarity, see Robertson, 'Glocalization,' 30; 34.

within, or tied to, the Atlantic World, in part because of that substance's ability to acquire validity, become situated, and weave itself into the fabric of everyday therapeutic practice elsewhere. Both knowledge of the bark and its global movement were, so this book contends, local – as in, related to place, and peculiar to it.

Chapter 5 reminds readers, at parting, once more of how plant trade, therapeutic exchange and epistemic brokerage are not extricable from space. Written in the style of a lengthy coda, it is concerned with how the bark's prevalence, wide fame and general usefulness in therapeutic practice among disperse societies, affected its natural habitat in the central and northern Andes. The last chapter argues that the bark's very mobility, and the popular demand that arose for it, altered the area's landscape of possession, commerce and demographics; the distribution and abundance of vegetation; and the livelihood, health and fate of the men and women implicated in harvesting, processing and conveying the bark. Consumption, and the imaginaries, therapeutic practices and medical understandings attendant to it, it contends, invariably begins with changes to the material world, to physical nature and to society.[147]

---

[147] Klingle, 'Spaces of Consumption in Environmental History,' 94.

1   Origin Stories

The grandest and most valuable acquisitions in medicine and surgery
have been discovered by accident, and often by the most illiterate and
unobserving part of mankind.
                    – Thomas Collingwood, *Observations on the Peruvian Bark*, 1785.

By the late 1700s and early 1800s, there were several stories in general
circulation of how the wonderful, even divine medicinal properties of
cinchona bark had first come to the knowledge of mankind. Hispanic
treatises on the bark, by naturalists like José Celestino Mutis and
Hipólito Ruiz López, usually related that the Indians of Loja had known
of the bark's medicinal virtues long before the arrival of the Spanish, and
around the year 1636 apprised the province's mayor, ill with an
intermittent fever, of them. The latter then passed the intelligence on
to the Viceroy and his wife, who had long suffered from that same
ailment. The Vicereine, after experiencing the bark's 'wonderful effects'
(*maravillosos efectos*), communicated her knowledge of its virtues to
members of the Jesuit Order.[1] Accounts from beyond the Spanish
Empire – French travel writings, Boston almanacs, Swedish advice
manuals and Italian pharmacopoeias – had by the decades around
1800 come to favour a slightly more elaborate and dramatic version of
that origin story. Most of them drew on the narrative put into circulation,
at least in its general outline, by the French naturalist Charles-Marie de
La Condamine with his 1738 report 'On the Cinchona Tree' (*Sur l'arbre
du quinquina*) to the Royal Academy of Sciences in Paris (FIGURE I.I).[2]
According to an ancient tradition, La Condamine's story went, the
veracity of which he would not vouch for, 'the natives of America
(*les Américains*) owed the discovery of this remedy to lions (*dûrent la
découverte de ce reméde aux Lions*)' – animals prone to a kind of

---

[1] Ruiz López, *Quinología*, 4; Hernández de Gregorio, *El arcano de la quina*, iii.
[2] According to Matthew Crawford, La Condamine was the first to add information on how
the Indians had presumably discovered the virtues of cinchona bark. Crawford, *Empire's
Experts*, 17–18.

Figure 1.1 *Dessein d'une branche de l'arbre du Quinquina, avec ses feuilles, ses fleurs & les fruits, en leurs divers états* (Sketch of a cinchona tree branch, with its leaves, flowers & fruits, in their various states), 1737. Charles M. de la Condamine, 'Sur l'arbre du quinquina,' *Mémoires de l'Academie Royale* MDCCXL.
Credit: Wellcome Collection.

intermittent fever. It was said, La Condamine wrote, that 'the people of the country, having noticed that the animals ate the bark of the cinchona tree, used it when they [themselves] succumbed to intermittent fevers (*fiévres d'accès*) [...] and [thus] came to understand its salutary virtue'. La Condamine concluded his tale with a passage presumably familiar to much of his readership, from one of the earliest and most influential written narratives of the bark's passage into Old World materia medica, Sebastianus Badus's 1663 *Anastasis Corticis Peruviae.*[3] According to Badus, who relied, in turn, on a letter purportedly written to him by the Genoese merchant Antonius Bollus, the Indians had long known of the bark's virtues but did 'not make the remedy known to the Spanish' (*ne Hispanis hominibus innotesceret Remedium*) because they were 'bitter'

---

[3] Sebastianus Badus, *Anastasis corticis Peruviae; seu chinae chinae defensio, Sebastiani Badi Genvensis patrij vtriusque Nosochomij olim medici, et Publicae Sanitatis in Ciuitate Consultoris. Contrà ventilationes Joannis Jacobi Chifletii, gemitusque Vopisci Fortunati Plempii, Illustrium Medicorum* (Genoa: Typis Petri Joannis Calenzani, 1663), 21–24.

(infensi) toward them.[4] To the rest of the world, La Condamine continued – and from thence his account coincided with the Hispanic version propagated by Mutis and Ruiz López –

the usefulness of that remedy only became known on occasion of an obstinate tertian fever of which the Countess of Chinchón, Vicereine of Peru, could not recover for several months. [...] the remedy [...] emerged from obscurity when the mayor (*corregidor*) of Loja [...] dispatched some cinchona to the Viceroy, assuring him that he would answer for the Vicereine's recovery, if she were given that febrifuge. [...] the Vicereine took the remedy and was cured. She immediately ordered from Loja a quantity of the same bark and [...] sent it to the friars of the Jesuit Order, who [...] on occasion of a visit of the procurator of the Province of Peru to Rome, sent [...] it to Cardinal Lugo of the same Society.[5]

This chapter exposes and examines the movement of narratives, and cultural imaginaries, about the bark across the Atlantic World, emphasizing its limits, idiosyncrasy and contingencies. The chapter is concerned in particular with the circulation of origin myths around 1800, about how the wonderful, admirable and divine medicinal properties of cinchona bark had first come to the knowledge of mankind. Earlier historians with an interest in the origin myths surrounding cinchona bark have tended to focus on exposing errors in the various historical traditions surrounding the bark. Much of the scholarship has assessed the, indeed, doubtful veracity of the legend of the Countess of Chinchón, of contemporaneous, rival, but afterwards largely forgotten, Iberian narratives by Gaspar Caldera de Heredia (b. 1591) and Pedro Miguel de Heredia (d. 1661), as well as later accounts put into circulation by travellers like Joseph de Jussieu (1704–1779), and, indeed, La Condamine.[6] Rather than highlighting inconsistencies and inaccuracies in the narratives, or seeking to uncover the actual circumstances of how cinchona was first attributed healing properties, this chapter is concerned

---

[4] Ibid., 21–22.

[5] Charles-Marie de la Condamine, 'Sur l'arbre du quinquina,' *Mémoires de l'Academie Royale* MDCCXL (1738 (1737)), 233–34.

[6] Jaramillo-Arango, 'A Critical Review'; Haggis, 'Fundamental Errors.' For a concise summary of the debate about 'errors', see Jarcho, *Quinine's Predecessor*, 2–8. See also José María López Piñero and Francisco Calero, introducción to *De Pulvere Febrifugo Occidentalis Indiae (1663), de Gaspar Caldera de Heredia y la introducción de la Quina en Europa*, ed. José María López Piñero and Francisco Calero (Valencia: Instituto de Estudios Documentales e Históricos sobre la Ciencia / Universidad de Valencia, 1992); Boumediene, *La colonisation du savoir*, 196–200; Fernando Ortiz Crespo, *La corteza del árbol sin nombre. Hacia una historia congruente del descubrimiento y difusión de la quina* (Quito: Fundación Fernando Ortiz Crespo, 2002), 11–43. For assessments of the later, eighteenth-century accounts, see, for instance, Honigsbaum and Willcox, 'Cinchona,' 25–26.

with the truths that the various untruths in general currency by the late 1700s and early 1800s reveal, through their very popularity and prevalence, about contemporary conceptions of the medicinal plant knowledge of America's Indians. In a first section, the chapter scrutinizes the story element of the Indians' alleged secrecy and bitterness towards the Spanish, exposing it as a widespread and long-lived topos. The second section examines the narrative of how the Indians had arrived at their knowledge of the bark's virtues by instinct, by the observation and emulation of nature. The third section shows that the rendering of native knowers as secretive and reliant on instinct served to heighten the bark's value and appeal, at a time when an Atlantic Enlightenment[7] both fashioned and fetishized the knowledge of 'wild Indians'. Laying the ground for subsequent chapters on the bark's wide distribution, the chapter argues that origin myths, in their appeal to contemporary sensibilities, made sense of and ultimately promoted ideas about cinchona's extraordinary efficacy and wonderful properties, while at the same time reifying ideas about indigeneity, epistemic hierarchies and geographies of knowledge. The chapter holds that the production of meaning about the bark[8] contributed to the bark's popularity, and wide spread in the Atlantic World – as much as, or more so than, the discerning appreciation of its supposed medical effects[9] or the support of prominent regal and religious advocates[10] historians have tended to privilege in their accounts of the bark's rise.

---

[7] On the 'Atlantic Enlightenment', as a 'framework intrinsic to the articulation of the modern world', 'perceived and described' by eighteenth-century intellectuals: Susan Manning and Francis D. Cogliano, 'Introduction. The Enlightenment and the Atlantic,' in *The Atlantic Enlightenment*, ed. Susan Manning and Francis D. Cogliano (London: Routledge, 2008), 1. Manning and Cogliano focus, 'by design, [on] a North Atlantic Enlightenment', 'protestant in denomination and Anglophone in orientation'. See ibid., 9. Recent investigations have, however, opened the field toward the southern Atlantic. See, for instance, James Delbourgo and Nicholas Dew, *Science and Empire in the Atlantic World* (New York: Routledge, 2008); Neil Safier, *Measuring the New World. Enlightenment Science and South America* (Chicago: University of Chicago Press, 2008).

[8] Paula Findlen reminds us to retrace the itinerary of the representations of things to understand 'the production of meaning that frequently has a separate life from the object in question'. Paula Findlen, 'Early Modern Things. Objects in Motion, 1500–1800,' in *Early Modern Things. Objects and their Histories, 1500–1800*, ed. Paula Findlen (London: Routledge, 2013), 20.

[9] See the section 'An Appraisal of the Historiography' in the introduction to this volume.

[10] According to Saul Jarcho, 'to the extent that a single factor can be isolated, it is the influence of royalty and nobility that produced the enduring acceptance of the bark'. Jarcho, *Quinine's Predecessor*, 74. There are several studies on Louis XIV and other French royal advocates who helped popularize the bark; see Stanis Perez, 'Louis XIV et le quinquina,' *Vesalius* IX, no. 2 (2003); Stanis Perez, 'Les médicins du roi et le quinquina aux XVIIe et XVIIIe siècles,' in *La santé des populations civiles et militaires. Nouvelles approches et nouvelles sources hospitalières, XVIIe–XVIIIe siècles*, ed. Élisabeth

Unalienable Truths

While most almanacs, advice manuals and medical treatises that referred to the discovery of cinchona's medicinal properties by the late 1700s and early 1800s followed the general contours of La Condamine's story, authors more often than not introduced minor variations, further detail or alterations according to their circumstances, creed or caprice. Many were rather imaginative and accordingly differed on the presumed reasons for the Indians' betrayal of the secret the latter had, by all accounts, 'sworn to keep among themselves, because of their hatred of their enemies':[11] the secret of the only remedy that could save these enemies from death due to the intermittent fevers endemic in these parts.[12] While George Motherby (1732–1793), the author of the 1775 London *New Medical Dictionary*, cited an Indian's compassion for a sick Spaniard,[13] others, like the Edinburgh-trained physician Thomas Collingwood (1751–1822), suggested that 'the fraudulent Spaniards' had extorted the secret of cinchona's effects from the natives by force.[14] Again, others, like the Paris medical student Julien Dufau (1782–1859), believed that the Spanish owed their knowledge of cinchona's medicinal properties to some kind of betrayal by an Indian unfaithful to his – or, indeed, her – oath.[15] As a matter of fact, it was commonly suspected that a Peruvian woman, 'enamoured with a Spaniard, and afraid to see her lover dying from intermittent fever, had revealed to him (*lui découvrit*) the means of finding relief with the aid of cinchona'.[16] The theme of the Peruvian woman betraying her knowledge of the bark's virtues – on account of love, friendship or compassion – even found its way into the realm of literature and the theatre. In Félicité de Genlis's (1746–1830)

Belmas and Serenella Nonnis-Vigilante (Villeneuve-d'Ascq: Presses Universitaires du Septentrion, 2010). On the role of religious supporters, particularly the Jesuit order, in the popularization of the bark in Europe, South and East Asia and across the Americas, see also Sabine Anagnostou, *Jesuiten in Spanisch-Amerika als Übermittler von heilkundlichem Wissen* (Stuttgart: Wissenschaftliche Verlagsgesellschaft, 2000).

[11] Francois Citte, 'De l'usage du quinquina et des règles de son application dans les fièvres intermittentes,' in *Collection des thèses soutenues a l'école de médecine de Montpellier pendant l'an XII*, ed. L'École de Médecine de Montpellier (Montpellier: G. Izar et A. Richard, 1804), 4.

[12] Julien Dufau, *Essai sur l'application du quinquina dans le traitement des fièvres intermittentes*, *Collection des theses soutenues a l'ecole de médecine de Paris* (Paris: Didot Jeune, 1805), 10.

[13] George Motherby, *A New Medical Dictionary; Or, General Repository of Physic Containing an Explanation of the Terms, and a Description of the Various Particulars Relating to Anatomy, Physiology, Physic, Surgery, Materia Medica, Pharmacy &c. &c. &c.* (London: J. Johnson, 1775), entry CORTEX PERUVIANUS.

[14] Thomas Collingwood, 'Observations on the Peruvian Bark,' *Medical Commentaries* X (1785), 265–66.

[15] Dufau, *Essai sur l'application du quinquina*, 10.   [16] Citte, 'De l'usage du quinquina,' 4.

1818 novel 'Zuma, or the Discovery of Cinchona' (*Zuma, ou la découverte du quinquina*), which was later adapted for the stage and the opera in Britain, France and the Netherlands, Zuma, 'the most beautiful of Indian women', and a faithful protégé of the dying Countess, dispensed the 'precious powder' to save her benefactress' life.[17]

Not all societies adopted the trope of the Indians' secrecy and betrayal as eagerly, or directly, as that, in ways that point to idiosyncrasies and contingencies in the circulation of stories, their translation, and their adaptation. Whereas some societies beyond Europe and its colonial, evangelizing and commercial entrepôts espoused the bark without attending to fanciful narratives about its origins,[18] others adopted altered or distorted versions of them. The account by the Ottoman medical author Bursalı Ali Münşî (d. 1733), for instance, though it featured a Spanish governor's sick wife, the Jesuit order and the Spanish conquerors' cruelty towards the inhabitants of the New World, diverged from the established narrative in that, according to him, the Spanish had kept the bark a secret from the natives in the hope of seeing the latter perish from fevers, leaving their land free for the taking.[19] Though presumably a simple misreading of his French, English and Italian sources, the variation also suggests that, to authors like Ali Münşî, the association between secrecy and indigeneity was not a self-evident or obvious one. Hispanic writers, though for different reasons, were also less eloquent regarding the Indians' secrecy and hatred of their enemies than scholars, novelists and playwrights from Britain, the Netherlands and France.[20] Indeed, few authors were quite as explicit, and imaginative, as those from the British, Dutch, and French territories, where the Black Legend – the myth of Spain's unique brutality in the conquest of the New World that began in sixteenth-century Protestant hostility[21] – ensured the smooth

---

[17] Stéphanie Félicité du Crest de Saint-Aubin, Countess de Genlis, *Zuma, ou la découverte du quinquina* (Paris: Maradan, Libraire, 1818), 10–11; 21. For examples of theatrical and operatic adaptations, see, for instance, the comic opera *Zuma, or the Tree of Health*, first performed February 21, 1818, London, Covent Garden. Margaret Ross Griffel, 'ZUMA, or, the Tree of Health,' in *Operas in English: A Dictionary*, ed. Margaret Ross Griffel (Plymouth: Scarecrow Press, 2013). For a Dutch adaptation, see Jan de Quack, *Zuma, of De ontdekking van den kina-bast: tooneelspel in vier bedrijven* (Amsterdam: J.C. van Kesteren, 1819).

[18] There is no evidence of the transfer of 'origin stories' to China, for instance. Unschuld, *Medicine in China*, 166.

[19] Aydüz and Yildirim, 'Bursalı Ali Münşî ve Tuhfe-i Aliyye,' 91.

[20] Boumediene, *La colonisation du savoir*, 200.

[21] Walter D. Mignolo, Margaret R. Greer and Maureen Quilligan, introduction to *Rereading the Black Legend. Discourses of Religious and Racial Difference in the Renaissance Empires*, ed. Walter D. Mignolo, Margaret R. Greer and Maureen Quilligan (Chicago: University of Chicago Press, 2007).

and willing acceptance of the idea that America's native inhabitants had guarded their knowledge of the bark's virtues out of their 'revulsion against the barbarism of their conquerors' (*révoltés de la barbarie de leurs vainqueurs*).[22] So would British, Dutch, Prussian and French diatribes against the inferiority of living nature in the Americas, and the cowardice and hostility of its 'savage' inhabitants, in the context of the notorious Enlightenment Dispute of the New World.[23] It was only, tellingly perhaps, in the writings of creole, that is, Hispanic, American-born authors that secretive Indians occasionally made their appearance by the late 1700s and early 1800s. As José Hipólito Unanue (1755–1833) put it in his 1791 essay on the nascent discipline of botany in the viceroyalty of Peru, the excellent 'empirical [...] medicine' and knowledge of the virtues of plants of the 'primitive inhabitants of Peru' was, to him and his kind, virtually lost. For, in the wake of the upheaval of the conquest, the Indians' 'temper' (*genio*) had become 'mysterious, stubborn and mistrustful' and led them to conceal their knowledge from the Europeans. Neither flattery nor bribery nor threats of violence would induce them to betray their secrets now and it was only sometimes, through the imprudence or treason of one of them, that their knowledge about the virtues of medicinal plants was ever unveiled to outsiders.[24]

Like his northern European counterparts, Unanue had not met in person the secretive Indians he described. In line with some of his contemporaries, the evidence he cited for his assertion that the Indians concealed their knowledge from the Europeans was a 1568 letter from a Lima-based soldier – Pedro de Osma – transcribed in part two of the

[22] Dufau, *Essai sur l'application du quinquina*, 10. See also Collingwood, 'Observations on the Peruvian Bark,' 265–66.
[23] The classic work on the subject is still that of Antonello Gerbi. Antonello Gerbi, *The Dispute of the New World: The History of a Polemic, 1750–1900*, trans. Jeremy Moyle (Pittsburgh, Pa.: University of Pittsburgh Press, 1973 (1955)), 5. On Spanish America in particular, see also David Brading, *The First America. The Spanish Monarchy, Creole Patriots, and the Liberal State 1492–1867* (Cambridge: Cambridge University Press, 1991), 422–64. For reflections on the subject of that dispute's source base, see Jorge Cañizares Esguerra, *How to Write the History of the New World: Histories, Epistemologies, and Identities in the Eighteenth-Century Atlantic World* (Stanford, Calif.: Stanford University Press, 2001), 235.
[24] José Hipólito Unanue, 'Botánica. Introducción a la descripción científica de las plantas del Perú, Vol. II, núms. 43 y 44 (29 de mayo y 2 de junio de 1791),' in *El Mercurio Peruano: 1790–1795. Antología*, ed. Jean-Pierre Clément (Frankfurt am Main: Vervuert, 1998 (1791)), 96. For a study of that essay as one attempt at assimilating and promoting a Linnaean taxonomy in the Viceroyalty of Peru, see Antonio Gonzàlez Bueno, 'Plantas y Luces: la Botánica de la Ilustración en la América Hispana,' in *La formación de la Cultura Virreinal. El siglo XVIII*, ed. Karl Kohut and Sonia V. Rose (Madrid: Iberoamericana, 2006), 197–28. On discourses about the Indians' mysterious, mistrustful and secretive character in Unanue's work and among Peru's creole elites more broadly, see Natalia Majluf, "The Creation of the Image of the Indian in 19th-Century Peru: The Paintings of Francisco Laso (1823–1869) (unpublished PhD dissertation, University of Texas, 1996), 345–359.

1574 'Medicinal History of the New World' (*Historia Medicinal de las Cosas que se traen de nuestras Indias Occidentales que sirven en Medicina*) authored by Nicolás Monardes (1493–1588), a Seville physician and trader who had never crossed the Atlantic.[25] De Osma, supposedly a reader of the first part of Monardes's medicinal history, who had followed the physician's exhortation to investigate native medicinal knowledge in the New World, allegedly encountered much reluctance among the Indians to divulge their admirable understanding of the virtues of medicinal plants and animal substances. When he was seeking to understand how to extract bezoar stones from vicuñas, de Osma lamented, he could obtain no intelligence from the Indians, 'because they are bitter towards us and do not want us to know their secrets'. It was an Indian boy who revealed to de Osma and his companions the method for extracting the bezoar stones, and in 'that instant, his countrymen threatened to slay him'.[26] 'Regardless of kind words and gifts, and of fierce words and threats', de Osma wrote, the Indians, 'being wicked and our enemies, will not disclose one secret, or one virtue of a herb, even if they were to see us dying.' Other than from children, de Osma told Monardes in the 1568 letter, the conquerors only learned about the medicinal virtues of America's plants from those 'Indian women, who get involved with the Spaniards and reveal and tell them everything they know'.[27] De Osma's letter was not the first intimation of American Indians' secrecy. Already the earliest natural history of the New World, Gonzalo Fernández de Oviedo y Valdés's (1478–1557) 1535 'General, and Natural History of the Indies' (*Historia general y natural de las Indias*)', had made reference to how the Indians were 'covetous (*avara*)' about their knowledge of medicinal herbs and plants,

---

[25] The letter was transcribed in part II of Nicolás Monardes's 'Medicinal History (*Historia Medicinal*)'. See Nicolás Monardes, *Primera y segunda y tercera partes de la historia medicinal, de las cosas que se traen de nuestras Indias Occidentales, que siruen en Medicina* (Sevilla: Alonso Escriuano, 1574), 73–77. For similar contemporary texts, see the passage on 'the botanical science in Peru' in Joseph Skinner's 'The Present State of Peru', which bears a striking affinity with the corresponding passage in Unanue's 1791 article, referring to the 'primitive inhabitants'' 'empiric practice of medicine' and their 'mysterious, tenacious and mistrustful disposition'. Skinner, too, referred his readership to Pedro de Osma's 1568 letter to document his assertion that the Indians were 'mysterious, tenacious and mistrustful'. Joseph Skinner, *The Present State of Peru: Comprising its Geography, Topography, Natural History, Mineralogy, Commerce, the Customs and Manners of its Inhabitants, the State of Literature, Philosophy, and the Arts, the Modern Travels of the Missionaries in the Heretofore Unexplored Mountain Territories, &c. &c. the Whole Drawn from Original and Authentic Documents, Chiefly Written and Compiled in the Peruvian Capital* (London: Richard Phillips, 1805), 43–44. For biographical notes on Monardes and the importance of his residence in Seville, see Marcy Norton, *Sacred Gifts, Profane Pleasures. A History of Tobacco and Chocolate in the Atlantic World* (Ithaca: Cornell University Press, 2008), 107–12.

[26] Monardes, *Historia medicinal*, 74.   [27] Ibid., 77.

and 'the secrets of nature', 'especially about whatever could be useful to the Christians'.[28]

The veracity, and accuracy, of Fernández de Oviedo y Valdés's account is doubtful, and so is that of de Osma's, or rather Monardes's, assertions. As a matter of fact, historians of Iberian science have cited the 1568 letter with some reserve, since there is no evidence to prove it was genuine. De Osma may well have been a creature of Monardes's imagination.[29] The early and ready acceptance of the idea that the Indians were secretive, despite its foundation on rather unstable ground, would appear inconsistent, were it not for the possibility that in this, as in so many other instances, familiar, ancient Old World paragons exercised a strong hold over early modern observers' minds and their representations of the New World.[30] Indeed, the secretive Indians' resemblance to other illiterate knowers in late medieval and early modern medical thought is conspicuous, and likely eloquent. From the earliest days of colonial rule in the Americas, rural Indians – not any Indian, for the category encompassed, into the early nineteenth century, Hispanized, noble or urban persons[31] – had been likened to Europe's peasantry.

[28] Gonzalo Fernández de Oviedo y Valdés, *Primera parte de la historia natural y general de las indias, yslas e tierra firme del mar oceano* (Sevilla: Iuam Cromberger, 1535), book XI, chapter V. On the original manuscript's slow way into print, see Henry Lowood, 'The New World and the European Catalog of Nature,' in *America in European Consciousness, 1493–1750*, ed. Karen Ordahl Kupperman (Chapel Hill: University of North Carolina Press, 1995), 308–09.

[29] On historians' reserve, see, for instance, Daniela Bleichmar's remark on 'the letter's laudatory tone and the way it dramatizes the importance and utility' of Monardes's work. Daniela Bleichmar, 'Books, Bodies, and Fields: Sixteenth-Century Transatlantic Encounters with New World *Materia Medica*,' in *Colonial Botany. Science, Commerce, and Politics in the Early Modern World*, ed. Londa Schiebinger and Claudia Swan (Philadelphia: University of Pennsylvania Press, 2005), 92. Other historians have, however, stressed the relevance of Monardes's epistolary networks in ways that suggest the letter may have been genuine. Norton, *Sacred Gifts, Profane Pleasures*, 117. On elements of imaginativeness in Fernández de Oviedo y Valdés's chronicle, see, for instance, Carlo Klauth, *Geschichtskonstruktion bei der Eroberung Mexikos: Am Beispiel der Chronisten Bernal Díaz del Castillo, Bartolomé de las Casas und Gonzalo Fernández de Oviedo* (Hildesheim: Georg Olms Verlag, 2012), 184.

[30] On the method of 'working from affinities with recognizable' Old World 'forms' in natural history, see Lowood, 'The New World,' 298. See also Roland Greene, 'Petrarchism among the Discourses of Imperialism,' in *America in European Consciousness, 1493–1750* (Chapel Hill: University of North Carolina Press, 1995), 135. On the same predicament, see also the classic works: Anthony Grafton, *New Worlds, Ancient Texts. The Power of Tradition and the Shock of Discovery* (Cambridge, Mass.: Harvard University Press, 1992), 6; Anthony Pagden, *The Fall of Natural Man. The American Indian and the Origins of Comparative Ethnology* (Cambridge: Cambridge University Press, 1982), 1–6; 11–12.

[31] On ethnic categories in South America, see David Cahill, 'Colour by Numbers: Racial and Ethnic Categories in the Viceroyalty of Peru, 1532–1824,' *Journal of Latin American*

Both were natural men, society-less, poor,[32] uncultured creatures 'close in condition, if not in kind, to the animals among which they worked'.[33] A university-trained physician versed in scholastic medicine like Monardes or a Romance humanist like Fernández de Oviedo y Valdés would have been familiar with the notion that many useful remedies and therapeutic practices could be learned from laypersons – rustics and peasants, simpletons, or women. They would have known also that these laypersons' knowledge was primarily based on the 'accurate but un-theorized observation of phenomena accessible to the senses', unlike that of learned scholars, who inquired into causes or first principles in the physical operation of natural substances.[34] They would have been aware, too, that women in particular communicated their knowledge only to their close associates – to other women – orally and in the vernacular, and that they would conceal it from men, since it represented one important way of exerting power over them. Men like Monardes and Fernández de Oviedo y Valdés, as well as their readership, would have known also that one of the most important tasks of literate physicians and scholars consisted of retrieving the secrets of women and other artisanal practitioners and committing them to Latin script to ensure their permanence and relative publicity within an elite scientific community.[35] The 'stereotype of the inscrutable

Studies 26, no. 2 (1994). In the Spanish American viceroyalties alike, herbalists usually were 'simple folks' – men and women who lived 'far-off', away from the city, in 'Indian villages'. Miruna Achim, 'From Rustics to Savants: Indigenous *Materia medica* in eighteenth-century Mexico,' *Studies in History and Philosophy of Biological and Biomedical Sciences* 42 (2011), 82.

[32] The image of Indian poverty, illiteracy and simplicity was fashioned particularly by the Franciscan order. Julia McClure, *The Franciscan Invention of the New World* (Basingstoke: Palgrave Macmillan, 2016), 178.

[33] The meaning of 'natural man' changed over the centuries, from that of a society-less creature, 'something less than human', in the sixteenth and seventeenth centuries, to the Enlightenment's 'natural man', someone whose mind is unfettered by the moral and intellectual constraints of civil society. Pagden, *The Fall of Natural Man*, 8–9; 97–98; 21–22.

[34] Katherine Park, *Secrets of Women. Gender, Generation, and the Origins of Human Dissection* (New York: Zone Books, 2006), 84–85. On how sixteenth-century physicians took seriously what they learnt from laypersons – 'empirics', artisans or women – see Michael Stolberg, 'Learning from the Common Folks. Academic Physicians and Medical Lay Culture in the Sixteenth Century,' *Social History of Medicine* 27, no. 4 (2014), 649–67. On 'observation' in natural inquiry before the sixteenth century, see Katherine Park, 'Observation in the Margins, 500–1500,' in *Histories of Scientific Observation*, ed. Lorraine Daston and Elizabeth Lunbeck (Chicago: University of Chicago Press, 2011), 15–44; 36. On the 'mechanic classes' as bearers of valuable empirical knowledge', see also Steven Shapin, *A Social History of Truth. Civility and Science in Seventeenth-Century England* (Chicago: The University of Chicago Press, 1994), 396.

[35] Park, *Secrets of Women*, 84.

Indian',[36] one of the most enduring elements of conventional wisdom about the Indians, was likely the continuance and transfer of a European topos of long standing to the Americas. It was, quite possibly, like the man-eating Indian[37] and the non-acquisitive, idle Indian, who needed to be coerced into labour,[38] one of a plethora of classical and humanist legacies that the Old World bequeathed to the New.

Monardes's *Medicinal History* was translated into Latin by Carolus Clusius in 1582. It soon after became available in French, German, Italian and English translations, which were re-edited several times during the sixteenth and seventeenth centuries.[39] As historians of Iberian medicine have often noted, it was through Monardes that many of early modern Europe's physicians, apothecaries and naturalists were first introduced to the medicinal plants of the Americas.[40] It was presumably along this same vein, among others, that the secretive Indian, the reluctant informer about the virtues of America's medicinal plants, put down roots in early modern medical discourse. Indeed, secretive Indians made their appearance in various seventeenth-century naturalist treatises shortly thereafter: in Sebastianus Badus's 1663 *Anastasis Corticis Peruviae*, most importantly, but also the Jesuit friar Bernabé Cobo's (1580–1657) 1653 *Historia del Nuevo Mundo*. Cobo, certain that the Indians knew many herbs to cure their illnesses, had heard from several of his correspondents about instances of secrecy, in which curious gentlemen had 'promised the Indian[s] such good pay, with flattery and kindness' to learn their medical secrets, but were invariably put off by them.[41] So had the Dutch physician Willem Piso (1611–1678), who lamented in his 1648 *Historia naturalis Brasiliae* that he had to 'wrest from the barbarians' – 'that sacrilegious people not enlightened by letters' – their

[36] Majluf, 'The Creation of the Image of the Indian in 19th-Century Peru': 345–358.
[37] Anthropophagy first made its literary appearance in the writings of Herodotus. The cannibal epithet has been applied to different human groups over the centuries – more prominently, America's Aztec and Tupinambá societies. See, for instance, the classic work by William Arens, *The Man-Eating Myth: Anthropology and Anthropophagy* (Oxford: Oxford University Press, 1979), 10–13.
[38] 'Idleness', long attributed in humanist thought to the 'poor' and to beggars, was transferred to an entire society – the Indians – in the wake of the conquest. See Nicolás Sanchez-Albornoz, 'El trabajo indigena en los Andes: teorías del siglo XVI,' *Revista ecuatoriana de historia económica* 2 (1987), 173.
[39] Even before the author's death, the 'Medicinal History' saw eighteen editions outside Spain: six in Italian, five in Latin, three in French, three in English and one in German. López Piñero, 'Los primeros estudios científicos.'
[40] Achim, 'From Rustics to Savants,' 278.
[41] Bernabé Cobo, *Inca Religion and Customs*, ed. John Howland Rowe, trans. Roland Hamilton (Austin: University of Texas Press, 1990 (1653)), 222.

'secret medicines'.[42] Hans Sloane, too, related in his 1707 *Voyage to the Islands Madera, Barbadoes, [...] and Jamaica* how the Indians of Guiana only revealed the virtue of contrayerba, an antidote, to the Spanish when they 'threatened to wound [an Indian prisoner] by one of their poison'd Arrows'.[43]

Secrecy was not solely a quality attributed to the Indians by the late 1700s and 1800s. It was also a genuine form of conduct adopted by many healers, regardless of their provenance, across the Atlantic empires. Early modern apothecaries, physicians and charlatans in the British, Portuguese and Habsburg empires had a tradition of secret remedies, rooted in the belief that the mysteries of medical knowledge should be available only to an elite few, and in the fact that proprietary knowledge of medicines, or procedures, would allow them to stand out within a bustling medical marketplace.[44] Indeed, before the late 1700s, when the Enlightenment, and later Romanticism, would make openness the ideal, and *the* rhetoric, of modernity,[45] dissimulation, secrecy and inscrutability were considered requisite virtues in various professions: not only in medical practitioners, but also in diplomats, courtiers and artisans.[46] Native healers in Spanish America would have had even more reason

---

[42] These quotes are taken from the Portuguese translation of the *Historia naturalis Brasiliae*. Guilherme Piso, *História natural do Brasil Ilustrada*, trans. Alexandre Correia (São Paulo: Companhia Editora Nacional, 1948 (1648)), 46. For a previous discussion of Piso, native secrecy and poisonous plants, see also Júnia Ferreira Furtado, 'Tropical Empiricism. Making Medical Knowledge in Colonial Brazil,' in *Science and Empire in the Atlantic World*, ed. James Delbourgo and Nicholas Dew (New York: Routledge, 2008), 36.

[43] Hans Sloane, *A voyage to the islands Madera, Barbados, Nieves, S. Christophers and Jamaica, with the Natural History of the Herbs and Trees, Four-footed Beasts, Fishes, Birds, Insects, Reptiles, &c. of the last of those islands; to which is prefix'd, an Introduction, Wherein is an Account of the Inhabitants, Air, Waters, Diseases, Trade, &c. of that Place, with some Relations concerning the Neighbouring Continent, and Islands of America*, 2 vols., vol. 1 (London: B.M., 1707), lv.

[44] Elaine Leong and Alisha Rankin, 'Introduction: Secrets and Knowledge,' in *Secrets and Knowledge in Medicine and Science, 1500–1800*, ed. Elaine Leong and Alisha Rankin (Farnham: Ashgate, 2011), 12. See also William Eamon, *Science and the Secrets of Nature. Books of Secrets in Medieval and Early Modern Culture* (Princeton: Princeton University Press, 1994), 319–50. The Atlantic empires also housed medical traditions that were, though competitive, not secretive, partly because, as Pablo F. Gómez has argued for Afro-Caribbean healers, 'secrecy limited public recognition'. Pablo F. Gómez, *The Experiential Caribbean: Creating Knowledge and Healing in the Early Modern Atlantic* (Chapel Hill: University of North Carolina Press, 2017), 87.

[45] For a qualification of the idea that the Enlightenment made 'openness' the ideal, and rhetoric, see Paola Bertucci, 'Enlightened Secrets: Silk, Intelligent Travel, and Industrial Espionage in Eighteenth-Century France,' *Technology and Culture* 54, no. 4 (2013).

[46] Koen Vermeir and Dániel Margócsy, 'States of Secrecy: An Introduction,' *British Journal for the History of Science* 45, no. 2 (2012); Jon R. Snyder, *Dissimulation and the Culture of Secrecy in Early Modern Europe* (Berkeley: University of California Press, 2009).

than their Portuguese or English contemporaries to hide their medicines. Native healing, associated with idolatry and native resistance, was considered illicit and was thus repressed under Spanish colonial rule.[47] Though punishments for *curandería* were usually minor,[48] particularly during the early years of the Spanish conquest, secrecy was a form of conduct it would have been perfectly prudent to adopt as a native herbalist. While native men and women would sometimes have chosen to hide their knowledge, however, there can be no denying the fact that several also shared their herbal skill amicably, compassionately or collaboratively, as is evident from the natural histories composed in collaboration with indigenous intellectuals[49] by Francisco Hernández (1517–1587),[50] Bernardino de Sahagún (1500–1590),[51] and Pedro Montenegro (1663–1728),[52] among others. It would seem, at the very least, that not every Indian's temper had become as mysterious and mistrustful in the wake of the upheaval of the conquest as Unanue would have us believe. The belief in a closely kept herbal skill, peculiar to the

[47] On how Indian healing would often have occurred in secrecy because of its association with idolatry and consequent repression, see Suzanne Austin Alchon, 'Tradiciones médicas nativas y resistencia en el Ecuador,' in *Saberes andinos. Ciencia y tecnología en Bolivia, Ecuador y Perú*, ed. Marcos Cueto (Lima: IEP, 1995), 27; 36. Miruna Achim has likewise argued that 'church censorship [...] did send [native] healers into hiding'. Achim, 'From Rustics to Savants,' 279. On accusations of witchcraft, and idolatry, against folk healers, see also Marcos Cueto and Steven Palmer, *Medicine and Public Health in Latin America* (Cambridge: Cambridge University Press, 2015), 24.

[48] Linda A. Newson, 'Medical Practice in Early Colonial Spanish America: A Prospectus,' *Bulletin of Latin American Research* 25, no. 3 (2006), 376. On 'reprimands' as the most common penalty for *curanderismo* in New Spain, see Sherry Fields, *Pestilence and Headcolds. Encountering Illness in Colonial Mexico* (New York: Columbia University Press, 2008), 74.

[49] On the concept of 'indigenous intellectuals', see Gabriela Ramos Cárdenas and Yanna Yannakakis, introduction to *Indigenous Intellectuals: Knowledge, Power, and Colonial Culture in New Spain and the Andes*, ed. Gabriela Ramos Cárdenas and Yanna Yannakakis (Durham: Duke University Press, 2014).

[50] On Francisco Hernández's six-volume 'Natural History of New Spain', see Jorge Cañizares Esguerra, *Nature, Empire, and Nation. Explorations of the History of Science in the Iberian World* (Stanford, Calif.: Stanford University Press, 2006), 8; López Piñero, 'Los primeros estudios científicos,' 128; Arndt Brendecke, *Imperium und Empirie. Funktionen des Wissens in der spanischen Kolonialherrschaft* (Köln: Böhlau Verlag, 2009), 263; Simon Varey, Rafael Chabrán and Dora B. Weiner, eds., *Searching for the Secrets of Nature. The Life and Works of Dr. Francisco Hernández* (Stanford, Calif.: Stanford University Press, 2000).

[51] On native informants and materia medica in Bernardino de Sahagún's 1577 'General history of the things of New Spain' (*Historia general de las cosas de nueva España*), see Achim, 'From Rustics to Savants,' 278.

[52] On the Jesuit friar Pedro Montenegro's materia médica misionera, see Carmen Martín Martín and José Luis Valverde, *La farmacia en la América Colonial: el arte de preparar medicamentos* (Granada: Universidad de Granada, 1995), 11; Anagnostou, *Jesuiten in Spanisch-Amerika*, 78; 281–83; 94–97; 305; 11.

Indians, was no doubt often corroborated by observation. It was also, however, a potent cultural discourse that would have structured and influenced observation.

Botanists by Instinct

La Condamine's decision, in his 1738 report on the cinchona tree, to add intelligence on how the Indians had arrived at their discovery by the observation and emulation of lions to the seventeenth-century narrative baseline – which had encompassed the Indians' secrecy, the countess's illness and the Jesuit order, but had given no account of how the Indians had learned of the bark's virtues – evidently appealed to his readership. Like the narrative about the Indians' secrecy, it was swiftly adopted and elaborated upon by other authors, who introduced alterations or further detail. By the late 1700s and early 1800s, there were several variations upon La Condamine's narrative in circulation. In one common variant, the Indians had arrived at the happy discovery by accident, with an Indian, 'in a delirious fever', quenching his thirst from a pond that had 'imbibed the virtues of the Bark' from cinchona trees upon its margin or, as other authors would have it, from cinchona trees that had fallen into its water as the consequence of an earthquake.[53] It was, at any rate, generally believed by the end of the eighteenth and the beginning of the nineteenth century that the Indians had arrived at their knowledge of the bark's virtues either by the observation and emulation of nature, or by an accident of nature.

Like the Indians' secrecy and hatred of their enemies, narratives about the place of nature in their ways of knowing and learning were a cultural discourse – other or rather than an observation. The various accounts of how the medicinal properties of cinchona bark had first

[53] See the entry 'On the Origin of the Use of Peruvian or Jesuits' Bark', in Robert Bailey Thomas, *The Farmer's Almanack, Calculated on a New and Improved Plan, for the Year of our Lord 1801. Fitted to the Town of Boston, but will serve for any of the adjoining states* (Boston: Manning & Loring, 1800), 41. See also Collingwood, 'Observations on the Peruvian Bark,' 265–66; Motherby, *A New Medical Dictionary*, entry CORTEX PERUVIANUS; Luigi Castiglioni, *Storia delle piante forastiere le piu importanti nell'uso medico, od economico* (Milano: Stamperia di Giuseppe Marelli, 1791), 45; Murray, *Vorrath von einfachen, zubereiteten und gemischten Heilmitteln*, 1, 1115. According to John Roberts, the 'natives' were compelled to drink the 'dirty water' of that lake because of a drought. John Roberts, *Observations on Fevers, wherein the Different Species, Nature and Method of Treating those Diseases, are Represented in New and Interesting Points of View. The Whole written in a simple and concise Manner, divested of the Terms of Art, and adapted to common Capacities, on the plain Principles of Common Sense. Being designed for the Use of those, who may be afflicted with those Diseases, as well as the Medical Reader* (London: Rozea et al., 1781), 34–35.

come to the knowledge of mankind, whether they followed the imita-
tion-of-lions or the accident-of-nature hypothesis, invariably echoed
and encouraged the then common perception and conventional
wisdom that the Indians – like other 'savages', incapable of forming
abstract ideas[54] – had learned of the properties of medicinal plants by
virtue of their unmediated, instinctive experience of and peculiar intimacy with nature and the American environment. References to the role
of nature, or an accident of nature, in the various late eighteenth- and
early nineteenth-century narratives bear witness to how the notion that
empirical, instinctive, first observations of nature were the certain prerogative of the lower orders, and of animals, had persisted throughout
the seventeenth century.[55] They also suggest that it regained currency,
and acquired complexity and salience in the Enlightenment, presumably along with the rise of the belief that civilized man had become
denatured, that is, come to diverge from a natural state and lost his
knowledge of nature.[56] Indeed, by the late 1700s and early 1800s, the
inimitability and singular efficaciousness of the medicinal plant
knowledge of the world's 'simple folks' was an established discourse.
Curiosity about the plant lore of vernacular folk and a belief in the
excellence of the empirical and use-oriented medicines of rustics and
other illiterate knowers – of which women, still in the eighteenth
century, were one subcategory – were widespread among botanists,
physicians and naturalists in Sweden and Prussia alike.[57] Both presumably extended, with European imperial expansion and exploration, not

[54] The mind of a 'true savage', according to Rousseau, was unable to form abstract ideas.
Jean-Jacques Rousseau, *A Discourse upon the Origin and Foundation of the Inequality among
Mankind* (London: R. and J. Dodsley, 1761 (1755)), 81. Cited in Terry Jay Ellingson,
The Myth of the Noble Savage (Berkeley: University of California Press, 2001), 82.
[55] Laurence Brockliss and Colin Jones, *The Medical World of Early Modern France* (Oxford:
Clarendon Press, 1997), 277. On the Portuguese-speaking world, see Beltrão Marques,
Natureza em Boiões, 61–62. See also Eamon, *Science and the Secrets of Nature*, 56; 187.
Seventeenth-century Dutch writers likewise wrote that 'cats and dogs' knew remedies.
Alix Cooper, 'The Indigenous versus the Exotic: Debating Natural Origins in Early
Modern Europe,' *Landscape Research* 28, no. 1 (2003), 54.
[56] Richard H. Grove, *Green Imperialism. Colonial Expansion, Tropical Island Edens and the
Origins of Environmentalism* (Cambridge: Cambridge University Press, 1995), 251. See
also Lorraine Daston and Fernando Vidal, 'Introduction: Doing What Comes
Naturally,' in *The Moral Authority of Nature*, ed. Lorraine Daston and Fernando Vidal
(Chicago: University of Chicago Press, 2004), 8.
[57] Carl Linnaeus, for instance, who was a 'keen ethnobotanist', studied 'the plant lore and
materia medica of tribal people [...], the Scandinavian peasantry', and 'wise women'.
Lisbet Koerner, 'Women and Utility in Enlightenment Science,' *Configurations* 3, no. 2
(1995), 250–51. See also Achim, 'From Rustics to Savants,' 280. Even Romantic
thought still 'related the female gender to the natural, and more specifically, to the
botanical'. Grove, *Green Imperialism*, 236.

only to American natives but also to Uzbek herders, South Asian non-Brahminic low-castes and African field slaves.[58]

The conception that plant knowledge was passed on from nature, or wild animals, to natives, and thence to civilized man, was likewise a common trope in the period.[59] By the late 1700s and early 1800s, there were several accounts in general circulation, from the Viceroyalty of Brazil to the island of Jamaica, that, like the narratives about the bark's happy discovery, detailed how the natives had found out about antidotes, poisons or herbs by chance observations of natural phenomena or by emulation of 'the conduct of brute animals' – snakes or wild boars.[60] Europeans had long cherished the idea that the barbarian, the primitive and the 'savage' were closer to nature than those who claimed civilization and, later, modernity for themselves.[61] Many Enlightenment thinkers placed man in a state of nature, the 'true savage' – of which American natives constituted a paradigmatic instance at the time[62] – 'at an equal distance from the stupidity of brutes and the [...] enlightenment of civilized man'[63] and deduced from this position the 'savage's' aptness to mediate between the two. Indeed, the period saw a revival of the idea of the chain of being (*scala naturae*), which predicated a functional relationship between the world's organisms in a linear, continuous and

---

[58] Historians have argued that, although Brahmin interpretations and texts shaped almost every realm of European perceptions and discourses, medico-botanical knowledge tended to privilege non-Brahminic, low-caste epistemologies. Richard H. Grove, 'Indigenous Knowledge and the Significance of South-West India for Portuguese and Dutch Constructions of Tropical Nature,' *Modern Asian Studies* 30, no. 1 (1996), 128–33. On Uzbek herders, and field slaves, see Koerner, 'Women and Utility in Enlightenment Science,' 251.

[59] Londa Schiebinger, 'Prospecting for Drugs. European Naturalists in the West Indies,' in *The Postcolonial Science and Technology Studies Reader*, ed. Sandra Harding (Durham, N.C.: Duke University Press, 2011), 114.

[60] Schiebinger, *Plants and Empire*, 82; Schiebinger, 'Prospecting for Drugs,' 114; Kathleen S. Murphy, 'Translating the Vernacular: Indigenous and African Knowledge in the Eighteenth-Century British Atlantic,' *Atlantic Studies* 8, no. 1 (2011). On the Iberian world, see Beltrão Marques, *Natureza em Boiões*, 61–62.

[61] Shepard Krech, *The Ecological Indian: Myth and History* (New York: W.W. Norton, 1999), 16–17; John G. A. Pocock, *Barbarism and Religion*, vol. 4, *Barbarians, Savages and Empires* (Cambridge: Cambridge University Press, 2008), 160; Pagden, *The Fall of Natural Man*, 21.

[62] Ellingson, *The Myth of the Noble Savage*, xiii.

[63] Jean-Jacques Rousseau, *A Discourse on Inequality*, trans. Maurice Cranston (London: Penguin, 1984 (1755)), 115. Cited in Ellingson, *The Myth of the Noble Savage*, 82. It is important to note that Rousseau distinguished the 'savage peoples known to us' from ethnographic travel from his hypothetical evolutionary construction of 'true' savages in a state of nature, since the 'savage peoples known to us' had already departed – whether far or little – 'from the state of nature'. Ibid. Georges-Louis Leclerc, Comte de Buffon (1707–1788), for instance, wrote that men in the Americas led 'an animal-like existence'. Gerbi, *The Dispute of the New World*, 8.

hierarchical series. That relationship was functional not only in that organisms fed upon one another, but also, as historians of botany have argued, in that knowledge of nature was imagined to pass from the lowest to the highest organism: from brutes, through the 'savage's' intercession, to civilized man.[64] Tellingly, whereas in the European context early modern naturalists in the wake of the seventeenth century fashioned themselves very much as discoverers of the secrets of nature,[65] naturalists in Spanish America frequently substituted that discourse for a rhetoric of approximation. As Miruna Achim has argued, especially creole experts in the natural world saw their peculiar privilege, and province, in the mediation, not between the enlightened sciences and nature, but between the enlightened sciences and the Indian – the communication of the knowledge of the inscrutable, withdrawn, natives to a wider world.[66]

Hence, the remarkable circumstance that the enlightened understanding of the empirical and use-oriented medicines of rustics and other illiterate knowers as singularly efficacious should be in no apparent contradiction to contemporary observers' increasingly severe reservations about these men and women's general understanding, and reliability, by the late 1700s and early 1800s.[67] For it was the illiterate knowers' supposed primitiveness, rudeness and simplicity, their closeness to nature, that endowed them, like animals, with the faculties to be able to derive herbs necessary for their preservation straight from 'the hands of their Creator' and that rendered their medicines so very inimitable, and efficacious, and them 'botanists by instinct'.[68] As Thomas Collingwood

---

[64] Schiebinger, 'Prospecting for Drugs,' 114. On the 'chain of being', see Mark Barrow Jr., *Nature's Ghosts. Confronting Extinction from the Age of Jefferson to the Age of Ecology* (Chicago: University of Chicago Press, 2009), 20–23.

[65] Bleichmar, *Visible Empire*, 48. The idea of the secrecy of nature is often attributed to Heraclitus of Ephesus, who around 500 BC deposited his manuscript at the temple of Artemis. Among its surviving phrases are 'Nature loves to hide'. Carolyn Merchant, *Reinventing Eden. The Fate of Nature in Western Culture* (New York: Routledge, 2013), xiii–xiv.

[66] Achim, *Lagartijas medicinales*, 75. For the term and concept of the 'rhetoric of approximation' see Majluf, 'The Creation of the Image of the Indian in 19th-Century Peru,' 308. On creole scholars as trans-Atlantic 'brokers', and intermediaries, see also Stefanie Gänger, 'Disjunctive Circles: Modern Intellectual Culture in Cuzco and the Journeys of Incan Antiquities, c. 1877–1921,' *Modern Intellectual History* 10, no. 2 (2013).

[67] On accusations of untruthfulness and caste stereotypes in Spanish American courts after the seventeenth century, see Susan Kellogg, *Law and the Transformation of Aztec Culture, 1500–1700* (Norman: University of Oklahoma Press, 1995), 68; 77. See also Lauren Benton, *Law and Colonial Cultures. Legal Regimes in World History, 1400–1900* (Cambridge: Cambridge University Press, 2002), 86.

[68] Edward Long, *The History of Jamaica. Or, General Survey of the Antient and Modern State of that Island: With Reflections on Its Situation, Settlements, Inhabitants, Climate, Products, Commerce, Laws, and Government*, 3 vols., vol. 2 (London: T. Lowndes, 1774), 380–81.

put it in his *Observations on the Peruvian Bark*, the Indians' knowledge of plants and herbs arose from a natural instinct, an intimacy with the environment that was the sole province, and privilege, of the 'most illiterate and unobserving part of mankind',[69] or, as Pierre Louis Moreau de Maupertuis (1698–1759) phrased it, of 'savage nations' (*nations sauvages*).[70] Indeed, skill in nature was to remain the Indians' particular province and prerogative in European and North and Spanish American scholarship and literature, even during the heyday of scientific racism in the late nineteenth and early twentieth centuries.[71]

Iberian expansion into the Americas, and the concomitant encounter of societies, resulted in the invention and reordering of epistemic hierarchies and geographies of knowledge. It allocated creole scholars the charge of brokerage, the enlightened natural sciences the prerogative of universality, transferability and scientificity, and the Indians the reverse: knowledge that was unalienable, for one thing – tied to the context and the lives of the people who had generated it – and, for another, reliant on instinct: the unthinking, unmediated observation of natural phenomena and lowly organisms that they, in their capacity as 'savages', were closer to than civilized man.[72]

Illiterate Saviours

Historians have long argued that the very fact that the 'simple' folks' knowledge needed to be drawn out, extracted, systematized and domesticated for its value to be realized was a way of denigrating the Indian and celebrating the European.[73] Yet the rendering of native knowledge as instinctive, secret and immediate also sometimes served to heighten its

---

[69] Collingwood, 'Observations on the Peruvian Bark,' 265–66.

[70] Pierre Louis Moreau de Maupertuis, *Lettre sur le progrès des sciences* (Berlin: 1752), 89. See also Boumediene, *La colonisation du savoir*, 253.

[71] Krech, *The Ecological Indian*.

[72] These supposed 'characteristics' of 'indigenous knowledge' persist to this day. See Arun Agrawal, 'Dismantling the Divide between Indigenous and Scientific Knowledge,' *Development and Change* 26 (1995). Various scholars have argued that the very notion of 'indigenous knowledge' is a 'creation of the encounter between the West and the rest'. David M. Gordon and Shepard Krech, 'Indigenous Knowledge and the Environment,' in *Indigenous Knowledge and the Environment in Africa and North America*, ed. David M. Gordon and Shepard Krech (Athens: Ohio University Press, 2012), 1. See also Marwa Elshakry, 'When Science Became Western: Historiographical Reflections,' *Isis* 101, no. 1 (2010).

[73] On British colonial naturalists' understanding of the expertise of their native informants as 'raw materials', see Murphy, 'Translating the Vernacular.' A similar sense pervades Londa Schiebinger's argument about the revival of the 'Great Chain of Being' (*scala naturae*). Schiebinger, 'Prospecting for Drugs,' 114.

value and enhance its appeal and authority.[74] Indeed, by the late 1700s and early 1800s, contemporaries were quite agreed not only that the Indians were excellent herbalists but also that their observations were superior to those of professional physicians[75] – that they often 'mock[ed] the Art and its wise Professors'.[76] The supremacy of native medicinal plant knowledge was a common discourse throughout the Atlantic World at the time. From the Spanish American Viceroyalties of Peru and New Spain to the British North American colonies,[77] across the Portuguese Empire[78] and the colonial Caribbean, physicians and sufferers alike eulogized the marvellous cures of the Indians and 'Negro Doctors' who had more knowledge, and greater 'skill in simples, and the virtue of plants', 'than the whites', and who 'put to shame physicians coming from Europe'.[79] As a matter of fact, in the period, apothecaries, drug merchants and physicians often included narratives about secretive Indians, or natural men in advertisements, medical treatises and advice literature because it was apparently an effective way of popularizing and marketing plant-based remedies.[80] The empirical medicines discovered

[74] Historians working on 'secrets' in medicine have argued that, in early modern Europe, the rendering of – medical, mechanical or spiritual – knowledge as concealed, confidential and secret potentially heightened its perceived value and authority. Leong and Rankin, 'Introduction: Secrets and Knowledge,' 10.

[75] Unanue, 'Botánica,' 96.   [76] Cited in Achim, 'From Rustics to Savants,' 280.

[77] See John Tobler's 1765 'South-Carolina and Georgia Almanack', which contained an 'Indian Remedy for Inveterate Ulcers'. Cited in Kevin J. Hayes, A Colonial Woman's Bookshelf (Eugene, Ore.: Wipf & Stock, 1996), 91.

[78] Eighteenth-century Brazilian surgeons like Luís Gomes Ferreira (1686–1764) similarly trusted and relied on the plant knowledge and experience of the inhabitants of the Sertão, in particular the Carijós. Júnia Ferreira Furtado, ed. Luís Gomes Ferreira, Erário mineral (Rio de Janeiro: Editora FIOCRUZ, 2002 (1735)), 28–29.

[79] Richard B. Sheridan, Doctors and Slaves. A Medical and Demographic History of Slavery in the British West Indies, 1680–1834 (Cambridge: Cambridge University Press, 1985), 81. See also Schiebinger, 'Prospecting for Drugs,' 113–19. For more examples, see John R. McNeill, Mosquito Empires. Ecology and War in the Greater Caribbean, 1620–1914 (Cambridge: Cambridge University Press, 2010), 82–86.

[80] On the fetishization of indigenous, and African medical knowledge, particularly by mestizo and creole sectors of society in Spanish America, see Steven Palmer, Doctors, Healers, and Public Power in Costa Rica, 1800–1940 (Durham, N.C.: Duke University Press, 2003), 31. On North America, see, for instance, Francisco Guerra's work on advertisements for the 'secret' medicines of slaves in The Virginia Almanack. Francisco Guerra, 'Medical Almanacs of the American Colonial Period,' Journal of the History of Medicine XVI, no. 3 (1961), 247–48. Irina Podgorny, in her work on charlatans, has argued that some healers claimed to have lived among indigenous groups who had introduced them to the secret healing powers of America's nature, while others renamed their balsams and tinctures with 'indigenous names'. Irina Podgorny, 'From Lake Titicaca to Guatemala: The Travels of Joseph Charles Manó and His Wife of Unknown Name,' in Nature and Antiquities. The Making of Archaeology in the Americas, ed. Philip L. Kohl, Irina Podgorny, and Stefanie Gänger (Tucson: University of Arizona Press, 2014).

Figure 1.2 Isaac Paling (1630–1719), *Savants hollandais en discussion*
(Dutch savants in discussion). The 'Dutch savants' are grouped around
a piece of cinchona bark, with one individual (sitting fourth from the
right) pointing his finger at it. n.p., n.d.
© *Bibliothèque de l'Académie nationale de médecine*, Paris.

by the Indians and other rustics promised to be 'of [...] singular and
extraordinary efficacy', in the contemporary imagination, and as such,
they appealed broadly and sold widely.

Cinchona was certainly among the most renowned and celebrated of
the remedies supposedly discovered by 'savages' and rustics at the time.
The bark's passage into the Old World medical repertoire during the
later 1600s and early 1700s had still been attended with many difficulties
and obstacles. Scepticism about the desirability and value of exotic,
foreign herbs in the diet and medical practice of Dutch, English or
French societies more broadly had at first caused hesitancy towards the
bark.[81] Cinchona's intellectual dissonance with various elementary
tenets of humoral theory had likewise caused controversy among Galenic
practitioners (FIGURE 1.2), especially the fact that it did not expel any
humours, and that its bitter, astringent taste indicated a warm quality,
when it was supposed to remove the heat of fevers.[82] By the late 1700s
and early 1800s, however, Scottish physicians, creole botanists and

[81] Cooper, 'The Indigenous versus the Exotic.'
[82] On the bark's early history, see, in particular, Jarcho, *Quinine's Predecessor*. On the bark's
  gradual acceptance over the seventeenth century, see also Maehle, *Drugs on Trial*,
  chapter 4; Wouter Klein and Toine Pieters, 'The Hidden History of a Famous Drug:
  Tracing the Medical and Public Acculturation of Peruvian Bark in Early Modern
  Western Europe (c. 1650–1720),' *Journal of the History of Medicine and Allied Sciences*

French writers alike were agreed that the bark was 'the most Excellent'[83] and powerful[84] of all remedies, 'vital to the preservation of mortals'[85] and essential to 'combat the many illnesses that weighed down on an ailing mankind' (*une foule de maux qui pèsent sur l'humanité souffrant*).[86] It had become, to them, the last in a series of miraculous remedies that had come out of the New World since the sixteenth century: guaiacum, sarsaparilla, contrayerba, mechoacán or, indeed, tobacco, which had been hailed as potent antisyphilitics, febrifuges, antidotes or panacea during the early modern period, ever since the earliest Iberian accounts of American nature by Nicolás Monardes, Francisco Hernández and José de Acosta (1540–1600) brought them to the attention of medicine.[87] Indeed, though the bark held 'the first place among the[m]'[88] by the late 1700s and 1800s, long-standing ideas about the New World and its native peoples as a source of powerful medicines presumably assisted and facilitated the bark's passage into the very heart of European medical discourse and practice.

Their principal characteristics in the various late eighteenth- and early nineteenth-century accounts – their secrecy and their closeness to

71, no. 4 (2016); Cook, 'Markets and Cultures,' 208–09; Boumediene, *La colonisation du savoir*, 221–26.

[83] Cockburn, *The Present Uncertainty in the Knowledge of Medicines*, Preface I. A1; Thomas de Salazar, *Tratado del uso de la quina* (Madrid: Viuda de Ibarra, 1791), 8; Ruiz López, *Quinología*, 38.

[84] Francisco José de Caldas, *Memoria sobre el estado de las quinas en general y en particular sobra la de Loja*, ed. Federico Gonzalez Suarez, Anales de la Universidad Central de Quito (Quito: Tipografía y Encuadernación Saletiana, 1907 (1809)), 13.

[85] Hernández de Gregorio, *El arcano de la quina*, 3. See also de la Condamine, who concluded his tale by stressing the importance of this 'singular trade, so useful to the preservation of mankind' (*si utile à la conservation du genre humain*). Condamine, 'Sur l'arbre du quinquina,' 235.

[86] Jean Baptist Timothee Baumes, *Traité des fièvres rémittentes et des indications qu'elles fournissent pour l'usage de quinquina*, 2 vols., vol. 2 (Montpellier: Sevalle, 1821), 3.

[87] Acosta's 'Natural and Moral History of the Indies' came out in 1589; Hernández's manuscripts were written in the 1570s and published in 1651 as 'Nova Plantarum, animalium et mineralium mexicanorum historia'. On the earliest writings about Spanish America's medicinal plants, see López Piñero, 'Los primeros estudios científicos.'; Varey, Chabrán, and Weiner, *Searching for the Secrets of Nature*. See also Antonio Barrera-Osorio, *Experiencing Nature. The Spanish American Empire and the Early Scientific Revolution* (Austin: University of Texas Press, 2006), chapter 1. Ulrich von Hutten in his 1519 *De guaiaci medicina et morbo gallico* and Gonzalo Fernández de Oviedo in his 1526 *Sumario* and in the 1535 *Historia general y natural de las Indias* most prominently championed guaiacum's use as an anti-syphilitic in the early-sixteenth century. On the trade in guaiacum and sarsaparilla, see López Terrada and Pardo Tomás, 'Las primeras noticias y descripciones de las plantas americanas (1492–1553),' 44–45; Pierre Chaunu and Huguette Chaunu, *Séville et l'Atlantique (1504–1650). Première Partie: Partie Statistique. Le mouvement des navires et des marchandises entre l'espagne et l'amérique de 1504 à 1650* (Paris: S.E.V.P.E.N., 1956), 1027–29.

[88] Tissot, *Anleitung für das Landvolk*, 288–89.

nature – were presumably ascribed to American natives on account of their supposed illiteracy, destitution and condition of degradation, which associated them with the empirical, illiterate knowers of the Old World. The uncommon confidence in their marvellous cures and the singularity of their remedies afforded them intellectual kinship with one particular topos within that discursive field: that of the illiterate saviour. Historians have long argued that early modern medical thought, in its reliance on the truths spoken by illiterate knowers, drew on a genre of Christian piety that placed the best hope of salvation for mankind in the poor and ordinary people, who were closer not only to nature but also to God than were the mighty.[89] From the earliest days of Christianity, the divine fool – simple, illiterate and humble – had told truths and offered hope of redemption for mankind.[90] That cinchona, at a time of increasing admiration for its singular medicinal properties, should have been supposed to owe its discovery to 'wild Indians'[91] – supposedly uncultured, natural men – was but logical and cogent in an age in which the knower's simplicity amounted to a promise of redemption. It was consistent, in the period's reasoning, to expect nothing less than one's best hope of salvation, the most excellent remedy, and the revelation of nature's most coveted secrets, from 'savages': the simplest, most illiterate and unobservant of men.

The various origin myths in general circulation by the late 1700s and early 1800s, published in Boston almanacs, Milan pharmacopoeias and Bogotá treatises alike, speak to how an Atlantic Enlightenment, with its dispersed, cosmopolitan audience, both fashioned and fetishized native knowers on the fringe of the world.[92] Fantastical stories about the bark's

---

[89] Cook, *Matters of Exchange*, 34.
[90] The discursive tradition of 'divine idiocy', of 'holy folly', grew out of the crossings of numerous traditions – early Christian thought, the Enlightenment and Romanticism, among others. Dana Heller, 'Holy Fools, Secular Saints, and Illiterate Saviors in American Literature and Popular Culture,' *CLCWeb: Comparative Literature and Culture* 5, no. 3 (2003), 4; 7.
[91] Cockburn, *The Present Uncertainty in the Knowledge of Medicines*, Preface I. A1.
[92] From the time of the Hippocratic Corpus, pharmacological writers had employed geographical epithets, and fantastical stories about remedies, for purposes of precision and 'exoticism'. Laurence M. V. Totelin, 'The World in a Pill. Local Specialties and Global Remedies in the Graeco-Roman World,' in *The Routledge Handbook of Identity and the Environment in the Classical and Medieval Worlds*, ed. Rebecca Futo Kennedy and Molly Jones-Lewis (London: Routledge, 2016), 158–61. On portrayals of 'an eastern world' through incense at the end of the fifth century BC, see Nicholas Purcell, 'Unnecessary Dependences. Illustrating Circulation in Pre-modern Large-Scale History,' in *The Prospect of Global History*, ed. James Belich et al. (New York: Oxford University Press, 2016), 67. On Enlightenment exoticism, see also G. S. Rousseau and Roy Porter, 'Introduction: Approaching Enlightenment Exoticism,' in *Exoticism in the Enlightenment*, ed. G. S. Rousseau and Roy Porter (Manchester: Manchester University

discovery in the wake of the mid-1700s served a purpose not so much of precision than of allure. Authors would rarely forgo an opportunity to linger on the Indians' unlettered simplicity, closeness to nature and secrecy, because to them that was a way of both making sense of and promoting the bark's extraordinary efficacy and wonderful properties. The bark's users, traders and advocates – from Scotland to the Viceroyalty of New Granada – would have been conversant with and, in some measure, tempted into its consumption, commerce or advocacy by the widely shared tale of the wonders of the bark's discovery.

★★★

By the late 1700s and early 1800s, cinchona had come to be considered a most excellent and powerful remedy across political entities and regions due to a variety of reasons. The support of prominent advocates – the French aristocracy and the Jesuit order – certainly assisted the bark's acceptance and popularity during the late 1600s and early 1700s, from Versailles to Beijing. As might, a century later, the discerning appreciation of its medical effects. The fact that cinchona would on occasion have afforded sufferers relief may well have mattered at a time when simple clinical observations, experiments and statistics to evaluate treatments were beginning to alter the grounds of medicine. Another important set of reasons for the bark's extraordinary popularity in the decades around 1800, however, lay in the cultural imaginary surrounding it and its hold over the minds of users, traders and advocates, at a time of growing readiness for adopting medical novelties.[93] The various narratives in general circulation of how the admirable, and divine, medicinal properties of cinchona bark had first come to the knowledge of mankind tell us little of certainty about the actual circumstances of how cinchona was first attributed healing properties. They are indisputably revealing, however, of the appeal of foreign herbs and the medical knowledge of 'wild Indians' about them. The widely shared tale of the bark's discovery contributed to fashioning the enduring topos of the Indian as excellent herbalist and botanist by instinct, in ways that made sense of and promoted cinchona's extraordinary efficacy and wonderful properties. The allure of the illiterate saviour[94] was not universal. Ottoman and Chinese consumers and proponents appear to have adopted the bark devoid, for

---

Press, 1990), 4–5. On the practice of attaching meanings to natural objects based on their 'origins', see Alix Cooper, *Inventing the Indigenous. Local Knowledge and Natural History in Early Modern Europe* (Cambridge: Cambridge University Press, 2007), 6.

[93] Norton, 'Tasting Empire,' 691.

[94] Heller, 'Holy Fools, Secular Saints, and Illiterate Saviors.'

the most part, of that cultural baggage. Indeed, the secretive Indian knower of nature was a topos contingent on the epistemic hierarchies, geographies of knowledge and ideas about indigeneity held among the Atlantic empires, a discourse woven from the epistemic threads of the Atlantic World. It was a trope that would have enticed primarily the bark's French, Iberian, English, Dutch and Italian promoters, buyers and movers, who were, in turn, primarily responsible for that substance's worldwide distribution and propagation. This is not to say that prominent advocacy or medical efficacy played no part, or a lesser part, than the cultural imaginary. Rather, one presumably augmented the effect, or likelihood, of the other. The Peruvian bark attained its unusual popularity at these crossings.

2   The Demands of Humanity

> The cinchona, placed in the dominions of his Majesty, [...] is, on account of its virtue, a specific of the first necessity. Its extraction ought to attract tribute to Spain from all the nations (*debe atraer a España el tributo de todas las Naciones*): but it must be handled abiding by constant rules that preserve it in the abundance, goodness, and at the moderate price that humanity demands.
>
> – Antonio Caballero y Góngora to José Gálvez, October 19, 1786.

On May 14, 1800, Luis François de Rieux (1768–1840), chief physician at the Royal Hospital in Cartagena de Indias, wrote to Miguel Cayetano Soler y Rabassa (1746–1808), then Minister of Finance of the Spanish Empire, to advise him about the importance of extensive, structured and equitable world trade in cinchona bark. Cinchona being, he wrote,

> the most important, and the most usual remedy that medicine possesses (*siendo la quina el Medicamento más importante, y el mas usual que posee la Medicina*); humanity demands, and justly so, that this medicament of first necessity reach our hands without the fraud it has so often suffered hitherto, and at the most equitable price, in order that not even the poor, the most numerous in all the countries and the worthiest of a monarch's piety, whose resources are scarce, be deprived of a remedy on the administration and faithfulness (*fidelidad*) of which the lives of men depend so many times.[1]

Luis François de Rieux, 'full of ideas [...] about the rights of man [...] acquired in France, where he had his cradle', was by no means the only contemporary to aver that his efforts on behalf of cinchona originated in his concern for humanity.[2] On the contrary, the presumption to speak and act not only in the name, but for the betterment, of a universal humanity is one we find recurrently, in the treatises and decrees of Spanish colonial officials, British physicians and French naturalists

---

[1] Rieux, 'Carta a Miguel Cayetano de Soler,' 346 v.
[2] For biographical references to Luis de Rieux, see Manuel Salvador Vázquez, 'Las quinas del norte de Nueva Granada,' in *Enfermedad y muerte en América y Andalucía (siglos XVI–XX)*, ed. José Jesús Hernández Palomo (Sevilla: CSIC, 2004), 421.

alike – writers accounting for their endeavours in regulating or intensifying cinchona commerce, like Rieux, and also in investigating its therapeutic properties or locating inexpensive substitutes outside the Spanish American harvest areas.[3] Part of a wider Enlightenment discourse, which consistently claimed both to address and represent the ideal collectivity of humanity,[4] the equitable distribution of faithful, and inexpensive cinchona was, though by no means the only project of enlightened medicine,[5] one that to many physicians, naturalists and colonial officials 'merited the utmost attention'.[6]

Probing the enlightened discourse that cinchona was in the service of humanity, this chapter endeavours to situate, measure and define the ideal collectivity of beneficiaries invoked by cinchona's advocates during the late 1700s and early 1800s. It exposes and examines how cinchona and knowledge of it travelled across and between societies within, or tied to, the Atlantic World by setting out the structure, volume and reach of trade in cinchona. Though scholarship on the general volume of Spanish bark trade,[7] as well as the Crown's inability to generate significant profits

---

[3] On September 24, 1786, Antonio Caballero y Góngora wrote about the relevance cinchona had for 'humanity'. Antonio Caballero y Góngora, 'El arzobispo Virrey da cuenta,' *Archivo General de Indias*, Indiferente 1554, Santa Fé, 1786-09-24. Hipólito Ruiz López likewise stressed the importance of a better understanding of the various cinchona species for the sake of 'all of humankind' (*todo el género humano*). Ruiz López, *Quinología*, 2. The physician Richard Dancer likewise referred to 'humanity' (*hominum generi*) in his Edinburgh doctoral dissertation on cinchona. Richard Dancer, *Dissertatio Medica Inauguralis de Cinchona* (Edinburgh: C. Stewart, 1809), 2. For an example from the French colonial context, see Joseph Gauché's appeal to the Royal Society of Medicine (La Société Royale de Médecine) to study 'that bark's efficacy' 'for the good of humanity': Gauché, 'Description d'un Quinquina indigène á St. Domingue.' The French quest for substitutes was often justified by means of its benefits for 'humanity'. Mallet, *Sur le Quinquina de la Martinique, connu sous le nom de Quinquina-Piton*, 1–2. Accusations of 'adulteration' of the bark were also often phrased around the writers' concern for 'humanity'. Padréll et Vidal, 'Dissertation sur l'usage et l'abus du quinquina,' 4–5.
[4] On the discourse of 'humanity' in the Enlightenment, Daniel Brewer, *The Enlightenment Past. Reconstructing Eighteenth-Century French Thought* (Cambridge: Cambridge University Press, 2008), 21.
[5] On the discourse of 'humanity' in relation to medicine, especially smallpox inoculation, see Martha Few, *For All of Humanity. Mesoamerican and Colonial Medicine in Enlightenment Guatemala* (Tucson: University of Arizona Press, 2015), 4.
[6] José Celestino Mutis, 'Carta testimonial de Mutis al dr. Francisco Martínez Sobral,' *in José Celestino Mutis*, ed. Federico Gredilla (Bogotá: Academia Colombiana de Historia, 1982 (1789)), 119.
[7] The most important works on the subject are those by Antonio García-Baquero González and Miguel Jaramillo Baanante. Antonio García-Baquero González, *Cádiz y el Atlantico* (1717–1778), 2 vols., vol. 1 (Cádiz: Diputación Provincial de Cádiz, 1988); Miguel Jaramillo Baanante, 'El comercio de la cascarilla en el norte peruano-sur ecuatoriano: evolución de impacto regional de una economía de exportación, 1750–1796,' in *El Norte en la Historia Regional. Siglos XVIII–XIX*, ed. Scarlett O'Phelan Godoy and Yves Saint-Geours (Lima: IFEA-CiPCA, 1998).

from it,[8] is quite extensive, the geography of that commerce beyond Spain and within the various consumer societies around 1800 has hitherto received little comprehensive attention.[9] Drawing on regional accounts of bark consumption as well as on trade statistics, medical treatises and the extensive Spanish administrative record on the subject, this chapter charts the bark's material availability by means of commerce and contraband as well as concomitant, non-commercial forms of distribution – charitable giving, medical relief programmes and diplomatic gift exchange – between 1751 and 1820. Preparing the ground for subsequent chapters concerned with the movement of consumption practices and expertise in indications, the chapter outlines, in the first part, the catchment and volume of Atlantic trade in the bark. In the second part, the chapter sketches the trade's routes and its reach, from the Spanish and Portuguese American possessions to North America and the Caribbean, Europe, northern and coastal Africa and the eastern Mediterranean, as well as the eastern reaches of the Spanish, Portuguese and Dutch empires in South and East Asia. The third part of the chapter is concerned with the trade's social reach within these societies – with the inner contours of the 'ailing mankind'[10] that benefitted from that remedy. As various historians have argued, though medicine trade across the Atlantic basin was lively and extensive around 1800, a set of substances – rhubarb,[11] opium[12] and cowpox lymph[13] – stood out for their exceptional importance, that is, their higher consumption and their

---

[8] Matthew J. Crawford's *The Andean Wonder Drug* highlight's Spain's inability to generate significant profits from the bark trade and to defend its 'natural' monopoly from Dutch, Portuguese, British and French trade and, especially, contraband.

[9] Even the most recent contributions in the field do 'little to explore [the bark's] medical utility outside Spain' or 'to trace the smuggling routes' and other unofficial conduits that distributed 'the Peruvian bark', as one of the book's reviewers put it. David Sowell, review of *The Andean Wonder Drug: Cinchona Bark and Imperial Science in the Spanish Atlantic, 1630–1800*, by Matthew James Crawford, *American Historical Review* 122, no. 3 (2017). Some world areas are particular 'blind spots'. As Saul Jarcho put it in 1993, 'the spread of Peruvian bark into [...] the Orient and the Americas, has failed to attract, even incidentally, the attention of [...] historians. Jarcho, *Quinine's Predecessor*, 94.

[10] Baumes, *Traité des fièvres rémittentes et des indications qu'elles fournissent pour l'usage de quinquina*, 2, 3.

[11] On rhubarb, see Clifford M. Foust, *Rhubarb. The Wondrous Drug* (Princeton, N.J.: Princeton University Press, 1992).

[12] On opium in the eighteenth century, see, for instance, Maehle, *Drugs on Trial*, chapter 3.

[13] Benjamin A. Elman, *A Cultural History of Modern Science in China* (Cambridge, Mass.: Harvard University Press, 2006), 109. On smallpox inoculation in Spanish America, see, Few, *For All of Humanity*. On East Asia, see Ann Jannetta, *The Vaccinators: Smallpox, Medical Knowledge, and the 'Opening' of Japan* (Stanford, Calif.: Stanford University Press, 2007).

wider reach.[14] The Peruvian bark was commonly the most valued and used of them from the Americas.[15] It was thus, the chapter holds, though perhaps not quite the single 'most important, and usual remedy that medicine possesse[d]', yet one of the best-known and most peripatetic medicinal substances to run through the warp and weft of Atlantic trade in pharmaceuticals during the late 1700s and early 1800s. Its journeys offer, as such, an important and rare window into the reach and workings of plant trade, epistemic brokerage and therapeutic exchange in that period.

World Bark Trade

According to the trade series compiled by Antonio García-Baquero González in his classical *Cádiz y el Atlántico* (Cádiz and the Atlantic), Spain imported a total of more than 223,932 *arrobas* – equalling 2,575 tons, on average 83 tons per annum – of cinchona from all of its Spanish American ports between 1747 and 1778, alongside various other medicines: 64 tons of various purges, 5 tons of copal, 3.9 tons of balsams, 7 kilograms of contrayerba and 4 tons of sarsaparilla.[16] At the time, most of the bark came to Cádiz – Spain's key port for the monopoly trade with the Americas from 1717 to 1778 – from the Viceroyalty of Peru, from whence it would have sailed the route around Cape Horn.[17] The exploitation of America's natural resources by the Spanish Crown had reached unprecedented heights by the second half of the eighteenth century, principally under the rule of Charles III (1716–1788, r. 1759–1788),[18] and the Crown's efforts to increase cinchona imports had by then already borne fruit. According to García-Baquero, between 1717 and 1738 Spain had only imported a total of some 24,293 *arrobas* – 279 tons, some 13 per annum – of cinchona.[19] In recent years, various historians,

---

[14] Much of the rapidly expanding literature on early modern global and Atlantic drugs trade has emphasized the prevalence of a small number of – five to ten – medicinal substances that dominated the market, with cinchona invariably being numbered as one of them. See, for instance, Walker, 'The Medicines Trade in the Portuguese Atlantic World,' 5; Benjamin Breen, 'Empires on Drugs: Materia Medica and the Anglo-Portuguese Alliance,' in *Entangled Empires: The Anglo-Iberian Atlantic, 1500–1830,* ed. Jorge Cañizares-Esguerra (Philadelphia: University of Pennsylvania Press, 2018), 602; Wallis, 'Exotic Drugs and English Medicine,' 31–33; Bleichmar, *Visible Empire,* 145–46.

[15] García-Baquero González, *Cádiz y el Atlantico,* 1, 340; Wallis, 'Exotic Drugs and English Medicine,' 31–33; Stefanie Gänger, 'World Trade in Medicinal Plants from Spanish America, 1717–1815,' *Medical History* 59, no. 1 (2015), 47.

[16] García-Baquero González, *Cádiz y el Atlantico,* 1, 340–41.   [17] Ibid., 275.

[18] On Charles III and his pursuit of natural science, for the benefit of the empire, see Paula De Vos, 'Natural History and the Pursuit of Empire in Eighteenth-Century Spain,' *Eighteenth-Century Studies* 40, no. 2 (2007), 217.

[19] García-Baquero González, *Cádiz y el Atlantico,* 1, 336–51.

adducing additional documentary evidence, have adjusted García-Baquero's figures upwards, suggesting that Cádiz imports may have been, at least for some periods of time, significantly higher than the amounts established to date. Studies have suggested that 314 tons reached Cádiz in 1755,[20] a year for which García-Baquero had counted 291 tons, 23.4 tons in 1769,[21] a year for which García-Baquero had calculated 5.4 tons, and more than 259 tons per annum between 1761 and 1775,[22] a 14-year period for which García-Baquero had assumed an average of 63.5 tons per annum. These figures may be precipitate, but they suggest that García-Baquero's are, at the very least, conservative figures. In the era of free trade between 1778 and 1796, when other peninsular ports were allowed to trade with Spanish America, cinchona commerce flourished along with an overall growth in imports from Spanish America.[23] Spanish merchants, according to recent historical studies, handled an average of more than 321 tons of cinchona per annum between 1775 and 1779, some 137 tons in the 5 years between 1780 and 1784, an average of 451 tons per annum in the years 1785 to 1789, and 224 tons per annum between 1790 and 1794.[24] Warfare would have caused serious temporary disruptions in the flow of trade, as the historian Miguel Jaramillo Baanante has argued, and may well account both for the decline in imports over the 5-year period between 1780 and 1784 – Spain entered the Anglo-French War (1778–1783) in 1779 – and their steep rise between 1785 to 1789, owing to stocks that would have accumulated during the war.[25] As Jaramillo Baanante has suggested, the normal volume would have oscillated around the 224 tons of the subsequent 5-year period (FIGURE 2.1).[26] Peninsular ports received the largest share of Spain's bark imports. Other lesser, official

---

[20] Luz del Alba Moya, *Auge y Crisis de la Cascarilla en la Audiencia de Quito, Siglo XVIII* (Quito: Facultad Latinoamericana de Ciencias Sociales, Sede Ecuador, 1994), 41.

[21] Juan Riera Palmero, 'Quina y malaria en la España del siglo XVIII,' *Medicina & Historia. Revista de estudios históricos de las ciencias médicas* 52 (1994), 21.

[22] According to Dora León Borja, between 1761 and 1775, as much as 259 tons of cinchona left the port of Callao for Cádiz annually, at a time when further exports would have reached Spain from the ports of Veracruz and Cartagena. Leon Borja, 'Algunos datos,' 101–02.

[23] John Fisher, *Commercial Relations between Spain and Spanish America in the Era of Free Trade* (Liverpool: Centre for Latin American Studies, University of Liverpool, 1985), 60–64.

[24] Jaramillo Baanante, 'El comercio de la cascarilla,' 661. Other historians' calculations roughly confirm these figures. According to Juan Riera Palmero, more than 20,000 *arrobas* – some 230 tons – were handled by Spanish merchants in 1788, some 438,026 *libras* or 201 tons in 1789, some 15,000 *arrobas* or 172.5 tons in 1791 and 250,000 *libras* or 115 tons in 1794. Riera Palmero, 'Quina y malaria en la España del siglo XVIII,' 21.

[25] Jaramillo Baanante, 'El comercio de la cascarilla,' 66.   [26] Ibid.

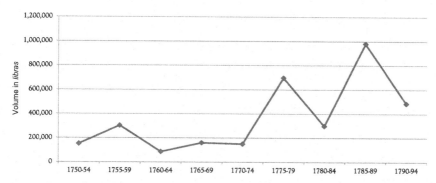

Figure 2.1 Volume of official cinchona exportations to peninsular ports, with 5-year averages.
Adapted from Miguel Jaramillo Baanante, 'El comercio de la cascarilla en el norte peruano-sur ecuatoriano: evolución de impacto regional de una economía de exportación, 1750-1796,' in *El Norte en la Historia Regional. Siglos XVIII-XIX*, ed. Scarlett O'Phelan Godoy and Yves Saint-Geours (Lima: IFEA-CiPCA, 1998), graphic 1.

transportation routes, however – along the continent's shores to ports in the Viceroyalties of New Spain, to the Spanish Philippines, or the United States[27] – would have added to the roughly 220 to 230 tons of cinchona traded legally and formally every year during the late 1700s.

There is agreement among historians that the amounts of cinchona handled by Spanish merchants were but a fraction of the overall volume of bark trade. Though the exact routes and the volume of illegal trade in cinchona elude us – the difficulty of following cinchona onto a contra-bandist's vessel or through the bustle of a marketplace renders any mapping or quantification necessarily fragmentary – the paper trail in Spain's archives leaves little doubt that foreign merchants handled a significant volume of contraband and that Spain faced 'much difficulty' in closing the trade's many 'gateways and entries'.[28] Portuguese, British,

---

[27] By the first decade of the nineteenth century, the United States directly imported from the port of La Guaira in Venezuela bags of cinchona bark. Jacques A. Barbier and Allan J. Kuethe, *The North American Role in the Spanish Imperial Economy: 1760–1819* (Manchester: Manchester University Press, 1984), 167. According to Alexander von Humboldt, 1030 *libras* of cinchona reached the port of Veracruz, in the Viceroyalty of New Spain, in 1802. Alexander von Humboldt, *Ensayo político sobre el reino de la Nueva España*, trans. Vicente Gonzalez Arnao, vol. 4 (Paris: Rosa, 1822), 74.

[28] As José García de León y Pizarro, the president-regent of the Kingdom of Quito, remarked in 1782, 'many were the gateways and entries of this commerce [in cinchona] today; and to close them one faces much difficulty and one requires an infinite number of guards'. 'Expediente y cartas de José García de Leon y Pizarro.' *Archivo General de Indias*, Indiferente 1554, Quito, 1782-08-18. On the importance, and elusiveness, of contraband, see also Jaramillo Baanante, 'El comercio de la cascarilla.'

Dutch and French contrabandists, overcoming the Spanish government's efforts to restrict trade with its colonies, frequently acquired cinchona directly from the harvesters in the Viceroyalties of New Granada and Peru. The inhabitants of the Dutch colonies of Surinam, Berbice and Essequibo in Guyana, for instance, were often accused of entering the Orinoco River 'under the pretext of fishing', and of trafficking cinchona in exchange for clothes with the harvesters.[29] For the latter, the attraction of selling their yield to smugglers and other unauthorized buyers presumably lay both in their more liberal pay and lesser concern with quality. As the governor (corregidor) of Loja, Pedro Xavier de Valdivieso y Torre (d. 1786, r. 1773–1784), put it, Loja bark cutters could always be sure to sell their yield to contrabandists who, 'unconcerned with [the bark's] quality (sin reparar en calidad)', paid 'three, and four pesos for an arroba',[30] where Crown officials would confiscate barks on account of their supposed worthlessness, and 'pretend to burn them'.[31] Several of the foremost cinchona merchants in Cuenca and Loja were also repeatedly accused of engaging in contraband with foreigners.[32] The Archbishop and Viceroy of New Granada, Antonio Caballero y Góngora (1723–1796, r. 1782–1789), was convinced that ample contraband in cinchona passed through the territories occupied by the Guajiros – one of many areas in the Americas the Spanish Crown controlled but nominally – along the River Hacha.[33] The Guajiros, 'the most troublesome tribe in the viceroyalty of New Granada' in the eyes of contemporary observers, were 'numerous and bold', and as 'a commercial tribe' were known to have 'considerable intercourse with the British and Dutch, who provide them with goods, slaves, and fire-arms'.[34] Another gateway for cinchona smugglers that Spanish officials found it difficult to

---

[29] 'Testimonio de los Autos, sobre que se establezcan el estanco de la Quina, o Cascarilla, en virtud de la real orden de S. M.,' Archivo General de Indias, Indiferente 1554, Cuenca, 1776-07-29 / Villa Orellana, 1776-08-18, 937.

[30] Pedro Xavier de Valdivieso y Torre, in a report to José García de Leon y Pizarro, dated as of February 21, 1782. 'Sobre la conservacion de Montes de Cascarilla de la Prov.a de Loxa y proveymiento de este Genero para la Real Botica,' Archivo Nacional de la Historia, Quito, Fondo General, Serie Cascarilla, Caja 2, Expediente 11, Loja, 1782-02-21 / 1783-04-27, 2–3.

[31] José Garcia de Leon y Pizarro, 'El Conde de Casavalencia informa difusamente sobre la representación del corregidor de Loxa y del Botanico encargados del acopio de Quina,' Archivo General de Indias, Indiferente 1556, Madrid, 1795-11-14, 178.

[32] Moya, Auge y Crisis de la Cascarilla, 179.

[33] Antonio Caballero y Góngora, 'El Arzobispo Virrey de Santa Fé Informa à V.E. el testimonio indirecto que ha tenido de las clandestinas extracciones de Quina que hacen los extranjeros en las costas septentrionales del Reyno,' Archivo General de Indias, Indiferente 1554, Carta 315, Santa Fé, 1782–1789.

[34] John Pinkerton, Modern Geography. A Description of the Empires, Kingdoms, States, and Colonies; with the Oceans, Seas, and Isles; in all the Parts of the World: Including the most

close was the border with the Portuguese Viceroyalty of Brazil. Officials frequently reported that Portuguese contrabandists shipped cinchona out of the Andes via the Marañon River in quantities sufficient to 'fill the entirety of Europe with it'.[35] With its source in the Andes, the Marañon runs northwest along the eastern base of the Andes before it turns eastward to meet the Ucayali River, together forming the Amazon River, which flows into Portuguese territory. Other instances of contraband apparently happened outside the harvest areas, at the stopovers along cinchona's shipping routes. Theft of legal exports was common, with the bark's lightness and low volume – it was usually shipped in the shape of fine, dried chippings – making it an easy target for small-scale smuggling. Spanish pharmacists often complained that boxes of cinchona reached them half empty, with evident traces of having been opened elsewhere – their nails removed or their leather straps cut – and with the cinchonas in them 'replaced with other rotten, dirty' barks.[36] In the late 1700s, Spanish officials persistently sought to avoid the exportation of cinchona through the empire's Caribbean ports. The Caribbean, with its many isolated beaches and coves, and with the major European colonial powers controlling multiple non-contiguous territories, was at the time the ultimate place for merchants to transgress the policies and statutes that impeded their ability to transact exchanges with one another.[37] The exportation of cinchona via Ocaña, on the Magdalena River northwards to the Caribbean port of Santa Marta, for instance, though it was comparatively inexpensive, was usually discouraged because it had the disadvantage of attracting contraband from 'foreign islands'.[38] So was the bark's extraction through the ports of Portobello or Cartagena.[39] The British colony on the island of Grenada in the south-eastern Caribbean – ceded to Britain by the treaty of Paris in 1763 – was generally thought to

---

Recent Discoveries and Political Alterations, 2 vols., vol. 2 (London: T. Cadell and W. Davies, Strand, 1817), 662.

[35] 'Expediente y cartas de José García de Leon y Pizarro.' Saul Jarcho also found evidence for early smuggling via the Marañon River. Jarcho, *Quinine's Predecessor*, 196–98.

[36] Joseph Diguja, 'El Excelentíssimo Señor Bailio Frey Don Julian de Arriaga, con fecha de siete de Mayo de este presente año, y de R. Orden, me previno,' *Archivo General de Indias*, Indiferente 1554, Quito, 1773-12-20. On replacement of cinchonas with false, or putrid, barks, see also Crawford, *Empire's Experts*, 127.

[37] Alan L. Karras, 'Transgressive Exchange. Circumventing Eighteenth-Century Atlantic Commercial Restrictions, or the Discount of Monte Christi,' in *Seascapes. Maritime Histories, Littoral Cultures, and Transoceanic Exchanges*, ed. Jerry H. Bentley, Renate Bridenthal and Kären Wigen (Honolulu: University of Hawai'i Press, 2007), 122.

[38] Caballero y Góngora, 'Copia de Carta Reservada.'

[39] Moya, *Auge y Crisis de la Cascarilla*, 33; Jaramillo Baanante, 'El comercio de la cascarilla,' 58; 62.

serve as a key entrepôt for contraband trade with cinchona,[40] as was Curacao, the most important Dutch transhipment port for illicit trade.[41] In other cases, shipments of cinchona coming from Spanish America were taken by British galleons. A Spanish shipment was taken just outside Cádiz in 1804,[42] and the same occurred with a shipment destined for Bordeaux in 1793, the contents of which went to public auction in Liverpool.[43]

Though the magnitude of illegal smuggling and theft of legal imports is, by nature, elusive, there is evidence to suggest that it by far surpassed the volume of Spain's official trade. 'From the newspapers' and trade bulletins that were published in 'London, Amsterdam, and elsewhere', from the 'observations of foreign commerce confirmed by men very versed in it', and from the 'large amounts of the most select cinchona that were extracted furtively' and traded 'by the British, the Dutch, the French, and the Danish', Spanish officials estimated that from the total volume of trade with 'a plant that grows only in the dominions of His Majesty', Spanish merchants handled, by the late 1700s, possibly half, possibly one-third, or as little as one-sixth.[44] Contemporary estimates of the overall harvest yields in some measure support these proportions. José Ignacio de Pombo (1761–1815), a spokesman of Cartagena's merchant guild, estimated that by the year 1800 the overall quantity of cinchona harvested in the Viceroyalties of New Granada and Peru amounted to some 2.5 million *libras* – that is, 1,150 tons –[45] five times the 220 to 230 tons handled legally, and officially, by Spanish merchants in the late 1700s and early 1800s. Contemporary observers' estimates

[40] 'Testimonio de los Autos.'
[41] In 1758, for instance, the ship *Curaçao Visser* shipped a cargo of 354 seroons of cinchona bark – some 35 tons – to Amsterdam. For this and other cinchona shipments, see Rutten, *Dutch Transatlantic Medicine Trade*, 29; 59.
[42] Bergen, *Versuch einer Monographie der China*, 310.
[43] 'Op Woensdag den 19 Juny 1793, 's morgens ten 11 uren precies, zal men, te Leverpool, publyk veilen en verkoopen: de GANTSCHE LADING, van het Schip Le Federatif, van St. Domingo komende en naar Bordeaux gedestineert gewest zynde,' *Rotterdamsche Courant N° 70*, 1793-06-11.
[44] 'Informe de la Contaduría de 9 de Julio de 1774 y respuesta del Sor Fiscal de 30 de Agosto del mismo año,' *Archivo General de Indias*, Indiferente 1554, Madrid, 1774-07-09 / 1774-08-30, 835–36. Other contemporary observers contended that 'for one pound that entered Spain the foreigners took two or three'. Miguel de Jijon y León, 'Recomendaciones para la explotación y comercialización real de la cascarilla,' *Revista del Archivo Nacional de Historia*, Sección del Azuay 6 (1986 (1776)). Also cited in Moya, *Auge y Crisis de la Cascarilla*, 41.
[45] José Ignacio de Pombo, 'Carta de José Ignacio de Pombo a José Celestino Mutis explicando su informe sobre las quinas,' *Archivo del Real Jardín Botánico*, Real Expedición Botánica del Nuevo Reino de Granada (1783–1816), José Celestino Mutis, Correspondencia, RJB03/0001/0001/0288, Cartagena (Colombia), 1805-10-30.

may have been a long guess, but other figures corroborate their sense that the volume of illicit trade in cinchona was a multiple of the legal volume. In 1776, a year in which, according to García-Baquero, 395 tons of bark reached Cádiz, just the 38 most important Cuenca merchants, forced to declare their possessions by royal decree, were found to be holding 618 tons of bark harvested in the Cuenca area – a significant portion, but surely not equal to the overall amounts extracted from the two Vice-royalties.[46] Calculating even with the by all accounts moderate estimate of foreign, illicit trade surpassing Spanish trade by a one-to-two ratio, the total bark trade volume would have oscillated around 450 tons per annum in the late 1700s and early 1800s. That estimate matches the calculations of Manuel Hernandez de Gregorio, the King's Apothecary, who wrote in 1804 that free trade in the bark then amounted to between 800,000 and 1 million *libras* – that is, some 368 to 460 tons – and that it appeared to him 'to increase further proportionate with an upsurge in [the bark's] uses'.[47] Calculating with less moderate estimates, the overall bark trade volume may well have amounted to, or surpassed, de Pombo's 1,150 tons. At any rate, as with other export sectors on the Atlantic seaboard of the New World – tea, cotton or tobacco – with cinchona, 'contraband dwarfed legal exchange'.[48]

Towards the early 1800s, the Spanish American territories were threatened with losing even their natural monopoly on the bark. While the Caribbean cinchonas that resulted from the British and French commercial quest for substitutes were for the most part quickly dis-carded,[49] the cinchona species discovered on Brazilian territory in 1805 as the result of a two-decades-long quest were more auspicious. During the 1790s in particular, the Portuguese Crown had not only naturalists but also colonial bureaucrats, clergymen and militia sergeants search Brazil's 'hinterlands' (*sertões*) and make inquiries among the 'persons of greater discernment' about trees resembling cinchona in appearance – prints of cinchona trees were issued for that

---

[46] Moya, *Auge y Crisis de la Cascarilla*, 40–41.

[47] Hernandez de Gregorio, 'Memoria,' 1034.

[48] Wim Klooster, 'Inter-Imperial Smuggling in the Amercas, 1600–1800,' in *Soundings in Atlantic History. Latent Structures and Intellectual Currents, 1500–1830*, ed. Bernard Bailyn and Patricia L. Denault (Cambridge, Mass.: Harvard University Press, 2009), 179. By 1745, legal tea imports represented about a quarter of English consumption. Carole Shammas, 'Changes in English and Anglo-American Consumption from 1550 to 1800,' in *Consumption and the World of Goods*, ed. Roy Porter and John Brewer (London: Routledge, 1993), 184.

[49] On the French quest, see McClellan and Regourd, *The Colonial Machine*, 260–62. On Britain, see Maehle, *Drugs on Trial*, 277.

purpose – in taste, or in terms of usage, that is, plants that were employed as febrifuges.[50] By decree of the prince regent, Dom João VI (1767–1826, r. 1799–1826), tree bark taken for cinchona was dispensed in 'the hospitals, even the military ones, of these realms' in 1804 to gain 'a proper understanding of its virtues' (*o devido conceito das virtudes*), by having doctors and surgeons conduct 'exact, and repeated observations' by means of a general, extensive administration, to decide 'the use, and consumption' of the 'bitter Brazilian barks'.[51] Even though tree bark taken for cinchona from the captaincies of Pernambuco, Oeiras do Piauí, Maranhão and Bahia was gathered and shipped to the metropolis, the Portuguese court and the empire's military and naval hospitals from the 1780s onwards, both historians and contemporaries have tended to assume that none of the plants discovered before 1805 was a cinchona variety.[52] According to the historian Vera Regina Beltrão Marques, it was only in 1805 that two cinchona varieties – *Cinchona macrocarpa* and *Cinchona pubescens* – were found in Rio de Janeiro. They were classified in 1806 and reconfirmed in 1811 by a commission consisting of, among others, Bernardino António Gomes (1768–1823).[53] It appears that, at least temporarily, these Brazilian and, in some measure, also some Caribbean barks, albeit in small amounts and with a low profile, entered commerce – as

[50] See, for instance, D. José Luis de Castro, 'OFÍCIO do (vice-rei do Estado do Brasil), conde de Resende, (D. José Luis de Castro), ao (secretário de estado da Marinha e Ultramar), D. Rodrigo de Sousa Coutinho,' *Arquivo Histórico Ultramarino*, 017 – RIO DE JANEIRO – CATÁLOGO DE DOCUMENTOS MANUSCRITOS AVULSOS / Cx. 165, D. 12275, Rio de Janeiro, 1798-05-04; Vicente Gomes da Silva, 'OFÍCIOS (14) de oficiais da marinha e comandantes de embarcações, ao (secretário de Estado dos Negócios da Marinha e Ultramar), visconde de Anadia, (D. João Rodrigues de Sá e Melo Meneses e Souto Maior),' *Arquivo Histórico Ultramarino*, 076 – REINO RESGATE 20121023 / Cx. 302-A, Pasta 5, Rio de Janeiro, 1806-02-04; Caetano Pinto de Miranda Montenegro, 'OFÍCIO do governador e capitão-general da capitania de Mato Grosso Caetano Pinto de Miranda Montenegro ao (secretário de estado da Marinha e Ultramar) Rodrigo de Sousa Coutinho,' *Arquivo Histórico Ultramarino*, 010 – MATO GROSSO – CATÁLOGO DE DOCUMENTOS MANUSCRITOS AVULSOS / Cx. 34, D. 1791, Vila Bela, 1798-06-14. For more examples, see Beltrão Marques, *Natureza em Boiões*, 132.
[51] 'Decretos do príncipe regente.'
[52] Jose Mariano Velloso's 'Portuguese Quinography' partly served the purpose of assisting the search for Brazilian cinchona varieties by enlightening its readership about the properties of 'false cinchonas'. Beltrão Marques, *Natureza em Boiões*, 132–33.
[53] Ibid., 134. See also Bernardino António Gomes, 'OFÍCIO de Bernardino António Gomes ao (secretário de estado da Marinha e Ultramar), visconde de Anadia (João Rodrigues de Sá e Melo Meneses e Souto Maior),' *Arquivo Histórico Ultramarino*, 017 – RIO DE JANEIRO – CATÁLOGO DE DOCUMENTOS MANUSCRITOS AVULSOS / Cx. 243, D. 16604, Lisboa, 1807-03-20.

cinchonas, or as creditable substitutes – and added further to the between 450 and 1,150 tons of cinchona in circulation.[54]

For these trade volume figures to become meaningful, and to ascertain how many people contemporaries' 'humanity' comprised, it is critical to understand how much cinchona was generally administered to a sufferer. While vernacular manuals, domestic recipe collections and medical treatises almost invariably offered precise counsel on the individual doses of bark required and their rate of application, sources rarely fixed an overall requisite quantity of cinchona to be administered. Rather, recipe collectors, physicians and apothecaries in different parts of the world were in agreement that the quantity of bark necessary for a cure was 'very different in different cases'.[55] Several writers advised, in case the suggested dose did not cure the sufferer, to 'repeat the same procedure',[56] administer a second or third dose,[57] or simply continue the administration of the medicine at regular intervals for as long as necessary. As Lady Eleanor Dundas (d. 1837) of Carron Hall put it in her recipe book, one was to 'repeat the medicine' 'untill [the Patient] misses the fit', or as the authors of the *Edinburgh new dispensatory* phrased it, 'till the paroxisms cease[d]' and the sufferer's 'appetite, strength, and complexion return[ed]'.[58] The authors' method of adjusting dosage to the sufferer's condition not only responded to differences in the 'obstinacy' (*Hartnäckigkeit*), or rebelliousness of fevers, and other ailments. It also complied with the period's general emphasis upon the variability of

---

[54] Timothy Walker refers to Brazilian cinchona exports. Walker, 'The Medicines Trade in the Portuguese Atlantic World,' 27.

[55] Lewis and Rotheram, *The Edinburgh new dispensatory*, 143.

[56] A cinchona-based recipe for an ague remedy pasted into Miss Myddleton's collection recommended 'repeat[ing] the dozes again' 'if the fit return'. 'Receipts copied from Miss Myddleton's Book.' The Portuguese recipe collection compiled by João de Jesus Maria likewise recommended repeating the 'same procedure' if the remedy did not cure the sufferer. João de Jesus Maria, 'Colleção medica de receitas p[ar]a quasi todos os achaques a q[ue] está sugeita a natureza humana, e muitas uezes exprimentadas com bom sucesso pellos melhores medicos, cirurgiõis, e quimicos deste reino, e estrangeiros: e muitas dellas tidas por segredos quasi infaliveis de cazas illustrissimas, e medicos grandes deste reino em muitos annos,' *Biblioteca Nacional de Portugal,* Manuscritos Reservados, COD. 5077, Lisboa, 1760, 43. A French recipe collection likewise counselled sufferers to 'repeat the same remedy if the fever returned'. 'Collection of medical receipts, with a few household and veterinary receipts: in French,' *Wellcome Library,* Archives and manuscripts, Closed stores WMS 4, MS.4087, n.p., n.d., 35.

[57] Luís Gomes Ferreira, 'Erário mineral,' in *Erário mineral,* ed. Júnia Ferreira Furtado (Rio de Janeiro: Editora FIOCRUZ, 2002 (1735)), 516.

[58] Lady Eleanor Dundas, 'Collections of medical and cookery receipts in English, by several hands,' *Wellcome Library,* Archives and manuscripts, MS.2242, Falkirk / London, c. 1785, 22. The authors of *The Edinburgh new dispensary* advised their readership to continue the administration until the sufferer's 'appetite, strength, and complexion return[ed]'. Lewis and Rotheram, *The Edinburgh new dispensatory,* 143.

individual constitutions and the resultant necessity of tailoring dosage to the patient's individual needs.[59] To ascertain standard dosage, at least approximately, one can only rely on the authors' references to the maximum quantities administered, the average doses dispensed or the doses found most beneficial in the clinical observations and experiences conducted at the time to evaluate treatments. Most Spanish American medical treatises, advice literature and recipe collections, for instance, advised doses ranging from half a drachm – less than 2 grams – to 1 ounce – some 29 grams – that is, less than 16 grams on average.[60] In Europe, with bark of low quality or in particularly violent illnesses, physicians occasionally saw themselves forced to administer by their own standards extravagant doses to effect a cure.[61] Experienced medical practitioners implicated in putting different or newly discovered varieties of cinchona on trial to gain 'a proper understanding of their virtues',[62] however – at the Lyon Grand Hôtel-Dieu hospital, the Spanish Court or

---

[59] For references to the necessity of adapting dosage to constitution, see Murray, *Vorrath von einfachen, zubereiteten und gemischten Heilmitteln*, 1, 1131. See also Buchan, *Domestic Medicine*, 176. See also 'Pareceres de los médicos sobre los efectos de la Quina de Santa Fé,' *Archivo del Palacio Real*, Papeles del Almacén de la Quina, Caja 22283 / Expediente 1, Madrid, 1784-12-19 / 1784-12-18. On New Spain, see Juan de Esteyneffer, *Florilegio medicinal de todas las enfermedades, sacado de varios, y clasicos autores, para bien de los pobres, y de los que tienen falta de Medicos, en particular para las Provincias remotas, en donde administran los RR. PP. Missioneros de la Compañia de Jesus* (Mexico: Herederos de Juan Joseph Guillena Carrascoto, 1712), 296. For secondary sources on the variability of dosage, depending on the sufferer's constitution, see, for instance, Roy Porter, 'The Eighteenth Century,' in *The Western Medical Tradition*, ed. Lawrence I. Conrad, et al. (Cambridge: Cambridge University Press, 1995), 87; Emma Spary, 'Health and Medicine in the Enlightenment,' in *The Oxford Handbook of the History of Medicine*, ed. Mark Jackson (Oxford: Oxford University Press, 2011).

[60] According to Hipólito Ruiz López, it was common to administer from 'half a drachm to two or more drachms', that is, between 1.94 and 7.76 grams, on average some 4.85 grams. Ruiz López, *Quinología*, 40. A Mexican recipe collection advised 'six drams', some 23.28 grams. See recipe number 223 – 'For Tertian Fevers' (*Para Fiebres tersianas*) – in 'Secretos medicos, y chirurgicos,' *Wellcome Library*, Archives and manuscripts, Closed stores WMS / Amer. 22, Mexico, n.d. The author of a Lima recipe collection advised one-eighth of an ounce, some 3.6 grams. Anon., 'El Medico verdadero. Prontuario singular de varios selectisimos remedios, para los diversos males à que està expuesto el Cuerpo humano desde el instante que nace. Compuesto por un curioso, para el alivio de todos los que se quieran curar con èl,' in *La medicina popular peruana*, ed. Hermilio Valdizán and Angel Maldonado (Lima: Imprenta Torres Aguirre, 1922 (1777)), 446. The highest dose advised in the sources under consideration was 'one ounce' of powdered cinchona, that is, some 30 grams. Villalobos, *Método de curar tabardillos*, 100–01; William Buchan, *Medicina doméstica* (Madrid: Imprenta Real, 1785), 54.

[61] See, for instance, Johann Claudius Renard, *Die inländischen Surrogate der Chinarinde in besonderer Hinsicht auf das Kontinent von Europa* (Mainz: F. Kupferberg, 1809), 9. See also Lochbrunner, *Der Chinarindenversuch*, 36.

[62] 'Decretos do príncipe regente.'

London's Royal College of Physicians – generally recommended, like their Spanish American counterparts, doses ranging from half a drachm to one or two ounces.[63] So did medical practitioners in societies within or adjoining Europe's colonial, evangelizing and commercial entrepôts in North America, the eastern Mediterranean or western Africa. From surgeons in Portuguese Angola who prescribed an ounce and a half of the bark[64] to Ottoman doctors in Bursa who administered doses ranging from 8 to 16 drachms, that is, 25 to 50 grams,[65] to physicians in the West Indies who generally administered 2 ounces of the bark,[66] medical practitioners the Atlantic World over would have agreed that more than 2 ounces of bark was an unduly large dose and that less than half a drachm was too low a prescription. Assuming that doses fluctuated around an average of 30 grams in societies rimming the Atlantic basin, the around 450 to 1,150 tons of bark traded per annum could have been administered in 15 to 38 million sickness episodes every year in the decades around 1800. At a time when Europe had a population of some 80 to 90 million – England, the Netherlands, the Habsburg territories, France, the Italian peninsula and Spain had 71.7 million inhabitants in 1750 and 121.7 million in 1850 – and the world a population of between 771 and 954 million inhabitants,[67] the bark might, potentially at least, have reached a rather substantial portion of humanity.

[63] A circular letter published by the Royal College of Physicians, dated November 15, 1799, recommended a dose of 'half a drachm in substance' of 'the red Bark'. See the article, pasted into 'Miss Myddleton's Book': 'Receipts copied from Miss Myddleton's Book.' In their clinical trials with Santa Fé bark on a thirteen-year-old girl, the inmate of a poorhouse in Jaen, Spanish doctors found that less than 2 ounces sufficed to cure her from quartan fevers, while an eighteen-year-old fieldworker, suffering from 'tertian fevers', convalesced after three-quarters of an ounce. 'Pareceres de los médicos sobre los efectos de la Quina de Santa Fé.' Doctor Michael O'Ryan at the Lyons hospital held that 'one or two ounces' of yellow bark were equal to curing a fever, while he required twice its amount of red bark. The Lyons hospital kept 'the most complete pharmaceutical collection in Europe', and its doctors had the necessary experience with a range of barks and sufferers to correlate a bark's efficacy with its colour. Michael O'Ryan, *A Letter on the Yellow Peruvian Bark, Containing an Historical Account of the first Introduction of that Medicine into France, and a Circumstantial Detail of Its Efficacy in Diseases, Addressed to Dr. Relph, Physician to Guy's Hospital* (London: J. Nunn, 1794), 20. Portuguese physicians usually found doses of between 11 and 60 grams adequate for their needs. Velloso, *Quinografia Portugueza*, 93–94.
[64] Pinto de Azeredo, *Ensaios sobre algumas enfermidades d'Angola*, 64.
[65] Günergun and Etker, 'From Quinaquina to "Quinine Law,"' 45; Aydüz and Yildirim, 'Bursalı Ali Münşî ve Tuhfe-i Aliyye.'
[66] James Clark, a physician on the island of Dominica, tended to begin with 'an ounce of bark in substance'; the same quantity given in the second remission would generally prevent a third 'fit'. Clark, *A Treatise on the Yellow Fever*, 82.
[67] The world had 771 million inhabitants in 1750 and some 954 million inhabitants in 1800. Massimo Livi-Bacci, *A Concise History of World Population* (Oxford: Blackwell, 1992), 26; 69.

Geographies of Consumption

There is overwhelming evidence that the bulk of these hundreds of tons and millions of doses of cinchona would have gone into the broader European markets, with a shifting tangle of Amsterdam,[68] Marseille,[69] Hamburg,[70] London[71] and Genoese merchant houses – the latter supplied 'not only all of Liguria' with cinchona from Cádiz, but also the remainder of the Italian Peninsula, the Swiss Confederacy and part of the Levant[72] – redistributing the bark from its ports of entry across Europe. By the late 1700s and early 1800s, cinchona was part of the standard medical repertoire across Europe, from the electorate of Hannover[73] to the Kingdom of Portugal,[74] from the Dutch

---

[68] According to Saul Jarcho, Amsterdam was a 'world center of the cinchona trade' between the late seventeenth and the late eighteenth century. Jarcho, *Quinine's Predecessor*, 207.

[69] Patrick Boulanger, 'Droguistes marseillais à la fin du XVIIIe siècle,' in *Herbes, Drogues et Epices en Mediterranée*, ed. George J. Aillaud (Paris: Editions du Centre National de la Recherche Scientifique 1988), 49. Spanish officials saw Marseille, next to Rouen, as the key port for the French bark trade. Jijon y León, 'Recomendaciones,' 144.

[70] Frigates returning from Callao, Cartagena and Guayaquil between 1801 and 1808 carried large cargoes of cinchona to Hamburg: the frigate *Juliana*, returning from Callao in 1801, carried 98,237 pounds (*Pfund*) of cinchona; the frigate *Wilhelm und Albert*, returning from Cartagena, carried 298 *quintales* or 24 pounds (*Pfund*); the frigate *Wilhelmsburg*, which left Callao in 1807, carried 300,136.5 pounds (*Pfund*) of bark. The *Juan Paris* carried 90 tons (198,068 *Pfund*) of cinchona; the frigate *Cesar Peter* carried 193,572.5 pounds (*Pfund*), some 87 tons, in 1808. Hans Pohl, *Die Beziehungen Hamburgs zu Spanien und dem Spanischen Amerika in der Zeit von 1740 bis 1806* (Wiesbaden: Franz Steiner Verlag GmBH, 1963), 271; 77; 79–80.

[71] On London, and England, as an international entrepôt for drugs and other commodities, see Wallis, 'Exotic Drugs and English Medicine.'

[72] According to the Spanish ambassador in Genoa, by 1785, 900 boxes with cinchona lay in the port of Genoa ready to be shipped. Juan Cornejo, 'Informe de Don Juan Cornejo Ministro de V.M. en Genova,' *Archivo General de Simancas*, Legajo 961/2, Genoa, 1785. Genoese merchants played a particularly important role in re-distributing American produce from Cádiz to the Mediterranean and northern European ports. Catia Brilli, 'Mercaderes genoveses en el Cádiz del siglo XVIII. Crisis y reajuste de una simbiosis secular,' in *Comunidades transnacionales: colonias de mercaderes extranjeros en el mundo atlántico (1500–1830)*, ed. Ana Crespo Solana (Aranjuez: Ediciones Doce Calles, 2006).

[73] Gabriele Beisswanger, *Arzneimittelversorgung im 18. Jahrhundert. Die Stadt Braunschweig und die ländlichen Distrikte im Herzogtum Braunschweig-Wolfenbüttel* (Braunschweig: Deutscher Apotheker-Verlag, 1995), 260–64.

[74] On the availability of cinchona in Portuguese pharmacies, see the report by Manoel José de Souza, signed in Lisbon, April 17, 1799: 'Neste Secretaria da Junta do Proto-Medicato da Repartiçao de Medicina se achaõ enformes que deram os Medicos d'esta Corte, e Provincias, relativamente a os Queitos a que se lhes mandou responder por Ordem da mesma Junta dos quaes o seu theor he o Seguinte,' *Arquivo Nacional da Torre do Tombo*, Ministério do Reino / Negócios diversos do Físico-Mor, Maço 469/ Caixa 585, 4, Lisboa, 1799-04-17 / Tavira, 1799-05-06. Portugal imported some cinchona bark from neighbouring Spain. Eugenio D. Larruga, *Memorias políticas y económicas sobre*

Republic[75] to Habsburg Transylvania.[76] It was available not only in trading hubs and capitals, but also in more provincial towns and cities such as Portuguese Évora,[77] Bender, in the Ottoman principality of Moldavia[78] and Brunswick, in the Hannover Electorate.[79] The bark held pride of place among medicinal imports from the Americas in many areas. It made up 40 per cent of all direct American drug imports into England after 1720, for instance,[80] and was among the most common and renowned of all foreign remedies – arriving, other than from the Americas, from the Levant or the territories bounding the Indian Ocean – in European pharmacies and dispensaries like the *Hôtel-Dieu de Carpentras*[81] or the *Hospital Escolar da Universidade de Coimbra.*[82] Indeed, the value and weight of cinchona imports into important transhipment ports like Cádiz,[83] Hamburg[84] and London[85] were often greater than

los frutos, comercio, fábricas y minas de España, vol. XXXV (Madrid: Por don Antonio Espinosa, 1795), 55.
[75] See, for instance, Terne, *Verhandelingen.*
[76] On the administration of cinchona in Habsburg Transylvania, see Lochbrunner, *Der Chinarindenversuch*, 35–36.
[77] On Évora, see the report written by Manoel Francisco de Carvalho on April 28, 1799, 'Enformes que deram os Medicos d'esta Corte.'
[78] The traveller Balthasar von Campenhausen found the Bender pharmacy, run by a Jewish apothecary, to have 1 pound of 'fever bark (*Fieberrinde*)' in store. Balthasar von Campenhausen, *Bemerkungen über Rußland, besonders einige Provinzen dieses Reiches und ihre Naturgeschichte betreffend* (Leipzig: Friedrich Christian Dürr, 1807), 148.
[79] On cinchona in Brunswick, see Beisswanger, *Arzneimittelversorgung im 18. Jahrhundert*, 260–70.
[80] Wallis, 'Exotic Drugs and English Medicine,' 32–33.
[81] A 1763 inventory of the plant-based remedies kept at the Hôtel-Dieu de Carpentras in southern France documents that around half of its remedies were 'home-grown' Mediterranean plants, while the other half came from abroad: 28 per cent from the contiguous Levant, 11 per cent from the territories bounding the Indian Ocean and 11 per cent from the 'New World', with cinchona being principal among those. Colette Dubois, 'Le quotidien d'une pharmacie hospitaliere: la boutique de l'Hôtel-Dieu de Carpentras,' in *Herbes, Drogues et Epices en Mediteranée*, ed. George J. Aillaud (Paris: Editions du Centre National de la Recherche Scientifique 1988), 90.
[82] Cinchona was the 'most extensively consumed drug' (*a droga mais consumida*) in the Coimbra Teaching Hospital in the late eighteenth century. João Rui Pita and Ana Leonor Pereira, 'A arte farmacêutica no século XVIII, a farmácia conventual e o inventário da Botica do Convento de Nossa Senhora do Carmo (Aveiro),' *Ágora. Estudos Clássicos em Debate* 14, no. 1 (2012), 231–32.
[83] García-Baquero González, *Cádiz y el Atlantico*, 1, 340. See also Gänger, 'World Trade in Medicinal Plants,' 47.
[84] Ulrich Pfister and Christine Fertig, 'Coffee, Mind and Body: Global Material Culture and the Eighteenth-Century Hamburg Import Trade,' in *The Global Lives of Things. The Material Culture of Connections in the Early Modern World*, ed. Anne Gerritsen and Giorgio Riello (London: Routledge, 2016), 229.
[85] Cinchona 'stands out for its exceptional importance' among all North and South American and Mesoamerican drugs imported into England in the second half of the eighteenth century and was the second most important drug in value imported into

those of any other – or most, in the latter case – medicinal substances, and invariably larger than those of any other American medicine, be that ipecacuanha, 'Virginian snake-root' or guaiacum. In some parts of Europe 'not a single' foreign plant remedy was employed 'as often and in such large quantities' (*so oft und dabei in so grosser Menge*) as cinchona, as the Mainz doctor Johann Claudius Renard put it in 1809,[86] or, as the Milan author Luigi Castiglioni (1757–1832) phrased it, 'could compare to cinchona (*kina-kina*), [...] for the extensive use made of it to this day'.[87] Levels of cinchona consumption were high, even 'absurd' (*ungeheuer*) in some parts in the eyes of contemporaries.[88] In peninsular Spain, with a population of 10.4 million in the late 1700s,[89] domestic consumption amounted to at least 20 tons per annum by 1792,[90] enough for 1.3 million doses. The amounts of cinchona imported into England during the 1750s, in turn, provided between 300,000 and 1,186,000 doses, and between 112,000 and 449,000 doses during the 1770s,[91] at a time when England had but around 6 million inhabitants.[92]

Cinchona was widely available across the Spanish and Portuguese American possessions, too. It could be obtained from any well-stocked pharmacy in the Viceroyalties of New Granada,[93] New Spain[94] and

England from 1752 to 1754, and the seventh most important in value between 1772 and 1774. Wallis, 'Exotic Drugs and English Medicine,' 31–33.
[86] Renard, *Die inländischen Surrogate der Chinarinde*, 3.
[87] Castiglioni, *Storia delle piante forastiere le piu importanti*, 45.
[88] Renard, *Die inländischen Surrogate der Chinarinde*, 3.
[89] There were 10.4 million persons counted in the Kingdom of Spain in the summer of 1787. Livi-Bacci, *A Concise History of World Population*, chapter 1.
[90] According to contemporaries, in the year 1792, Spain imported 716,734 *libras* – 330 tons – of which 674,102 *libras* – 298 tons – were re-exported, and 42,633 *libras* – 20 tons – consumed domestically. Hernández de Gregorio, *El arcano de la quina*, v. Other contemporaries estimated that in the five years following 1771, one-tenth of the cinchona imports that reached Cádiz remained within peninsular Spain to cover the country's needs. Miguel de San Martin Cueto, 'Razon de las libras de cascarilla que se han extrahido para fuera del reyno,' *Archivo General de Indias*, Indiferente 1554, Cádiz, 1776-10-25. Cinchona consumption in eighteenth-century Spain has been researched comparatively in depth. Juan Riera Palmero, 'La Medicina en la España del siglo XVIII,' in *Medicina y Quina en la España del siglo XVIII*, ed. Juan Riera Palmero (Valladolid: Universidad de Valladolid, 1997); María Luisa de Andrés Turrión, 'Quina de la Real Hacienda para el ejército español en el siglo XVIII,' in *Guerra y milicia en la España del X Conde de Aranda. Actas del IV Congreso de Historia Militar*, ed. José A. Armillas Vicente (Zaragoza: Gobierno de Aragon, Departamento de Cultura y Turismo, 1998); Riera Palmero, 'Quina y malaria en la España del siglo XVIII.'
[91] Wallis, 'Exotic Drugs and English Medicine,' 34.
[92] Edward Anthony Wrigley and Roger S. Schofield, *The Population History of England: 1541–1871: A Reconstruction* (Cambridge: Cambridge University Press, 1989), 210.
[93] On the consumption of local cinchona varieties in Santa Fé, see also Salvador Vázquez, 'Las quinas del norte de Nueva Granada,' 55.
[94] Esteyneffer, *Florilegio medicinal*, 279.

Peru – in Lima, but also in provincial towns like Trujillo, Pisco, Ica, Huancavelica, Moquegua and Cuzco[95] – as well as in Brazil, where cinchona and cinchona-based compound medicines were sold from Rio de Janeiro to Bahia and from Pernambuco to Maranhão.[96] Contemporary estimates suggest an annual consumption of some 12,000 *libras*,[97] around 344,000 doses, for the entirety of Spanish America, at a time when these dominions had about 13.5 million inhabitants.[98] Internal consumption in the Viceroyalty of Peru would appear to have been comparatively low, too, amounting to between 0.4 and 1.4 tons of cinchona – between 25,000 and 86,000 doses per annum – that is, less than 1 per cent of the overall exports.[99] Actual consumption may have been higher than official figures suggest, however. Given that colonial officials often complained that the bark was sold on the streets and 'on Fridays in the marketplace' in cities like Santa Fé, and in the 'villages adjacent to the hills' where it was harvested,[100] we may venture to presume that cinchona was distributed through a variety of formal and informal channels in Spanish and Portuguese America by the late 1700s and early 1800s. It was likely employed in far more than the – 25,000,

---

[95] Pablo Macera, *Precios del Perú XVI–XIX. Fuentes* (Lima: Fondo Editorial / Banco Central de la Reserva, 1992), 131.

[96] Danielle Sanches de Almeida, 'Entre lojas e boticas: O comércio de remédios entre o Rio de Janeiro e Minas Gerais (1750–1808)' (Universidade de São Paulo, 2008), 84. See also *Documentos régios que authorizão a verdadeira Agoa de Inglaterra, da composição do doutor Jacob de Castro Sarmento, manipulada presentemente por José Joquim de Castro, na sua Real Fabrica, por decreto de sua Alteza Real o Principe Regente nosso Senhor. Com huma relação dos professores de medicina, e cirurgia deste reino de Portugal e seus Dominios, que tem attestado a excellencia da dita Agoa de Inglaterra* (Lisboa: Impressão Régia, 1809).

[97] Ruiz López, *Quinología*, 14.

[98] Nicolás Sánchez-Albornoz, 'The Population of Colonial Spanish America,' in *The Cambridge History of Latin America*, ed. Leslie Bethell (Cambridge: Cambridge University Press, 1984), 34.

[99] According to Cosme Bueno's estimates, the entirety of the Viceroyalty of Peru consumed some 'eight to nine quintales' – 414 kilograms – in 1785. Cosme Bueno, 'Carta al S. Visitador y Superior Intendente General,' *Biblioteca Nacional del Perú*, Expediente sobre el Estanco de la Cascarilla, Sección Manuscritos – C388, Lima, 1785-09-19. The consulate that same year estimated that consumption amounted to 2,400 *libras* – that is, 1,104 kilograms – while Sebastian Mena, then in charge of inspecting Lima's pharmacies, arrived at an annual consumption of 3,000 *libras* – 1,380 kilograms – for the Viceroyalty. For a discussion of these sources, see Jaramillo Baanante, 'El comercio de la cascarilla,' 52.

[100] Sebastián José López Ruiz, 'Representación.' *Archivo General de Indias*, Indiferente 1554, n.p., 1779-11-22, fls. 117–18; Sebastián José López Ruiz, 'Para los efectos que convengan remite copia de una representación que ha dirigido al Virrey en que refiere los abusos y desordenes que se cometen en la colectación beneficio y comercio de las quinas de aquel reyno,' *Archivo General de Indias*, Indiferente 1557, Santa Fé, 1804-08-13, 639.

86,000 or 344,000 – sickness episodes per annum that official trade figures would have allowed for.

A substantial share of British, Portuguese, French, Dutch and Spanish cinchona imports was re-exported to these Atlantic realms' imperial, commercial and evangelizing entrepôts – primarily in the Caribbean and North America, along the African coast and in South and East Asia.[101] The British, Dutch and French West Indies were important consumer markets for the bark.[102] So were the British and French North American colonies – or, after 1776, the United States – with New England pharmacies, Louisiana hospitals' provisions and plantation medicine chests in the antebellum South encompassing cinchona as a staple.[103] Portuguese merchants shipped bulk quantities of cinchona and cinchona-based medicines to its Lusophone enclaves along the African coast – to Mozambique, Benguela and Luanda, western Africa's largest slaving port – and to Timor, Goa and Macao, leased from China in 1557.[104] So did the Dutch West and East India companies (*Vereenigde Oost-Indische Compagnie*, or VOC; *Geoctroyeerde Westindische Compagnie* or WIC), which transported cargoes of cinchona to the Dutch Antilles, the Guyana colonies, the forts and lodges on the Gold Coast of West Africa and the Dutch East Indies.[105] British trade companies, according to contemporary statistics, re-exported in the five years from 1789 to 1793 some 123,779 pounds of cinchona – between

---

[101] In the years 1772 to 1774, for instance, London's drug re-exports – into other European markets and British colonies – amounted to 53 per cent of the value of imports. Wallis, 'Exotic Drugs and English Medicine,' 28.

[102] See Chapter 4.

[103] On New England, see Norman Gevitz, '"Pray Let the Medicine be Good": The New England Apothecary in the Seventeenth and Early Eighteenth Centuries,' in *Apothecaries and the Drug Trade. Essays in Celebration of the Work of David L. Cowen*, ed. Gregory J. Higby and Elaine C. Stroud (Madison, Wis.: American Institute of the History of Pharmacy, 2001), 15. On cinchona consumption in Louisiana under Spanish rule, see 'Colonie de la Louisiane,' *Archivo del Palacio Real*, Suministros de Medicinas y Papeles de la Real Botica 1671–1781, Caja 22284, Madrid, n.d. On plantation medicine chests in the South, see Sharla M. Fett, *Working Cures. Healing, Health, and Power on Southern Slave Plantations* (Chapel Hill: University of North Carolina Press, 2002), 68–69.

[104] On the availability of cinchona in Timor and Mozambique, see Walker, 'The Medicines Trade in the Portuguese Atlantic World,' 27. On cinchona consumption in Luanda, see Pinto de Azeredo, *Ensaios sobre algumas enfermidades d'Angola*, 64. On shipments of cinchona-based patent medicines to São Filipe de Benguala, see 'Doutor Juiz de Fora Prezidente, Vereadores e Procurador da Camera abaixo asignados da Cidade de São Filippe de Benguella, e Sua Capitania por Sua Alteza Real o PRÍNCIPE REGENTE Nosso Senhor, que Deus Guarde,' *Arquivo Nacional da Torre do Tombo*, Ministério do Reino / Negócios diversos do Físico-Mor / Maço 469, Caixa 585, 22, São Filipe de Benguela, 1800-09-02.

[105] Rutten, *Dutch Transatlantic Medicine Trade*, 12–13.

3.7 and 1.9 million doses – to Britain's 'colonial possessions in the East and West Indies'.[106] Spanish American exports of cinchona ointments, powders and extracts and of unprocessed red, white and orange bark from Acapulco via the Manila galleon,[107] in turn, though by all accounts small in comparison to Cádiz-bound freights, were presumably both destined for the Spanish settler population on the Philippines and resold to China via Cantonese merchants, whose intermediation allowed traders from Spain and Spanish America to participate in Asian commercial networks.[108]

Indeed, cinchona was widely consumed not only in Europe's colonial, evangelizing and commercial entrepôts in the Atlantic World but also far beyond. Its use was popular and prevalent wherever Iberian and English, Dutch or French imperialism, proselytizing and trade intersected with, or submitted to the rules of, other – Levantine, Mediterranean and Cantonese – trade networks. The bark came to the knowledge of Chinese physicians from the cosmopolitan, populous maritime entrepôt of Canton and from the 'barbarians at Macao'[109] – a term that presumably encompassed both Portuguese settlers and members of the Jesuit order who, before their expulsion from all Portuguese territory after 1759, had popularized cinchona from their pharmacy at St Paul's College

[106] According to the statistics compiled by John Relph in 1794, between 1789 and 1793 Britain imported a total of 634,783 pounds of ground cinchona, of which 123,779 were exported and 511,004 pounds stayed in the country. John Relph, *An Inquiry into the Medical Efficacy of a new Species of Peruvian Bark, Lately Imported into this Country under the Name of Yellow Bark: Including Practical Observations Respecting the Choice of Bark in General* (London: James Phillips, 1794), 2.

[107] Archival sources studied by Mexican historians reveal that, between 1772 and 1809, seven Manila galleons left the port of Acapulco with cinchona or cinchona-based medicines. Shipments sailed in 1772, 1775, 1799; two shipments in 1801; one in 1807; and another one in 1809. Reyna María Pacheco Olivera, 'Análisis del intercambio de plantas entre México y Asia de los siglos XVI al XIX' (unpublished master's thesis, Universidad Nacional Autónoma de México, 2006), 126. On the Manila galleon trade in cinchona, see also Leon Borja, 'Algunos datos,' 96.

[108] Katherine Bjork, 'The Link That Kept the Philippines Spanish: Mexican Merchant Interests and the Manila Trade, 1571–1815,' *Journal of World History* 9, no. 1 (1998). Bjork's argument is consistent with the position taken by Flynn and Giráldez, who argue that Europeans participated in existing Asian commercial networks. Dennis O. Flynn and Arturo Giráldez, 'Born with a "Silver Spoon": The Origin of World Trade in 1571,' *Journal of World History* 6, no. 2 (1995). On the history of the galleon trade see Dennis O. Flynn, Arturo Giráldez and James Sobredo, eds, *European Entry into the Pacific. Spain and the Acapulco-Manila Galleons* (Ashgate: Variorum, 2001). In 1757, the Qing court changed its policy towards Western trade, consigning it to a single port, Canton. Fa-ti Fan, *British Naturalists in Qing China. Science, Empire, and Cultural Encounter* (Cambridge, Mass.: Harvard University Press, 2004), 6.

[109] That quote is taken from Chao Hsüeh-min's (1719–1805) 'Addenda and corrigenda to the Pen-ts'ao kang-mu (Pen-ts'ao kang mu shih-i)', cited in Unschuld, *Medicine in China*, 166.

(*Colégio de São Paulo*) in Macao.[110] Other than through the veins of Portuguese medicine trade,[111] Chinese sufferers would also have procured the bark through Spain's Asian commerce in the bark, a sector that Spanish officials were eager to expand in the late 1700s.[112] Cinchona was also among the few goods imported into Nagasaki, Japan's only governmentally sanctioned point of entry for foreign merchants, by the Dutch East India Company. Following the expulsion of the Portuguese in 1638, only Chinese merchants and representatives of the Dutch East India Company were allowed to enter Japan to engage in trade, and the commodities brought were dictated by the – usually rather explicit – requests and regulations of the shogunate.[113] In the Maghreb, cinchona had long been a valued remedy, which its inhabitants obtained through the region's long-standing participation in Mediterranean trade and contraband with Genoa, Catalonia, Marseille and Venice – and later, Britain, Denmark and France – and also from the Spanish pharmacies in Melilla and the Peñón de Alhucemas, just off the Moroccan coast. Madrid's Royal Pharmacy often supplied Spanish apothecaries directly with the necessary medicaments – among them, at times, several hundred kilograms of cinchona.[114] Other 'oriental nations', according to

---

[110] Recipes retrieved from the 1766 'Collection of various recipes and particular secrets from the principal pharmacies of our order in Portugal, India, Macao, and Brazil' (*Colleccão de varias receitas e segredos particulares das principaes boticas da nossa Companhia de Portugal, da India, de Macáo e do Brasil*), a 633-page manuscript compilation, document that Jesuit pharmacies in Macao had long relied on cinchona for several of their recipes and manufactured numerous cinchona-based compound remedies. For transcriptions and excerpts from that collection, see Sabine Anagnostou, *Missionspharmazie. Konzepte, Praxis, Organisation und wissenschaftliche Ausstrahlung* (Stuttgart: Franz Steiner Verlag, 2011), 292; Ana Maria Amaro, *Introdução da medicina ocidental em Macau e as receitas de segredo da Botica do Colégio de São Paulo* (Macau: Instituto Cultural de Macau, 1992).

[111] Walker, 'The Medicines Trade in the Portuguese Atlantic World,' 19.

[112] Viceroy Antonio Caballero y Góngora repeatedly mentioned plans to increase Asian cinchona consumption 'through Acapulco and the Philippines' for the benefits of the Royal Treasury. Antonio Caballero y Góngora, 'Carta.' *Archivo General de Indias*, Indiferente 1555, Santa Fé, 1786-10-19; Antonio Caballero y Góngora, 'Carta al Marquès de Sonora,' *Archivo General de Indias*, Indiferente 1554, Santa Fé, 1786-10-19.

[113] The list of imports was drawn from a review of inventories from the second half of the eighteenth century, found in *Het archief van de Nederlandse factorij in Japan*, 'Dejima Dagregisters' (*Algemeen Rijksarchief*, The Hague), three sample inventories of which are reproduced in Martha Chaiklin, *Cultural Commerce and Dutch Commercial Culture. The Influence of European Material Culture on Japan, 1700–1850* (Leiden: CNWS, 2003), 78–191. See also Robert Liss, 'Frontier Tales: Tokugawa Japan in Translation,' in *The Brokered World. Go-Betweens and Global Intelligence, 1770–1820*, ed. Simon Schaffer et al. (Sagamore Beach: Watson Publishing International, 2009), 9.

[114] 150 *libras* of cinchona reached the Spanish pharmacies in 1784, 204 *libras* in 1785, 877 *libras* in 1786 and 200 *libras* in 1787, a total of 1,431 *libras*, or 658 kilograms. Andrés Turrión, 'Quina de la Real Hacienda para el ejército español en el siglo XVIII,' 422.

Spanish cinchona merchants' complaints, procured 'this excellent febrifuge' through English and other European exporters at the Smyrna market, in the Ottoman Empire.[115] According to Ottoman physicians, the bark had at first been known only in Constantinople,[116] but by the early 1700s, 'sailors and other travellers had popularized it far beyond [...], in other towns and lands'.[117] The city of Smyrna on the Aegean coast was not only, according to contemporary observers, itself an important consumer market for the bark, but also a cosmopolitan entrepôt whence cinchona was 'distributed (*se derrama*) throughout the Asian portion of Turkey (*la Turquía Asiática*), in very large quantities (*en muy gruesas quantidades*)'.[118] Smyrna had long attracted factors from Amsterdam, London, Marseille and Venice as well as Ottoman Armenian and Jewish merchants. It remained, by the late 1700s and early 1800s, an important entrepôt provisioning Anatolia and the Syrian and Persian markets with produce from western Europe.[119] The population of the sprawling Ottoman Empire – extending, at the time, from Bosnia in the west to the Mesopotamian provinces of Basra, Baghdad and Mosul in the east – also procured supplies of the bark through the port of Cairo. Hundreds of kilograms reached the Ottoman palace from that Mediterranean entrepôt every year in the early 1800s.[120] The volume of Chinese, Moroccan, Japanese and Ottoman cinchona commerce, in part because it largely consisted of contraband, is elusive, but there can be little doubt that consumption and demand, 'in the countries in Africa and Asia (*los países de África y Asia*)' where cinchona was known, and which European merchants frequented, was significant.[121] As Miguel de Jijon y León (1717–1794) put it in 1776, the bark was 'of the greatest necessity and, use, all over the world' (*de tan precisa necesidad y uso en todo el Mundo*), but 'particularly among the Asians' (*especialmente entre los Asiáticos*)[122] – a term

---

[115] 'Testimonio de los Autos,' 972. See also Jijon y León, 'Recomendaciones,' 131.
[116] Edhem Eldem, *French Trade in Istanbul in the Eighteenth Century* (Leiden: Brill, 1999), 86. On cinchona consumption in the Ottoman Empire, see also Günergun and Etker, 'From Quinaquina to "Quinine Law,"' 50.
[117] Aydüz and Yildirim, 'Bursalı Ali Münşî ve Tuhfe-i Aliyye,' 92.
[118] Jijon y León, 'Recomendaciones,' 131.
[119] Elena Frangakis-Syrett, *The Commerce of Smyrna in the Eighteenth Century (1700–1820)* (Athens: Centre of Asia Minor Studies, 1992), 35. On the history of Izmir and its merchant community, see Daniel Goffman, 'Izmir: From Village to Colonial Port City,' in *The Ottoman City between East and West. Aleppo, Izmir, and Istanbul*, ed. Edhem Eldem, Daniel Goffman and Bruce Masters (Cambridge: Cambridge University Press, 2001).
[120] Campenhausen refers to some 300 okas, or 384 kilograms, per annum. Campenhausen, *Bemerkungen über Rußland*, 192–93.
[121] Pombo, 'Carta de José Ignacio de Pombo a José Celestino Mutis.'
[122] Jijon y León, 'Recomendaciones,' 144.

that encompassed, by the late 1700s, not only the inhabitants of the Chinese and Mughal empires and Tokugawa Japan but also men and women of Arabic, Turkish or Persian extraction (FIGURE 2.2).[123]

Limits to Distribution

Cinchona not only reached geographically disperse societies. The social depth of its consumption was just as varied and wide. Among the world's ruling elites, cinchona was widely known by the late 1700s and early 1800s. The shogunate in Japan,[124] the Tsar and his kin in Russia[125] and the Kangxi Emperor in China[126] valued and kept supplies of the bark for their own use. So did some ruling families in British and Mughal India: Muhammed Ali Khan, the Nawab of Arcot (1717–1795, r. 1749–1795), for instance, had the British physician Paul Jodrell (1746–1803) administer the bark – to good effect – in the treatment of his youngest son.[127] The Moroccan 'Alawi court likewise prized cinchona. Gift exchange was assiduous between Charles III and the Sultan of Morocco, Mohammed Ben 'Abd Allāh al-Khatib (1710–1790, r. 1757–1790), with Spain and Morocco alternately making and breaking diplomatic arrangements between 1767 and the Aranjuez Convention in 1780.[128] When the sultan chose his gifts from Spain in 1771, his list encompassed books about astronomy and globes, sweet cinnamon pepper and nutmeg bark, and also a number of plant-based remedies from Spanish America that the Sultan – who relied on Spanish medicine for his family's health – apparently was accustomed to using: jalap root, various balsams and ointments 'to reduce fleshiness' and to close wounds, and cinchona.[129]

---

[123] On eighteenth-century European definitions of 'Asia' and 'Asians', see Jürgen Osterhammel, *Unfabling the East. The Enlightenment's Encounter with Asia* (Princeton, N.J.: Princeton University Press, 2019), 22.

[124] Liss, 'Frontier Tales: Tokugawa Japan in Translation,' 9.

[125] Jarcho, *Quinine's Predecessor*, 91; Bruce-Chwatt and Zulueta, *Rise and Fall*, 82. On substantial and consistent demand for four American drugs – cinchona, sassafras, sarsaparilla and guaiacum – at the Russian court in the late seventeenth century, see Griffin, 'Russia and the Medical Drug Trade in the Seventeenth Century', 17.

[126] Two French Jesuits, Jean de Fontaney and Claude de Visdelou, treated the emperor successfully with cinchona introduced from the French colony of Pondicherry in a much-evoked anecdote. See, for instance, Linda L. Barnes, *Needles, Herbs, Gods, and Ghosts. China, Healing, and the West to 1848* (Cambridge, Mass.: Harvard University Press, 2007), 108–09; Bruce-Chwatt and Zulueta, *Rise and Fall*, 102.

[127] Mark Harrison, *Medicine in an Age of Commerce and Empire. Britain and Its Tropical Colonies, 1660–1830* (Oxford: Oxford University Press, 2010), 133.

[128] Juan Bautista Vilar and Ramón Lourido Díaz, *Relaciones entre España y el Magreb: siglos XVII y XVIII* (Madrid: MAPFRE, 1994), 319–26.

[129] Duque de Losada, 'Nota de varios encargos que hace el Rey de Marruecos,' *Archivo del Palacio Real*, Real Botica, Reinados Carlos III / Legajo 197, 3, Aranjuez, 1771-04-20.

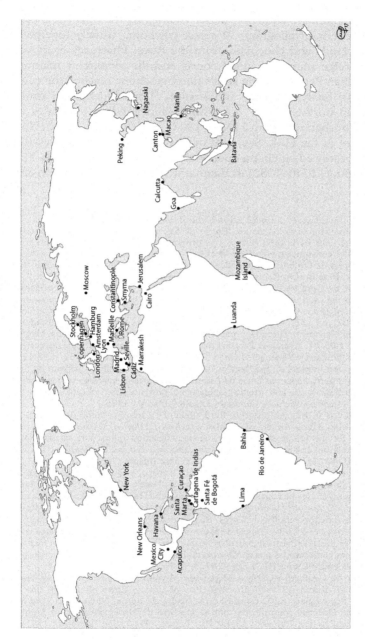

Figure 2.2 Some of the world's principal commercial entrepôts and urban consumer markets for cinchona in the late 1700s and early 1800s.

Following the time-honoured practice of sumptuary gift-exchange in
Eurasian diplomatic etiquette, under Charles III the Spanish court gave
away hundreds of kilograms of select cinchona, usually together with
tobacco, vanilla and chocolate, from the Royal Pharmacy every year as
gifts to foreign ministers, allied courts and the monarch's relations.[130]
On the Italian Peninsula, the king's finest cinchona reached the courts of
Naples and Sicily, Tuscany and Venice,[131] the Duchy of Parma and
Piacenza in northern Italy[132] – with their dependent territories under
Spanish Bourbon rule from 1732 to 1808 – and the Pope and the Spanish
ambassador at the Holy See, 'out of filial affection'.[133] Charles III also
regularly bestowed cinchona upon Holy Roman Empress Maria Theresa
(1717–1780, r. 1740–1780) in Vienna, who could not thank him enough

[130] In 1777, the treasurer counted 1,444 *libras* of select cinchona– 664 kilograms – for the
king's gifts. Miguel de Muzquiz, 'En 31 de Diciembre del año próximo anterior me
pasaron el Gefe de la Real Botica, y el primer ayuda de dicho Real Oficio la cuenta
adjunta del consumo que en todo el año de 1777 hubo de quina,' *Archivo del Palacio
Real*, Copias de Ordenes comunicadas por el Ministro de Hacienda al Señor Sumiller
de Corps. Real Botica, Reinados Carlos III / Legajo 197, 3, n.p., 1778-09-12.
[131] 'Notas de las Corachas de Cacao soconusco, Botes de Tabaco de Tabaco Havano, de
media arroba cada uno; de Sevilla y negrillo de a seis libras; y Quina que su majestad
envía de regalo a la Corte de Napoles en este año de 1773,' *Archivo General de Simancas*,
Legajo 907, n.p., 1773-07-20; 'Quina y Tabaco para los Regalos acostumbrados de las
Cortes de Napoles y Toscana, y para Venecia,' *Archivo General de Indias*, Legajo 907,
n.p., 1777-05-24; 'Quina y Tabacos para los Regalos que S.M. regala en este año a
Napoles, Toscana y Venecia,' *Archivo General de Simancas*, Legajo 907, n.p., 1783-07-
19; 'Relacion puntual de los Regalos de cacao soconusco, tabaco habano de Sevilla y
Quina, que S.M. envía por la via de Alicante, y Genova, a sus Magestades Sicilianas, y
demás particulares de la Corte de Napoles en el presente año de 1786, como se hizo en
el 1785,' *Archivo General de Simancas*, Legajo 907, n.p., 1786.
[132] The bark was usually addressed to the Duque de Parma, Fernando de Borbón-Parma,
and his wife, Maria Amalia of Austria, Duchess of Parma. Occasionally, it was also
destined for the Duque's vassals. 'Instancia solicitando quina para los vasallos de la
encomienda del Infante Duque de Parma,' *Archivo General de Simancas*, Legajo 907,
n.p., 1775-07-13; 'Tacabo y Quina para la Señora Infanta Duquesa de Parma,' *Archivo
General de Simancas*, Legajo 907, n.p., 1783; 'Tabaco, y Quina para la Señora Infanta
Duquesa de Parma,' *Archivo General de Simancas*, Legajo 907, n.p., 1786-01-22; Miguel
de Muzquiz, 'El Rey ha resuelto que se embie al Señor Infante Duque de Parma un cajón
de Quina,' *Archivo del Palacio Real*, Copias de Ordenes comunicadas por el Ministro de
Hacienda al Señor Sumiller de Corps. Real Botica, Reinados Carlos III / Legajo 197, 3,
San Lorenzo, 1772-12-22; Miguel de Muzquiz, 'El Rey quiere que se remitan a la Señora
Infanta Duquesa de Parma dos arrobas de quina selecta,' *Archivo del Palacio Real*, Copias
de Ordenes comunicadas por el Ministro de Hacienda al Señor Sumiller de Corps. Real
Botica, Reinados Carlos III / Legajo 197, 3, El Pardo, 1783-03-30.
[133] Marques de Grimaldi, 'Carta a Miguel de Muzquiz,' *Archivo General de Simancas*,
Legajo 907, Rome, 1779-10-21; Miguel de Muzquiz, 'Entre otras cosas que el Rey
envia al Papa quiere S.M. remitirle una arroba de quina,' *Archivo del Palacio Real*,
Copias de Ordenes comunicadas por el Ministro de Hacienda al Señor Sumiller de
Corps. Real Botica, Reinados Carlos III / Legajo 197, 3, San Ildefonso, 1771-07-27;
'Regalo al Papa,' *Archivo General de Simancas*, Legajo 907, n.p., 1774-08-30.

for a 'gift of such a precious medicine, so rare when one wants it of excellent quality'.[134] Maria Anna Sophia of Saxony (1728–1797), widow of the Bavarian Elector Maximilian III (1727–1777, r. 1745–1777), often reminded Charles III of his duties as her brother-in-law, if his regular remittances of cinchona and Spanish tobacco, both so 'vital for her health', were ever overdue.[135] Already under the Spanish monarch Ferdinand VI (1713–1759, r. 1746–1759), gifts of cinchona had reached the courts of the Netherlands[136] and Denmark.[137] French, Habsburg and Portuguese ambassadors at the completion of their tour of duty at the Spanish court usually received cinchona among their presents,[138] and when the Ottoman ambassador Ahmet Vâsif Efendi (c. 1730–1806) left the Spanish Court in 1788 to return to Constantinople, Charles III bestowed upon him diamonds, vicuña cloth, a golden chest adorned with diamonds on green enamel, fine crimson cloth and 'two arrobas of cinchona in four small boxes'.[139] In 1784, on the occasion of a Treaty of Neutrality signed with the Ottoman Empire, Charles III sent a small Spanish fleet to Sultan Abdülhamid I's (1725–1789, r. 1774–1789) court in Constantinople with a series of precious gifts.[140] 'Amongst other things for the Sultan', the gifts included gold tableware, silver artefacts, embroidered satin and velvet cloth, and chocolate, vanilla, tobacco and cinchona laid out in 'curious boxes'.[141] Charles III had long sought

---

[134] Simon de las Casas, 'Carta a Miguel de Muzquiz,' *Archivo General de Simancas*, Legajo 907, Vienna, 1772-10-03.

[135] Miguel de Muzquiz, 'Queriendo el Rey regalar a Su Hermana la Señora Electriz Viuda de Baviera veinte y cinco libras de quina de la mas selecta, lo participo a V.E. de orden de S.M.,' *Archivo del Palacio Real*, Copias de Ordenes comunicadas por el Ministro de Hacienda al Señor Sumiller de Corps. Real Botica, Reinados Carlos III / Legajo 197, 3, El Pardo, 1780-02-18.

[136] The King remitted some two *quintales* – about 92 kilograms, an amount sufficient for some 450 cases – of cinchona through Pedro Gil de Olondriz, his treasurer in The Hague, home to the Dutch States General, the government and the royal family. 'Nota del cacao de Soconurco, Polbillo de Oaxaca, Vaynillas Tavaco y Quina que se ha distribuido a los sujetos y parages que irán mencionados,' *Archivo General de Indias*, Indiferente 1552, Cádiz, 1753-10-09.

[137] Ferdinand VI remitted 'six pounds of the best cinchona' (*de la mejor que huviese*) together with 'other things from the Indies' – cacao, tobacco, vanilla – to the Marquis de Puente Fuerte, the Spanish ambassador in Copenhagen. Julián de Arriaga, 'Carta,' *Archivo General de Indias*, Indiferente 1552, Madrid, 1753-01-23.

[138] Riera Palmero, 'Quina y malaria en la España del siglo XVIII,' 22.

[139] José (Conde de Floridablanca) Moñino y Redondo, 'Regalo de Alhajas y Joyas hecho por S.M. y A.A. al embiado de la Puerta Otomana Ahmet Vasif Effendi,' *Archivo General de Simancas*, Secretaría de Hacienda / Legajo 906, El Pardo, 1788-02-29.

[140] On the negotiations between Spain and the Ottoman Empire in the late eighteenth century, see Hüseyin Serdar Tabakoğlu, 'The Impact of the French Revolution on the Ottoman-Spanish Relations,' *Turkish Studies* 3, no. 1 (2008).

[141] 'Regalos hechos a la Corte de Constantinopla con motivo de la Paz, concluida en este año,' *Archivo General de Simancas*, Secretaría de Hacienda / Legajo 906, n.p., 1785.

good relations with the Ottoman Empire to institute direct trade links with the Levant, avoiding British and French intermediaries, and he and his ministers may well have taken advantage of the occasion to showcase cinchona as a Spanish product. At the same time, however, they would have taken a great deal of care in selecting the gifts for the Ottoman sultan to avoid any blunder in a long-awaited diplomatic alliance. They must have been materially certain the sultan would be familiar enough with the bark to recognize it instantly as a precious medicinal substance and that it would delight him as much as the gold, chocolate or velvet they bestowed along with it. By the late 1700s and early 1800s, cinchona bark was a medicinal substance sufficiently renowned to be valued, and prized, by the upper strata of Ottoman, Habsburg and Mughal societies alike, but also one that, so it would seem, was not so abundantly available of excellent quality as to render a gift in them unnecessary or unwelcome.

Cinchona's close association with gold, satin and velvet was not imaginary or purely symbolic. Though Spain did not capitalize significantly on the bark – even between 1782 and 1796, cinchona amounted only to 1.4 per cent of all of Spain's imports in value[142] – the bark cost practitioners and sufferers a high price. Cádiz merchants sold the bark at 8 *reales* per *libra* between 1747 and 1762, at a time when stimulants like coffee cost 1.6 *reales* per *libra* in Cádiz and when chocolate sold for 2.5 *reales* per *libra*. Bark prices rose further over the later decades of the eighteenth century. Cinchona sold at 16 *reales* per *libra* in 1778, at 23 *reales* per *libra* in 1793, and at 28 *reales* per *libra* by 1794.[143] Prices soared on occasion, owing to temporary dislocations in supply and demand[144] or the sudden rise in popularity of particular bark varieties. In 1786, 'coloured cinchona bark' (*quinas coloradas*), harvested in the provinces of Cuenca, Riobamba, Guaranda, Alausí and Guayaquil, became so popular that English merchants paid up to 60 *reales* per *libra* for it in Cádiz.[145] Cádiz prices were a multiple of the amounts paid in the harvest areas, in

---

[142] Shipments with a total value of 152,472,717 *reales de vellón* entered Cádiz in the period, while cinchona worth 6,000 *reales de vellón* reached Barcelona. Fisher, *Commercial Relations between Spain and Spanish America in the Era of Free Trade*, 70.

[143] According to García-Baquero, between 1747 and 1762, Cádiz merchants sold the bark at 25 silver *pesos* per *arroba*, while coffee cost 5 *pesos* per *arroba* – 1.6 *reales* per *libra* – and chocolate sold for 8 *pesos* per *arroba* – that is, 2.5 *reales* per *libra*. García-Baquero González, *Cádiz y el Atlántico*, 1; 164–65. For the Cádiz prices from 1778, 1793 and 1794, see Jaramillo Baanante, 'El comercio de la cascarilla,' 68.

[144] See, for instance, Jaramillo Baanante, 'El comercio de la cascarilla,' 70.

[145] José Garcia de Leon y Pizarro, 'Informe,' *Archivo General de Indias*, Indiferente 1556, Madrid, 1706-01-07.

transfer sites like Piura and Paita and in Lima,[146] and merchants, apothecaries and itinerant barber-surgeons added further markups to the cost of cinchona bark when reselling it to medical practitioners and end consumers. Apothecaries in cities like Rome or Lisbon, according to contemporary observers, 'made double or triple profits' (*hacen una ganancia del doble, à del triple*), from reselling cinchona, or cinchona-based medicines.[147] Bark prices were affordable, and appeared 'very reasonable',[148] to the upper and middle echelons of various consumer societies at the time – the gentry, clergymen or civil servants, and also many craftsmen, merchants, physicians, lawyers and freehold farmers – but surely not to the poor – the men and women whose resources were scarce, as de Rieux had phrased it. In late eighteenth-century England, at a time when a shilling a day was a fair wage for a worker, customers paid from 18 pence to 9 shillings for a pound of cinchona – a 'variation founded upon a supposed comparative difference in their respective goodness'.[149] Similarly, in Portugal, the price of what physicians considered a curative dose of the best cinchona oscillated between 400 and 600 *réis*,[150] when skilled workers – carpenters, masons and painters – made 300 to 400 *réis* a day and labourers and farmhands between 120 and 200 *réis*.[151] In the electorate of Mainz, red cinchona bark, then the most esteemed by local physicians, cost 55 *Gulden* a pound, while bark of lesser quality still cost between 9 and 25 *Gulden*.[152] A worker's, even an artisan's, annual income then amounted to some 80 to 100

[146] By 1786, for instance, in the harvest areas, merchants paid less than 48 *reales* per *arroba*, at a time when 1 *arroba* equalled 25 *libras*. Jaramillo Baanante, 'El comercio de la cascarilla.'
[147] According to Nicolas de Azára, Spanish ambassador in Rome, by 1785 'apothecaries sold common cinchona for two paolos per ounce, and the one they pretended was select cinchona [...] for three, making a double or triple profit (*hacen una ganancia del doble, à del triple*)'. Nicolás de Azara, 'Carta a Pedro de Lorena,' *Archivo General de Simancas*, Legajo 961/2, Rome, 1789-10-09. In Portugal, 'monopolists' acquired a bottle of Castro's cinchona-based English Water for 1,600 *réis*, and resold it to the public at up to 4,000 to 4,800 *réis*. *Documentos régios*.
[148] Buchan, *Domestic Medicine*, 169.
[149] Relph, *An Inquiry into the Medical Efficacy of a new Species of Peruvian Bark*, 4.
[150] See the report written by the Évora physician Manoel Francisco de Carvalho on April 28, 1799: 'Enformes que deram os Medicos d'esta Corte,' 56. According to the Minas Gerais-born physician Francisco Melo Franco (1757–1823), 'an ounce of cinchona commonly costs 480 réis'. Cited in José Pedro Sousa Dias, *A água de Inglaterra: paludismo e terapêutica em Portugal no século XVIII* (Lisbon: Caleidoscópio, 2012), 62.
[151] See the data files on Lisbon wages compiled by a research team based at the Instituto de Ciências Sociais da Universidade de Lisboa. Jaime Reis and Conceição Andrade Martins, 'Prices, Wages and Rents in Portugal 1300–1910' (2009), http://pwr-portugal.ics.ul.pt/?page_id=56.
[152] Renard, *Die inländischen Surrogate der Chinarinde*, 18.

Gulden at most[153] and 'only the wealthy', according to the Mainz doctor Johann Renard, could at all afford red cinchona. 'Families without fortune, artisans, manufacturers, people with a small income or families with many children' could generally not, says Renard, afford any bark.[154] The authors of popular medical advice manuals and charitable pamphlets – John Haartman (1725–1788), whose 'Clear Advice' (*Tydelig Underrättelse*) addressed poor Finnish parishioners, or the Swiss Samuel Auguste André Tissot's (1728 – 1797) 1761 'Advice to the country folk, with regard to their health' (*Anleitung für das Landvolk in Absicht auf seine Gesundheit*) – unanimously recommended cinchona as the only secure remedy in fevers, but were well aware that the 'common people' (*das gemeine Volk*) would, for pecuniary reasons, often be unable or 'reluctant to undergo a treatment' (*wird sich nicht so leicht dieser Cur unterwerfen*) that resorted to the bark.[155]

In and beyond Europe's colonial, evangelizing and commercial entrepôts in North America and the Caribbean, coastal Africa, the eastern Mediterranean and South and East Asia, prices would commonly have been even higher than in the metropolis. High taxes and markups, as well as warfare and the low value placed on the New England currency, made medicines shipped to the British North American colonies far costlier than they were in London.[156] Surgeons of the Royal Navy could not afford to purchase the Peruvian bark in the West Indies – it sometimes cost 'two guineas a pound', around 42 shillings, four times the price paid back home in England – and petitioned for an 'allowance of bark from government, while upon that station'.[157] Portuguese traders in Angola complained they could not make enough money to pay for the medicines of which they were 'in daily need'.[158] Indeed, the price of bottled 'English Water' (*Água de Inglaterra*), a cinchona-based patent medicine

---

[153] Karl Härter, *Policey und Strafjustiz in Kurmainz. Gesetzgebung, Normdurchsetzung und Sozialkontrolle im frühneuzeitlichen Territorialstaat* (Frankfurt a. M.: Vittorio Klostermann, 2005), 600–01.

[154] Renard, *Die inländischen Surrogate der Chinarinde*, 8.

[155] Tissot, *Anleitung für das Landvolk*, 294. Haartman's *Tydelig Underrättelse om de mäst gångbara Sjukdomars kännande och Botande genom Lätta och Enfaldiga Hus-medel* (Clear advice about recognizing and curing the most common diseases with plain and simple medication), first published in 1759, advised the use of the Peruvian bark, but suggested substitutes, since Haartman knew that many of his patients could not procure or afford the bark. Huldén, 'The First Finnish Malariologist,' 5.

[156] Gevitz, '"Pray Let the Medicine be Good,"' 93; 97.

[157] John Hunter, *Observations on the Diseases of the Army in Jamaica: And on the Best Means of Preserving the Health of Europeans, in that Climate* (London: Printed for G. Nicol, Pall-Mall, Bookseller to His Majesty, 1788), 138.

[158] Cited in Joseph C. Miller, *Ways of Death. Merchant Capitalism and the Angolan Slave Trade, 1730–1830* (London: James Currey, 1988), 286.

popular throughout the Kingdom of Portugal and its overseas dominions, was higher in West African captaincies than back home in Portugal
or in Brazil. While the English Water was available at 1,000 *réis* per bottle
in Portugal from around 1772,[159] it was resold to the public in Portugal's
American overseas dominions at up to 4,000 to 4,800 *réis*, and at up to
6,400 *réis* in Angola and Benguala (Figure 2.3).[160] As a consequence, in
1809 the Portuguese government sought to regulate the price at 1,600
réis for every big bottle and 900 *réis* for every small bottle for 'private
commissioners in the captaincies of Rio de Janeiro, Bahia, Pernambuco,
Pará, and Maranhão etc.', and at 2,400 *réis* for every big bottle and 1,300
réis for every small bottle in the captaincies of Angola and Benguala.[161]
The English Water was presumably more expensive in the West African
captaincies since very few ships sailed there from Portugal directly. The
inhabitants of Angola and Benguela would have received most of their
supplies of English Water via Brazil.[162] Prices were comparable or even
more moderate than in the metropolis only in the southern Spanish
American Viceroyalties. Cinchona was costly in pharmacies – in Lima,
by the late 1700s, sufferers paid one *real* for a dose of powdered bark,[163]
at a time when even higher earners like chaplains and physicians made
but some 3,000 *reales* per annum[164] – but it was likely more affordable
from other suppliers. Given that cinchona was administered in the
potions of slave healers in Tucumán, in the Viceroyalty of Rio de la
Plata,[165] and used in all kinds of fevers by Indian healers in the Quito
Audiencia,[166] it was presumably available at a more favourable price from

---

[159] André Lopes de Castro (1734/35–1803) first started manufacturing the English Water,
which had hitherto been imported from England, in Portugal between 1772 and 1774.
José Pedro Sousa Dias, *A água de Inglaterra: paludismo e terapêutica em Portugal no século
XVIII* (Lisboa: Caleidoscópio, 2012), 56; 62; 74.

[160] *Documentos régios.*   [161] Ibid.

[162] Most of the imports for the town of Luanda were Brazilian-produced goods, since the
slave trade from Angola was primarily a two-way trade, between Angola and Brazil. In
1795, for example, of the 337 Portuguese vessels which left the harbour of Lisbon, only
14 went to Africa, as opposed to 51 that sailed for Brazil. Herbert S. Klein, 'The
Portuguese Slave Trade from Angola in the Eighteenth Century,' *The Journal of
Economic History* 32, no. 4 (1972).

[163] Anon., 'El Medico verdadero,' 446.

[164] By the year 1770, a hospital orderly made 2,400 *reales*, a porter 192 *reales*, a laywer 800
reales, and a chaplain, physician or apothecary more than 3,000 *reales* per annum.
Macera, *Precios del Perú XVI–XIX*, xxiv.

[165] Carlos Alberto Garcés, 'Místicos, curanderos y hechiceros: Historias de afroamericanos
en la sociedad del Tucumán colonial,' *Contra Relatos desde el Sur. Apuntes sobre África y
Medio Oriente* V, no. 7 (2010), 23.

[166] Eduardo Estrella, 'Ciencia ilustrada y saber popular en el conocimiento de la quina en
el siglo XVIII,' in *Saberes Andinos. Ciencia y tecnología en Bolivia, Ecuador y Perú*, ed.
Marcos Cueto (Lima: Instituto de Estudios Peruanos, 1995), 56–57.

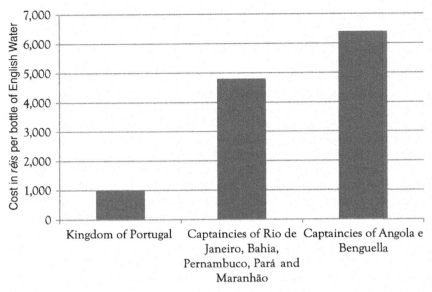

Figure 2.3 Approximate market prices in *réis* for bottles of 'English Water' (*Água de Inglaterra*) between 1772 and 1809, at a time when that medicine was fabricated in Portugal and exported to the West African captaincies of Angola and Benguella via Brazil.
Data from Sousa Dias, *A água de Inglaterra*, 45; *Documentos régios*; *Enformes que deram os Medicos d'esta Corte*, 56.

peddlers or herbalists, on the streets and in marketplaces, with ties to the harvest areas.

Even though those whose resources were scarce in places like New England, Finland and Angola could generally have ill afforded cinchona, those within the reach of religious orders, private charity and increasingly systematic medical relief programmes would still have had access to the bark. The Spanish Crown, in conjunction with a wider reform in the state's understanding of its responsibilities towards the population, developed a comprehensive system of free-of-charge medical attention during the late 1700s that encompassed the distribution of medicines from the Royal Pharmacy among religious convents, localities afflicted by epidemics on the Iberian peninsula, in its American empire and beyond,[167] and hospitals – at a time when hospitals were still fundamentally charitable

[167] On the Crown's eighteenth-century distribution of medicines among the army, convents and the court, see Andrés Turrión, 'Quina de la Real Hacienda para el ejército español en el siglo XVIII,' 419–21. See also María Esther Alegre Pérez, 'La asistencia social en la Real Botica durante el último cuarto del siglo XVIII,' *Boletín de la*

institutions, places of shelter for those who were poor and ill or near death.[168] Cinchona remittances regularly reached the Discalced Franciscans of Ciempozuelos, whose vows of poverty prevented them from purchasing cinchona for their confreres,[169] 'sick paupers' (*enfermos pobres*) in Santa Fé's San Juan de Dios Hospital,[170] and Jerusalem's Franciscan monasteries, to cure tertian fevers among the 'ailing friars that live in the convents of these holy sites'.[171] The Spanish Crown was not the only government to dispense free cinchona to sick paupers. The Portuguese Crown was likewise frequently called upon, and granted, cinchona or cinchona-based medicines to sickly localities, from Estremadura[172] to Angola.[173] Cinchona preparations were also distributed by the monarchs of Islamic societies, in which medical aid was likewise a recognized act of benevolence and charity.[174] When an epidemic was raging in the sultanate of Morocco in 1799 and 1800 – in Marrakesh,

Sociedad Española de Historia de la Farmacia 35, no. 139 (1984). On health care and poor relief under Bourbon rule more broadly, see also Pedro Carasa, 'Welfare Provision in Castile and Madrid,' in *Health Care and Poor Relief in 18th and 19th Century Southern Europe*, ed. Ole Peter Grell, Andrew Cunningham and Bernd Roeck (Aldershot: Ashgate, 2005), 100. On the earlier history of hospital and welfare work in Habsburg Spain, see Teresa Huguet-Termes, 'Madrid Hospitals and Welfare in the Context of the Habsburg Empire,' in *Health and Medicine in Hapsburg Spain: Agents, Practices, Representations*, ed. Teresa Huguet-Termes, Jon Arrizabalaga and Harold J. Cook (London: The Wellcome Trust Centre for the History of Medicine at UCL, 2009).

[168] Morris J. Vogel, *The Invention of the Modern Hospital. Boston 1870–1930* (Chicago: University of Chicago Press, 1980). On Spanish American hospitals, see Gabriela Ramos, 'Indian Hospitals and Government in the Colonial Andes,' *Medical History* 57, no. 2 (2013).

[169] Manuel de Laganes, 'El Guardián y Comunidad de este Convento de San Francisco de Descalzos de la V. de Ciempozuelos,' *Archivo del Palacio Real*, Copias de Ordenes comunicadas por el Ministro de Hacienda al Señor Sumiller de Corps. Real Botica, Reinados Carlos III / Legajo 197, 3, Ciempozuelos, 1775-07-06.

[170] Juan José Villaluenga, 'Copia del oficio de Juan José Villaluenga del 18 de julio de 1786 al arzobispo virrey Antonio Caballero y Góngora comentando una donación anual de quina a fray Juan Antonio Gago, prior del Hospital de San Juan de Dios, de Cassa Plaza, para su administración a enfermos pobres,' *Archivo del Real Jardín Botánico*, Real Expedición Botánica del Nuevo Reino de Granada (1783–1816). José Celestino Mutis. Documentación oficial. Oficios varios, RJB03/0002/0003/0097, Quito/ Turbaco, 1786-10-08.

[171] Fray Josef Fernandez Alexo, 'Fray Josef Fernandez Alexo Religioso observante de San Francisco y Comisario de los Santos Lugares de Jerusalén expone,' *Archivo General de Simancas*, Legajo 960, Jerusalem, 1781.

[172] 'OFÍCIO (3) ao (secretário de Estado dos Negócios da Marinha e Ultramar) Martinho de Melo e Castro,' *Arquivo Histórico Ultramarino*, 076 – REINO RESGATE 20121023 / Cx. 30-A, Pasta 5, Faro (Lisboa), Caldas da Rainha, 1784-08-24 / 1787-05-08.

[173] Patrick Figueiredo, 'A "Água de Inglaterra" em Portugal,' in *A Circulação do Conhecimento: Medicina, Redes e Impérios*, ed. Cristiana Bastos and Renilda Barreto (Lisboa: Imprensa do Instituto de Ciências Sociais, 2011), 123–24.

[174] Miri Shefer-Mossensohn, *Ottoman Medicine. Healing and Medical Institutions 1500–1700* (Albany: State University of New York Press, 2009), 110.

Tangier, Meknes and Tétouan, in particular – Sultan Mawlay Sulayman (1766–1822, r. 1792–1822) had it combated by means of cinchona-based preparations, courtesy of the Spanish monarch Charles IV (1748–1819, r. 1788–1808).[175] Religious hospitals also frequently expended large quantities of the bark: the charitable *Hôtel-Dieu de Carpentras* in southern France,[176] for instance, and Rome's 'Hospital of the Holy Spirit' (*Ospedale di Santo Spirito*), which used some 6 tons of cinchona bark between 1778 and 1785.[177] Religious orders were important pharmaceutical suppliers of the bark in the late 1700s and early 1800s, too, and several of them would have dispensed the bark charitably: pharmacies pertaining to Capuchin monasteries in Solothurn, in the Basel diocese,[178] to the Cistercian order in Eger, in the Habsburg-ruled Kingdom of Hungary,[179] and to the Jesuit order, the networks of which stretched from Büren to Macao, from Lima to Rome and from Milan to Goa.[180] Medical poor relief programmes in northern Europe, where, from the Reformation onwards, provision of health care and poor relief came to be seen as the responsibility of the community as a whole,[181] commonly avoided the more expensive foreign plant remedies and replaced them with cheaper,

[175] Braulio Justel Calabozo, 'El doctor Masdevall. Protomédico del sultán Marroquí Muley Solimán,' *Al-Andalus – Magreb* II (1994), 180; 221–22.
[176] Dubois, 'Le quotidien d'une pharmacie hospitaliere,' 94.
[177] According to the Spanish ambassador, the hospital used 18,099 *libras romanas*, that is, 6 tons. Azara, 'Carta a Pedro de Lorena.'
[178] See the cinchona-based remedy against a 'cold fever' contained in a recipe collection attributed to the Solothurn Capuchin monastery's pharmacy. 'Kräuterrezepte und Hausmittel,' *Archiv für Medizingeschichte*, Universität Zürich, Rezeptbücher, MS J 29, Solothurn (?), c. 1767.
[179] See, for instance, the report on a 1786 inspection of the pharmacy (*Bericht von der den 27. [...] ber 786 geschehenen Visitation der Apotheke deren P.P. Cisterciten in Erlau*) from the Hungarian National Archives (*Archivum. Magyar Nemzeti Levéltár*). I would like to thank Katalin Pataki for sharing this document, and her research, with me. Katalin Pataki has extensively studied monasteries and medical provisions in Hungary. Katalin Pataki, 'Healers, Quacks, Professionals: Monastery Pharmacies in the Rural Medicinal Marketplace', *Society and Politics* 12, no. 1 (2018), 32–49; Katalin Pataki, 'Medical Provision in the Convents of Poor Clares in Late Eighteenth Century Hungary', *Cornova* 6, no. 2 (2016), 33–58.
[180] On the Jesuit networks taking cinchona across the world, see Anagnostou, *Missionspharmazie*, 298; 321; 27. See also Sabine Anagnostou, 'The International Transfer of Medicinal Drugs by the Society of Jesus (Sixteenth to Eighteenth Centuries) and Connections with the Work of Carolus Clusius,' in *Royal Netherlands Academy of Arts and Sciences*, ed. Florike Egmond, Paul Hoftijzer and Robert P. W. Visser (Amsterdam: Koninklijke Nederlandse Akademie van Wetenschappen, 2007). The Jesuit pharmacy of San Pablo College in Lima was a particularly important distribution centre for American plant-based remedies. Luis Martin, *The Intellectual Conquest of Peru. The Jesuit College of San Pablo, 1568–1767* (New York: Fordham University Press, 1968).
[181] Andrew Cunningham, 'Some Closing and Opening Remarks,' in *Health Care and Poor Relief in 18th and 19th Century Southern Europe*, ed. Ole Peter Grell, Andrew Cunningham and Bernd Roeck (Aldershot: Ashgate, 2005), 11.

home-grown substitutes. Physicians were, however, frequently allowed and encouraged to administer cinchona and some other select remedies out of the public's purse to sick paupers. Hamburg's *Paupers' Pharmacopoeia*, for instance, a paragon of pharmaceutical knowledge for the German territories at the time, encompassed expensive foreign drugs if they were considered indispensable or significantly more effectual than local substitutes. Quassia wood, Peruvian balsam, copaiba balsams and cinchona were thus administered to the poor on a regular basis.[182] Since the lives of patients often depended upon it, as one contemporary physician phrased it, it was imperative that cinchona also be accessible to the poor.[183] In 1806, the charitable hospital in Mainz spent more than 1,000 *Gulden*, a small fortune at the time, for cinchona and cinchona-based remedies.[184] The world over, the distribution of free cinchona frequently also extended to men and women on whose utility or productivity masters and governments relied. Slaves in Ottoman households[185] and on plantations in the West Indies,[186] workers in Spanish mines[187] and servants on the Arabian Peninsula[188] were administered doses of the bark to restore their health and ability to work. From Marrakesh to Rome, from Jerusalem to Carpentras, in the late eighteenth and early nineteenth centuries, charity or economic expediency often entailed access to the bark even for those whose resources were too scarce to purchase it.

Rather than poverty or scarcity of resources as such, it was distance from the veins of cinchona's passage – from colonial entrepôts, urban centres or charitable hospitals – that excluded men, women and children

---

[182] Almuth Weidmann, *Die Arzneiversorgung der Armen zu Beginn der Industrialisierung im deutschen Sprachgebiet, besonders in Hamburg* (Stuttgart: Deutscher Apotheker-Verlag, 1982), 144; Robert Jütte, 'Hanseatic Towns: Hamburg, Bremen and Lübeck,' in *Health Care and Poor Relief in 18th and 19th Century Southern Europe*, ed. Ole Peter Grell, Andrew Cunningham and Bernd Roeck (Aldershot: Ashgate, 2005).

[183] Cited in Weidmann, *Die Arzneiversorgung der Armen*, 144.

[184] Renard, *Die inländischen Surrogate der Chinarinde*, 9.

[185] Aydüz and Yildirim, 'Bursalı Ali Münşî ve Tuhfe-i Aliyye,' 96.

[186] James Thomson, *A treatise on the diseases of negroes, as they occur in the island of Jamaica: with observations on the country remedies* (Jamaica: Alex. Aikman, jun., 1820), 14; 19; 27.

[187] The Royal Pharmacy in Madrid, for instance, supplied the miners in Almaden with cinchona almost yearly during the mid-1750s. Miguel de Muzquiz, 'En los años de 1752, 54 y 56 se sirvió el Rey mandar entregar varias porciones de Quina selecta para la curación de los terzados y enfermos del Real Hospital de las Minas de Almaden,' *Archivo del Palacio Real*, Copias de Ordenes comunicadas por el Ministro de Hacienda al Señor Sumiller de Corps. Real Botica, Reinados Carlos III / Legajo 197, 3, San Ildefonso, 1776-09-12.

[188] James Bruce (1730–1794), for instance, on his journey across the Arabian Peninsula, administered the bark to his 'rais', ill with a fever, to set him up again. James Bruce, *Travels to discover the source of the Nile, in the years 1768, 1769, 1770, 1771, 1772, and 1773*, vol. 5 (Edinburgh / London: J. Ruthven / J. Robinson, 1790), 300.

from the remedy's consumption. Urban populations generally had better access to remedies on account of a higher concentration of wealth[189] as well as a greater density of suppliers – the number of fixed shops specializing in the distribution of drugs grew even faster than the medical profession between 1780 and 1900[190] – and particularly so to foreign plant remedies like cinchona. Outside the colonial urban centres in the Viceroyalty of New Spain – that is, Mexico City, Puebla, Guadalajara, Veracruz, Valladolid and Oaxaca –[191] for instance, cinchona bark was barely available. Similarly, in Muscovy, though 'generally employed' for the cure of intermittent fevers in the capital, Moscow,[192] along the Caspian Sea, on the Caucasian plains and in the Crimea, where apothecaries were scarce, 'the poor and even many of the rich [were] unable to procure the bark',[193] as they were in the commercially more isolated regions of north-western Europe. Finnish sufferers' only source of supply for cinchona was Stockholm. Though severe 'intermittent fevers' (*växelfeber*) reigned in the south-western archipelago around Turku in the late 1700s – some 1,800 men, women and children died of these fevers between 1751 and 1773[194] – the district physicians who urged the administration of cinchona were well aware that most sufferers would be unable to procure the bark.[195] Even in states like Britain and in the Holy Roman Empire, in outlying rural areas where apothecaries were

[189] Mark S. R. Jenner and Patrick Wallis, 'The Medical Marketplace,' in *Medicine and the Market in England and Its Colonies, c. 1450–1850*, ed. Mark S. R. Jenner and Patrick Wallis (Basingstoke: Palgrave Macmillan, 2007), 11.

[190] Louise Hill Curth, 'Introduction: Perspectives on the Evolution of the Retailing of Pharmaceuticals,' in *From Physick to Pharmacology: Five Hundred Years of British Drug Retailing*, ed. Louise Hill Curth (Aldershot: Ashgate, 2006), 5.

[191] Cinchona was available 'in all of [New Spain's] pharmacies' (*en todas las Boticas curiosas*). Esteyneffer, *Florilegio medicinal*, 296. Licensed apothecaries in colonial New Spain were concentrated, however, in the colonial urban centres of Mexico City, Puebla, Guadalajara, Veracruz, Valladolid and Oaxaca. Paula De Vos, 'From Herbs to Alchemy: The Introduction of Chemical Medicine to Mexican Pharmacies in the Seventeenth and Eighteenth Centuries,' *Journal of Spanish Cultural Studies* 8, no. 2 (2007), 140.

[192] Robert Lyall, *The Character of the Russians, and a Detailed History of Moscow* (London: T. Cadell, in the Strand, and W. Blackwood, 1823), cvii.

[193] Edward Daniel Clarke, *Travels in Various Countries of Europe, Asia and Africa. Part the First: Russia, Tahtary and Turkey*, 4 ed., 2 vols., vol. 2 (London: T. Cadell and W. Davies, 1817), 206.

[194] This estimate is taken from F. J. Rabbe, who published *Några anteckningar om frossan i Finland samt botemedlen deremot* – which translates as 'Some Notes about the Ague in Finland and Remedies against it' – in the volume *Finska Läkaresällskapets handlingar* in 1856. Cited in Lena Huldén, Larry Huldén and Kari Heliövaara, 'Endemic Malaria: An "Indoor" Disease in Northern Europe. Historical data Analysed,' *Malaria Journal* 4, no. 19 (2005), 4.

[195] A Finnish cinchona-grinding mill was only established in Turku in 1813. Huldén, 'The First Finnish Malariologist,' 5.

scarce, supply was more restricted and vulnerable, and particularly foreign drugs were not always to be had when wanted, by both patients and professionals.[196] In the rural areas near Mainz, as Renard put it, even those who could afford it, if they fell ill on their landed estates, could not rely on finding cinchona in 'small village pharmacies nor in the pharmacies of the minor neighbouring towns', nor in the medical supplies of country doctors.[197] By the late 1700s and early 1800s, many of the world's royal courts, bazaars and port cities were part of a vibrant medical market that redistributed remedies like cinchona to 'the four corners of the Earth' (*las cuatro partes del mundo*).[198] More so even than a person's social stratum, religious creed, or political belonging, it was his or her relative nearness to, or distance from, the boundaries and confines of that market that determined 'what he or she could reasonably expect to have available as medical provision'.[199]

★★★

Officials, physicians and naturalists concerned with the bark's equitable distribution around 1800 were not the last to presume to speak and act in the name and for the betterment of a universal humanity. The British quest to smuggle and transplant cinchona seedlings during the 1850s, which put an end to the South American monopoly, was commonly portrayed as an act to rescue the tree from certain extinction at the hands of its 'ignorant', and 'barbarous' Andean keepers for the good of nature and mankind.[200] Like these later schemes, which were, as historians have argued, about the good of British imperial troops and administrative personnel rather than humanity, governmental, commercial and scholarly efforts around 1800 were likewise in the service of a particular sort and sector of humanity. The boundaries of the universal humanity propagated in the treatises and decrees of Spanish officials, British physicians and French naturalists in late 1700s and early 1800s were, however, unlike later ones, neither national and imperial nor strictly social,

---

[196] Steven King, 'Accessing Drugs in the Eighteenth-Century Regions,' in *From Physick to Pharmacology: Five Hundred Years of British Drug Retailing*, ed. Louise Hill Curth (Aldershot: Ashgate Publishing, 2006), 58.
[197] Renard, *Die inländischen Surrogate der Chinarinde*, 9.   [198] Caldas, *Memoria*, 13.
[199] Charles W. J. Withers, *Placing the Enlightenment. Thinking Geographically about the Age of Reason* (Chicago: University of Chicago Press, 2007), 54. Along a similar vein, historians have argued that men and women living in colonial territories had the advantage of better and more direct access to medical innovations like smallpox vaccines around 1800. Jürgen Osterhammel, *Die Verwandlung der Welt. Eine Geschichte des 19. Jahrhunderts* (München: C.H. Beck, 2011), 274–75.
[200] Philip, 'Imperial Science Rescues a Tree: Global Botanic Networks, Local Knowledge and the Transcontinental Transplantation of Cinchona,' 190–91.

religious or geographical. Millions of men and women around the Atlantic World and beyond, be they Ottoman courtiers, Hamburg paupers or Andean villagers, had by the late 1700s and early 1800s come to have access, and to assign a medical purpose, to dried shreds of cinchona tree bark. Rather, the contours of the bark's availability were constricted and bound by physical and cultural distance from the veins of its passage: from the commercial, imperial and diplomatic ties and relationships that entwined its English, creole, Levantine and Portuguese distributors and that formed the trade's volume, vigour and, above all, reach. The humanity propagated in the treatises and decrees of Spanish officials, British physicians and French naturalists encompassed a wide range of men, women and children, so long as they lived or moved in places tied to the wider world: in convents or at court, in hospitals or near seaports, beside marketplaces or in town. Cinchona reached societies from the North Sea Basin to the Gulf of Mexico and from the Gulf of Guinea to the Caribbean Sea, but it was commonly in the service of a rather particular sort, and sector, of humanity.

3 Community of Practice

Cure for an Ague from the Whitehall paper No. 1791: Take thirty-five grains of Salt of Wormwood, twenty-five grains of powdered snake-root, half an ounce of bark put it into a pint of red Port shake it well & take a wineglass full every three hours when the fit comes on.
— 'Receipts Copied from Miss Myddleton's Book,' 1785.

A compound tincture of cinchona common to many apothecary books, recipe collections and health advice manuals in the late 1700s and early 1800s was Masdevall's Antipyretic Opiate (*opiate antifebril*), a remedy named after its inventor, the Spanish court physician Joseph Masdevall (d. 1801), who allegedly devised it in the context of his efforts to combat an epidemic of putrid and malignant fevers in Aragón and Catalonia from 1783 to 1785.[1] Consisting of absinthe salt, ammoniac salt, stibiated tartar, emetic tartar and Peruvian bark, the opiate was to be administered in a 'sufficient quantity of absinthe syrup' and to be preceded by a purgative comprising various antimonial wines to 'purify the sufferer's blood', as well as 'viper water' (*aqua viperina*) and either cream of tartar, confection of hyacinth or Sal-Policrest, a sulphate salt.[2] Credited and celebrated for subduing the epidemic in Aragón and Catalonia, Masdevall's method had earned him instant fame within Spain and its empire. It was adopted by physicians and hospital orderlies as well as

---

[1] For biographical information on Joseph Masdevall, and an account of the epidemic, see Riera Palmero, 'La Medicina en la España del siglo XVIII,' 16; Juan Riera Palmero, *José Masdevall y la medicina española ilustrada. Enseñanza, epidemias y guerra a finales del siglo XVIII* (Valladolid: Secretariado de Publicaciones, 1980). On its spread, see also Juan Riera Palmero, 'Epidemias y comercio americano de la quina en la España del s. XVIII', in *Capitulos de la medicina española ilustrada. Libros, cirujanos, epidemias y comercio de quina* (Valladolid: Universidad de Valladolid, 1992), 85–87.

[2] Masdevall first outlined his 'method' in a 1786 report on the epidemic. Joseph Masdevall, *Relación de las epidemias de calenturas pútridas y malignas, que en estos últimos años se han padecido en el Principado de Cataluña; y principalmente de la que se descubrió el año pasado de 1783 en la ciudad de Lérida, Llano de Urgel y otros muchos Corregimientos y Partidos, con el método feliz, pronto y seguro de curar semejantes enfermedades* (Barcelona: Imprenta Real, 1786), 68–88.

householders from the Bay of Gibraltar[3] to viceregal Lima,[4] and from Chancay[5] to the City of Mexico.[6] Italian and German translations of Masdevall's 1786 'Report of an Epidemic of Putrid and Malignant Fevers' (*Relación de las epidemias de calenturas pútridas y malignas*) – the 1789 'Account of an Epidemic Suffered in the Principality of Catalonia in the year 1783' (*Relazione dell'epidemia sofferte nel Principato di Catalogna nell'anno 1783*),[7] and the 1792 'Account of the Epidemics of Putrid and Malignant Fevers' (*Bericht über die Epidemien von faulen und bösartigen Fiebern*)[8] – spread the word to the Italian Peninsula and the Holy Roman Empire. News of the 'curious medicine', as Joseph Townsend phrased it in his 1791 *Journey through Spain*, also swiftly reached Britain and its growing empire.[9] The first French edition of Masdevall's report, the 'Medicines, or a Digest of the Method of Mr. Masdevall' (*Medicamens, et précis de la méthode de Mr. Masdevall*), was published in New Orleans in 1796, then formally under Spanish rule, rather than in metropolitan France, presumably because epidemics of fevers were perpetually haunting the soggy terrain of Louisiana, and particularly the Lower

---

[3] On the opiate's administration in the Bay of Gibraltar, see Joaquin de Villalba, *Epidemiologia española, ó, Historia cronológica de las pestes, contagios, epidemias y epizootias que han acaecido en España desde la venida de los cartagineses hasta el año 1801. Con noticia de algunas otras enfermedades de esta especie que han sufrido los españoles en otros reynos, y de los autores nacionales que han escrito sobre esta materia, así en la península como fuera de ella* (Madrid: Fermin Villalpando, 1803), 167.

[4] See, for instance, its inclusion in the 'Domestic Treatise on Some Ailments Rather Common in this Capital' (*Tratado doméstico de algunas enfermedades bastante comunes en esta capital*), a medical compendium authored by the Lima doctor Tomás Canals, designed to 'enlighten' the city's inhabitants with 'some (medical) knowledge'. Tomas Canals, *Tratado doméstico de algunas enfermedades bastante comunes en esta capital* (Lima: Imprenta Real del Telegrafo Peruano, 1800), X–XI.

[5] Villalobos, *Método de curar tabardillos*, 100–01.

[6] For a reference to the application of Masdevall's 'method' in New Spain, see Juan Sánchez y Sánchez, *Disertacion quimico-medica sobre la opiata antifebril, inventada por el ilustre señor doctor D. Josef de Masdevall, Medico de Camara con Exercicio de S.M. Católica, & c.* (Malaga: D. Felix de Casas y Martinez, 1794), 15.

[7] Joseph Masdevall, *Relazione dell'epidemie sofferte nel principato di Catalogna principalmente nell'anno 1783. Scritta in lingua spagnuola dal nobile sig. dottor Giuseppe Masdevall, attuale medico di camera del re cattolico: In cui si espone il suo nuovo metodo specifico per guarire le febbri putride, maligne, ed altre simili malattie* (Ferrara: Per gli eredi di Giuseppe Rinaldi, 1789).

[8] Joseph Masdevall, *Bericht über die Epidemien von faulen und bösartigen Fiebern welche in den letzten Jahren im Fürstenthum Catalonien geherrscht haben nebst der glücklichen, geschwinden und sichern Heilmethode dieser Krankheiten*, trans. E. H. Spohr (Braunschweig: Schulbuchhandlung, 1792).

[9] Joseph Townsend, *A Journey through Spain in the Years 1786 and 1787, with particular attention to the Agriculture, Manufactures, Commerce, Population, Taxes, and Revenue of that Country* (London: C. Dilly, 1791), 137.

Mississippi Valley, in the late 1700s.[10] Thence, the news quickly spread to the United States.[11] Nor did the recipe halt at boundaries of creed or religious belief. When an epidemic was raging in the sultanate of Morocco in 1799 – particularly in Marrakesh, Tangier, Meknes and Tétouan – the Moroccan 'Alawi court under Sultan Mawlay Sulayman chose to combat the disease with the aid of Masdevall's method.[12] While the foreign-language editions of Masdevall's report generally contained literal, accurate translations of the recipe, manuscript recipe collections kept by men and women in the Swiss Confederacy, the Italian territories and the Kingdom of France sometimes encompassed abridged versions of, or alterations to, the original recipe, presumably on account of a sufferer's palate, persuasion or means. While a French householder's recipe for opiate 'against the fevers' contained betony syrup in addition to cinchona, absinthe salt, ammoniac salt and vitriolated tartar,[13] the Swiss army surgeon Johannes Hueber's recipe book contained a formula for 'opiate febrifuge' consisting of absinthe salt, cinchona, antimonial substances, absinthe syrup and confection of hyacinth, and also theriac, centaury salt, and gentian.[14] In other instances, amendments would have

---

[10] The New Orleans edition contained the recipe for the 'opiate'. Joseph Masdevall, *Medicamens, et précis de la méthode de Mr. Masdevall docteur médecin du Roi d'Espagne Charles IV, pour guérir toutes les maladies épidémiques, putrides & malignes, fièvres de différents genres &c. & pour en préserver. Divisés en paragraphes & en numéros correspondans, à l'usage des familles dépourvues de médecins* (New Orleans: Chez Louis Duclot, 1796), 15–16. A French translation released in Marseille was to follow in 1800, a Toulouse edition in 1810. Manuel Valera Candel, *Proyección internacional de la ciencia ilustrada española. Catálogo de la producción científica española publicada en el extranjero 1751–1830* (Murcia: Universidad de Murcia, Servicio de Publicaciones, 2006), 136–37.

[11] 'Review of "Medicaments, et Precis de la Methode de Mr. Masdevall, &c &c. – That is, Prescriptions, and a Sketch of the Method of Mr. Masdevall, Physician of Charles the Fourth, King of Spain, for curing all epidemic, putrid, and malignant Distempers, Fevers of different Kinds, &c. &c. with the Means of Prevention. Divided into Paragraphs, for the Use of Families who are unable to procure Physicians. New-Orleans. Duclot. 1796. pp. 47 8vo,"' in *The Medical Repository*, ed. Samuel L. Mitchill, Edward Miller and Elihu H. Smith (New York: T. & J. Swords, 1800).

[12] Owing to the difficulty of procuring all the ingredients, it appears that José Antonio Coll, the disciple of Masdevall's sent to assist in the Moroccan epidemic, came to prefer a simplified version of the remedy: emetic tartar, followed by cream of tartar and salt of fig, followed, in turn, by a cinchona-based decoction. Justel Calabozo, 'El doctor Masdevall. Protomédico del sultán Marroquí Muley Solimán,' 99–102; 80; Braulio Justel Calabozo, *El médico Coll en la corte del sultan de Marruecos (año 1800)* (Cádiz: Servicio de Publicaciones de la Universidad de Cádiz / Instituto de Cooperación con el Mundo Arabe, 1991), 197.

[13] 'Collection of medical receipts, with a few household and veterinary receipts: in French,' f. 2.

[14] Johannes Hueber, 'Handbüchlein angefangen den 6ten Mey 1727 in Coullioure ein Meerport in der Provinz Roussillon, in den Pireneischen Gebirgen. Von Johannes

responded to religious differences. The Moroccan 'Alawi court, for instance, insisted that the wine included in the recipe 'be replaced with vinegar'.[15]

This chapter examines medical recipes for 'preparations of the bark',[16] as they moved between and across societies in the late 1700s, and early 1800s – through practitioners' and sufferers' continuous exposure to the written word, medical practice and word of mouth. Like the historiography of cinchona more broadly, scholarship on bark recipes has commonly been confined, and defined, by familiar entities – the colony, a nation-state or a particular medical tradition[17] – in ways that have obscured and buried the reach of the epistemic exchange the bark's worldwide passage entailed.[18] Premised upon a study of various genres of popular print, pharmacopoeias and a selection of manuscript notebooks of medical recipes, this chapter exposes and examines how men and women from New Orleans to Tétouan came to share modes of cinchona preparation and administration. The chapter focuses, in the first section, on the purgative substances – antimonial wines and cream of tartar, to name but two – that commonly preceded the administration of cinchona, for the 'proper evacuations' thought necessary before a sufferer could 'safely use the jesuites bark'.[19] It turns, in the second part, to the range of substances – wine, syrup and absinthe salt – that frequently accompanied cinchona. Some of the most common formulae and popular arrangements of ingredients metamorphosed into patent, proprietary or eponymous medicines – that is, remedies named after a

Hueber Chirurgien in dem löblichen Regiment de D'Hemel Schweitzer, zu diensten dero Königl. Majeßtet in Franckreich,' *Archiv für Medizingeschichte*, Universität Zürich, Rezeptbücher, MS J 4, Collioure, 1727-05-06. For the Italian recipe, see 'Collection of medical receipts by several hands: in Italian,' *Wellcome Library*, Archives and manuscripts, Closed stores WMS 3, MS.4110, n.p., c. 1800.

[15] Cited in Justel Calabozo, 'El doctor Masdevall. Protomédico del sultán Marroquí Muley Solimán,' 175.

[16] Richard Kentish, *Experiments and observations on a new species of Bark, shewing its great efficacy in very small doses* (London: J. Johnson, 1784), 95.

[17] For references to preparations of the bark in the Ottoman Empire, for instance, see Günergun and Etker, 'From Quinaquina to "Quinine Law."' On the preparation of cinchona extracts in the Spanish Empire, see Matthew Crawford, 'An Empire's Extract: Chemical Manipulations of Cinchona Bark in the Eighteenth-Century Spanish Atlantic World,' *Osiris* 29 (2014). On Peruvian bark recipes in viceregal Lima, see Stefanie Gänger, 'In Their Own Hands. Domestic Medicine and "the Cure of all Kinds of Tertian and Quartan Fevers" in Late-Colonial Lima,' *Colonial Latin American Review* 25, no. 4 (2016). On Portuguese bark recipes, see Sousa Dias, *A água de Inglaterra*.

[18] Publications studying the passage of medical recipes, and formulae, are generally scarce, as has recently been noted by Marta Hanson and Gianna Pomata in 'Medicinal Formulas and Experiential Knowledge in the Seventeenth-Century Epistemic Exchange between China and Europe,' *Isis* 108, no. 1 (2017), 2.

[19] Buchan, *Domestic Medicine*, 167.

practitioner-inventor, like Masdevall's Antipyretic Opiate – and the third section examines these commercially available cinchona-based compound medicines and their worldwide distribution and reception. The chapter argues that the ingredients that preceded and accompanied the administration of the bark had, by the late 1700s and early 1800s, coalesced into a series of common formulae and classic arrangements like the opiate that would have been attainable across the societies that constituted the Atlantic World in the period: homemade in a Lima household, available from an Italian apothecary and popular at the Moroccan court. While men and women commonly tinkered with the particulars of these formulae because of their culinary habits, economic means or creed, the recipes also exhibited structural similarities in the composition that were owing to the material and therapeutic properties cinchona brought to bear in practices of administration: the presence of kindred febrifuges to assist the bark in the cure and, at a time when societies within or tied to the Atlantic World shared a habitual fondness for sugar and alcohol, a necessity for solvents and sweeteners to cover its, by all accounts, nauseous taste.[20] Indeed, men and women developed a taste for bittersweet, spirituous preparations of the bark, the chapter holds, in part through the social fabric of the Atlantic World – social relations and contingent 'act[s] of communication' between physician-authors and their readership, soldiers and enemy civilians, planters and slaves.[21]

'Proper Evacuations'

By the late 1700s and early 1800s, fevers and a wide array of other complaints were commonly seen to be the effect of accumulated and deleterious morbid matter or humour – taken in with food and air or occasioned by the supressed evacuation of waste matter – it was necessary to eliminate before cinchona could take effect.[22] Self-help manuals,

[20] On the bark's bitter, 'nauseous taste', see, for instance, Buchan, *Domestic Medicine*, 136. For similar remarks from Levantine, Mesoamerican or Iberian authors, see, for instance, Aydüz and Yildirim, 'Bursalı Ali Münşî ve Tuhfe-i Aliyye,' 93; Joseph Rosuela, 'Pertenece este al uso de Frai Joseph Rosuela,' *Wellcome Library*, Archives and manuscripts, WMS/Amer.22–24, San Diego, 1771-03-22; Francisco Tavares, *Pharmacopeia Geral para o reino, e dominios de Portugal, publicada por ordem da Rainha Fidelissima D. Maria I*, 2 vols. (Lisboa: Na Regia Officina Typografica, 1794), 91.
[21] This follows the argument by Marcy Norton, who contends that Europeans developed a taste for chocolate not (solely) on account of biological affinities or as a manifestation of ethos but because the social relationships that Spanish imperial organization entailed continually exposed them to it. Norton, 'Tasting Empire.' Norton offers a similar interpretation of tobacco's transculturation in Norton, *Sacred Gifts, Profane Pleasures*.
[22] Michael Stolberg, 'Medical Popularization and the Patient in the Eighteenth Century,' in *Cultural Approaches to the History of Medicine. Mediating Medicine in Early Modern and*

recipe collections and almanacs, accordingly, almost uniformly counselled their readership to initiate their treatment by 'evacuating' the stomach and the intestines by means of emetics and purgatives.[23] The most popular were laxative salts like Sal-Policrest, Glauber's and Epsom's,[24] and plant-based purges like rhubarb,[25] ipecacuanha,[26] senna and jalap root.[27] Antimonial drugs or mercury compounds were likewise valued for their apparent ability to expel undesirable matter from the body by promoting sweating, vomiting and purging.[28] Sufferers from Graubünden to Luanda, from Barcelona to Chancay and from London to Constantinople would have been familiar with courses of therapy preceding cinchona with mercury compounds such as cinnabar[29] or antimony,[30] a lustrous metalloid often found in ores together with either

Modern Europe, ed. Willem de Blécourt and Cornelie Usborne (Basingstoke: Palgrave Macmillan, 2004), 14.

[23] On the persistent popularity of purgatives, bleeding, emetics and sudorific substances more broadly in the eighteenth century, see Michael Stolberg, *Homo patiens. Krankheits-und Körpererfahrung in der Frühen Neuzeit* (Köln: Böhlau, 2003), 41; 129–30; 94–99.

[24] For manuals advising the administration of Glauber's or Epsom's salt, see, for instance, Thomson, *A treatise on the diseases of negroes*, 19; João Curvo Semedo, *Polyanthea Medicinal. Noticias galenicas, echymicas. Repartidas em tres Tratados, dedicadas a's saudisas memorias, e veneradas cinzas do eminentissimo senhor Cardenal de Sousa, Arcebispo de Lisboa, Capellam Mor do Serenissimo Senhor Rey Dom Pedro II & seu Conselheyro de Estado* (Lisboa: Antonio Pedrozo Galram, 1727), 548; Buchan, *Domestic Medicine*, 167.

[25] See, for instance, 'Neste Secretaria da Junta do Proto-Medicato,' 53; Buchan, *Domestic Medicine*, 167. See also Irving, *Experiments on the Red and Quill Peruvian Bark*, 173. On the medical history of rhubarb, see Foust, *Rhubarb*.

[26] For advice manuals advising to precede the bark's administration with that of ipecacuanha, see, for instance, Buchan, *Domestic Medicine*, 240; John Theobald, *Every Man His Own Physician: Being, a Complete Collection of Efficacious and Approved Remedies, for Every Disease Incident to the Human Body. With Plain Instructions for Their Common Use* (London / Boston: Griftin / Cox and Berry, 1767), 2; 'Pareceres de los médicos sobre los efectos de la Quina de Santa Fé.' On ipecachuanha, see M. R. Lee, 'Ipecacuanha: The South American Vomiting Root,' *Journal of the Royal College of Physicians of Edinburgh* 38 (2008).

[27] See, for instance, Buchan, *Domestic Medicine*, 167; Dundas, 'Collections of medical and cookery receipts in English.' On the medical history of jalap, see Rogelio Pereda-Miranda, Daniel Rosas-Ramírez and Jhon Castañeda-Gómez, 'Resin Glycosides from the Morning Glory Family,' in *Progress in the Chemistry of Organic Natural Products*, ed. A. Douglas Kinghorn (Wien: Springer, 2010), 81.

[28] Porter observes the same procedure – the 'standard recourse to purging' that would 'evacuate the system and thus pave the way for the truly effectual medicine', cinchona. Porter, 'The Eighteenth Century,' 422.

[29] Danielle Molinari, 'Ricete,' *Archiv für Medizingeschichte*, Universität Zürich, Rezeptbücher, MS J 8, Graubünden/ Chur [?], c. 1740–1762, 236.

[30] Pedro Maria Gonzalez, *Tratado de las enfermedades de la gente de mar, en que se exponen sus causas, y los medios de precaverlas* (Madrid: Imprenta Real, 1805), 204. See also Robert Robertson, *Observations on Fevers and Other Diseases: Which Occur on Voyages to Africa and the West Indies* (Cambridge: Cambridge University Press, 2011 (1792)), 144. The seventh of João Curvo Semedo's (1635–1719) medical secrets published in the Polyanthea medicinal 'Lusitanian Water' could also be preceded by an 'antimonial

sulphur or mercury. Some advice manuals would have their readership continue, if patients were 'robust and plethoric', by bleeding them[31] – indiscriminate phlebotomy had become controversial by the late 1700s[32] – and only then administer a dose of the bark. Judging by the entries in the notebooks of householders, sufferers in different parts of the world abided by these counsels. The rhubarb- and dandelion-based 'laxative tea' that 'Madame la presidente de Maliverny' – probably Marie Thérèse Maliverny (1703–1757), wife of Claude Maliverny who served as Président à Mortier in the Provence Parliament – dispensed before she administered her cinchona and absinthe-based remedy for tertian and quartan fevers would have answered that purpose.[33] So would the 'Jalap Cream of Tartar' Lady Eleanor Dundas ordered, alongside her ague-potions, 'the day the patient has not fitt'.[34] It was also not unusual for practitioners in the late eighteenth and early nineteenth centuries to administer potions that contained the bark and purgative substances

emetic'. Francisco Suarez de Ribera, *Ilustracion, y publicacion de los diez y siete secretos del Doctor Juan Curvo Semmedo, confirmadas sus virtudes con maravillosas observaciones* (Madrid: Imprenta de Domingo Fernandez de Arrojo, 1723), 125. So could Pinto de Azeredo's cinchona-based cures. Pinto de Azeredo, *Ensaios sobre algumas enfermidades d'Angola*, 61. Ottoman physicians in the late 1700s likewise recommended the use of these purgatives before the administration of cinchona. Günergun and Etker, 'From Quinaquina to "Quinine Law,"' 45.

[31] On the necessity of bleeding 'plethoric' patients suffering from intermitting fevers, see, for instance, Murray, *Vorrath von einfachen, zubereiteten und gemischten Heilmitteln*, 1, 1131; José Manuel de Dávalos, 'Specimen Academicum de morbis nonnullis Limae grassantibus, ipsorumque Therapeia,' *Journal de Médicine* (1787), 14; John Huxham, *An Essay of Fevers, and Their Various Kinds: As Depending on Different Constitutions of the Blood: with Dissertations on Slow Nervous Fevers; on Putrid, Pestilential, Spotted Fevers; on the Small-pox; and on Pleurisies and Peripneumonies* (London: S. Austen, 1750), 23; 'Medicina Primitiva ou Colecção de Remedios Escolhidos e aprovados por experiências constantes, Para o uso das Pessoas do campo, dos Ricos e Pobres, traduzida do inglez, de Wesley sobre a décima terceira edição, revista, e aumentada consideravelmente, e Agora traduzida do Franzes na Lingua Portugueza,' *Arquivo Nacional da Torre do Tombo*, Manuscritos da Livraria, PT/TT/MSLIV/0134, London, 1760-11-10, f. 53. On bleeding in the treatment of fevers in the Ottoman Empire, see Günergun and Etker, 'From Quinaquina to "Quinine Law,"' 43.

[32] Mark Harrison, 'Disease and Medicine in the Armies of British India, 1750–1830,' in *British Military and Naval Medicine*, ed. Geoffrey L. Hudson (Amsterdam: Rodopi, 2007), 96–97; Jean Luiz Neves Abreu, *Nos Domínios do Corpo. O saber médico luso-brasileiro no século XVIII* (Rio de Janeiro: Editora Fiocruz, 2011), 91–92; Huldén, 'The First Finnish Malariologist,' 2.

[33] 'Livre des remedes de Madame la presidente de Maliverny seulement de sus dont ie fait le pruve tan de sus que l'on ma donne que sus que ie pris aux livres,' *Wellcome Library*, Archives and manuscripts, Closed stores WMS 4, MS.3409, Aix[-en-Provence], 1719–1739, fs. 90–91.

[34] Dundas, 'Collections of medical and cookery receipts in English,' f. 8. For more examples, see recipe number 223 in 'Secretos medicos, y chirurgicos.'

blended together – cinchona with tartar, as in Masdevall's Antipyretic Opiate,[35] senna or rhubarb.[36]

When at their wits' end, householders would have occasionally summoned a titled physician or one of the variety of unlicensed practitioners that crowded their city's medical marketplace[37] – one of its barber-bloodletters, curanderos or charlatans – but, at least with intermittent fevers and other common ailments, the advantages that professional treatment offered in comparison to household medicine were by no means obvious.[38] As a matter of fact, by the late 1700s and early 1800s, the treatments that sufferers inflicted upon themselves differed little from those applied by university-trained physicians, ship surgeons or orderlies in convents and military hospitals. Practitioners in the Hanseatic League and the West Indies, in peninsular Spain and the Viceroyalty of Peru almost invariably initiated their treatment of intermittents and other fevers, like that of 'all ailments', as the Lima physician José Manuel Dávalos (1758–1821) put it, by 'evacuating' the stomach and the intestines by means of purges. They continued, if proper, by bleeding them and only then administered a dose of cinchona bark, sometimes together with yet another purge.[39] Some authors of medical treatises and

---

[35] See also the Peruvian medical notebook 'Recipes Known from Experience', for instance, which administered tartar alongside cinchona. Anon., 'Recitario eficaz para las familias. Medicamentos caseros,' in La medicina popular peruana, ed. Hermilio Valdizán and Angel Maldonado (Lima: Imprenta Torres Aguirre, 1922 (n.d.)), 308.

[36] See, for instance, Mrs Finger and Anna Maria Reeves, 'Collection of medical, cookery, and household receipts: with additions by several hands. With numerous inserted receipts and cuttings from newspapers, etc. pasted in,' Wellcome Library, Archives and manuscripts, MS.2363, n.p., c. 1750–1775, 4; 10.

[37] Jenner and Wallis, 'The Medical Marketplace,' 1. On the medical marketplace in colonial New Spain, see Fields, Pestilence and Headcolds, 40–88.

[38] On the centrality of domestic medicine, lay healing and self-dosing in eighteenth-century Europe and North America, see, for instance, Enrique Perdiguero, 'The Popularization of Medicine during the Spanish Enlightenment,' in The Popularization of Medicine 1650–1850, ed. Roy Porter (London: Routledge, 1992); Mary E. Fissell, 'Popular Medical Writing,' in The Oxford History of Popular Print Culture. Cheap Print in Britain and Ireland to 1660, ed. Joad Raymond (Oxford: Oxford University Press, 2011); Hilary Marland, 'Women, Health, and Medicine,' in The Oxford Handbook of the History of Medicine, ed. Mark Jackson (Oxford: Oxford University Press, 2011). For Spanish and Portuguese America, see, for instance, Neves Abreu, Nos Domínios do Corpo; Palmer, Doctors; Gänger, 'In Their Own Hands.'

[39] Dávalos, 'Specimen Academicum de morbis nonnullis Limae grassantibus,' 14. For an example from the Portuguese context, see Jesus Maria, 'Colleção medica de receitas.' For a Spanish medical treatise advising the same proceeding in 'benign tertian fevers', see Luis Joseph Pereyra, Tratado completo de Calenturas: fundado sobre las leyes de la Inflamacion, y Putrefaccion, que constantemente observaron los mayores, y mas ilustrados medicos del mundo (Madrid: Imprenta de Antonio Marin, 1768), 249. See also George Baker, 'Observations on the Late Intermittent Fevers to which is added a Short History of the Peruvian Bark,' Medical Transactions III (1785), 676. For an example from the German-language context, see the writings of Johann Jacob Rambach, who described

handbooks cautioned their readership about the impropriety, even harm-fulness, of 'evacuations', fearing that purgatives might 'counteract the effects of cinchona'.[40] Still, contemporaries' sense that in fevers and a wide array of other ailments their bodies were replete with deleterious matter they needed to rid themselves of before they could 'safely use the jesuites bark' was too deeply ingrained to yield easily. As Lady Eleanor Dundas noted in her collection of medical and cookery recipes, most agues proceeded 'from a Foulness of Stomack' and no Medicine would 'take Effect untill the patient ha[d] had a gentle Ematic'.[41] The language and logic of evacuation that men and women from Scotland to Illinois, from Provence to the Bosporus, and from the Lemanic region to coastal Peru could apparently so very easily call on, was still an intrinsic and deep-seated element of the widely shared 'cultural background against which disease ideas', and practices of medical consumption, were being worked out in the late eighteenth and early nineteenth centuries.[42]

Preparations of the Bark

Even though cinchona bark was not invariably believed to require a specific mode of preparation to reveal its curative properties – it was

how Hamburg physicians begun the 'cure' of the 'intermittent fevers (*Wechselfieber*)', so prevalent in the city, with 'vomits and purges' (*Brechmitteln und Abführungen*) and invariably 'closed with cinchona'. Rambach, *Versuch einer physisch-medizinischen Beschreibung von Hamburg*, 310–12. James Lind, who practiced as a surgeon of the Royal Navy in the Mediterranean, Guinea and the West Indies, advised 'when an European is taken ill of a fever', in any country 'between the tropics', first to consider whether the 'violence of the fever' admitted of bleeding, then to 'cleanse' the 'stomach and intestines [...] either by a vomit, a purge [...] or by an oily and purging clyster, and then to [...] administer the bark'. James Lind, *An Essay on Diseases incidental to Europeans in hot Climates with the Method of Preventing their fatal Consequences. To which is added, an Appendix Concerning Intermittent Fevers, and, a simple and easy Way to render Sea Water fresh, and to prevent a Scarcity of Provisions in long Voyages at sea* (London: Printed for J. Murray, 1788), 234. See also John Pringle, *Observations on the diseases of the Army* (London: A. Millar; D. Wilson; T. Durham, T. Payne, 1764), 209–10.
[40] Thornton, *New Family Herbal*, 122. Ralph Irving wrote that many practitioners joined to the bark 'rhubarb and other cathartics', but presumed that these were 'for the most part unnecessary, and often improper'. Irving, *Experiments on the Red and Quill Peruvian Bark*, 173. Tissot likewise doubted the propriety of administering purgatives after the administration of the bark. Tissot, *Anleitung für das Landvolk*, 294.
[41] Dundas, 'Collections of medical and cookery receipts in English.' For similar passages, see Johann Eberhard Friedrich Schall, *Handbuch für Leute die keine Aerzte sind zur Beförderung nützlicher und angenehmer Kenntnisse. Dritter und letzter Theil* (Riga: Johann Friedrich Hartknoch, 1781), 142.
[42] The phrasing is taken from the work of Conevery Bolton Valencius, who made that observation in relation to Mary E. Smith, a young Illinois girl studying at a boarding school in St Louis in the late 1830s. Conevery Bolton Valencius, *The Health of the Country. How American Settlers Understood Themselves and Their Land* (New York: Basic Books, 2002), 59.

'pretty generally believed that the bark [was] most effectual when given in substance, and in pretty large doses'[43] – by the late 1700s and early 1800s, across the various consumer societies, sufferers and practitioners alike accorded considerable importance to administering, or concocting, cinchona preparations that were 'elegant, palatable, and at the same time sufficiently efficacious'.[44]

Some cinchona preparations quite simply served the purpose of rendering the bitter, powdered bark palatable and bearable to sufferers who could not be made to swallow it pure and 'in substance'.[45] Sugar, alongside other sweeteners, was perhaps the most common and pervasive of vehicles employed to that end. Men, women and children in the various Atlantic consumer societies would have shared a – patterned and habitual – fondness for sweetness by the late 1700s and early 1800s,[46] and sugars played an important part in rendering the bitter-tasting Peruvian bark agreeable to their taste. Cinchona was given in honey,[47] extract of liquorice,[48] mashed sweet fruit like sugar apples[49] or in various kinds of syrups, made from juices, starches

[43] Irving, *Experiments on the Red and Quill Peruvian Bark*, 173. Or, as William Buchan phrased it, 'no preparation seem[ed] to answer better than the most simple form in which it [could] be given, viz. in powder'. Buchan, *Domestic Medicine*, 167. The *Edinburgh new dispensatory* forwarded the same opinion. See Lewis and Rotheram, *The Edinburgh new dispensatory*, 142. See also Thomas Skeete, *Experiments and Observations on Quilled and Red Peruvian Bark: Among Which are Included Some remarkable Effects arising from the Action of Common Bark and Magnesia upon each other, with Remarks on the Nature and Mode of Treatment of Fevers, Putrid Sore-Throat, Rheumatism, Scrophula, and other Diseases; in order to ascertain the Cases in which Bark may be administered – either alone, or combined with other Remedies – to the best Advantage* (London: J. Murray, 1786), 141. The same opinion was generally expressed in the Iberian world. See, for instance, Sánchez y Sánchez, *Disertacion quimico-medica sobre la opiata antifebril*, 8.

[44] Thomas Percival, 'Experiments on the Peruvian Bark,' *Philosophical Transactions* LVII (1767), 227.

[45] John Clark, *Observations on Diseases which prevail in Long Voyages to Hot Countries, Particularly on those in the East Indies, and on the Same diseases as they appear in Great Britain* (London: J. Murray, 1792), 185; Buchan, *Domestic Medicine*, 112.

[46] Sidney W. Mintz, *Sweetness and Power: The Place of Sugar in Modern History* (New York: Penguin Inc., 1985), xxv.

[47] See, for instance, Suarez de Ribera, *Ilustracion, y publicacion de los diez y siete secretos del Doctor Juan Curvo Semmedo*, 192–95.

[48] Irving, *Experiments on the Red and Quill Peruvian Bark*, 173. The *Edinburgh new dispensatory* likewise recommended liquorice as a vehicle. Lewis and Rotheram, *The Edinburgh new dispensatory*, 142. See also Skeete, *Experiments and Observations on Quilled and Red Peruvian Bark*, 153.

[49] See the recipe for a 'Black Antipyretic Electuary by Cicilio Folie' (*Eletuario Antipiretico nero di Cicilio Folie*) – presumably Cecilio Folio, or Caecilius Folius (1615–1650) – which comprehended 'cinchona calisaya', gentian root, arnica flowers, 'Jewish Bitumen (*Bitum(e) Giudaico*)', Venetian treacle, ammoniac salt and sugar-apples in 'Collection of medical receipts by several hands: in Italian.'

or saps.[50] Indeed, some pharmacies, like Juan Lucas Camacho's (fl. –
1759) in Lima, even sold ready-made 'cinchona syrup' (*xarabe de
cascarilla*) to their customers.[51] Practitioners and sufferers appear to
have continuously revised, enriched and adjusted their therapeutic
arsenal on the basis of advice literature and medical treatises, and
their choice of sweeteners often reflected both their reading and place
of abode, with its peculiar culinary lore. Physicians at the Ottoman
court adopted a formula comprising 'syrup of [...] red roses' from the
proliferation of recipes devised by Thomas Sydenham, presumably on
account of the popularity of rose flavouring in Ottoman palace cuisine.[52] The prevalence of Peruvian bark recipes employing capillaire
syrup, a sucrose syrup based on dried maidenhair fern, among French
sufferers may have owed as much to its presence – in the form of
ground cinchona – and capillaire syrup–'boluses' (*bols*) – in popular
advice manuals like Jean-Adrien Helvetius's (1661–1727)[53] as to that
syrup's popularity in French cuisine.[54] Sugar was an especially significant vehicle for rendering cinchona preparations appealing to the
youngest of sufferers. Though cures for children were structurally
similar to those employed for adults – they, too, would often have
endured purges before being administered the bark – an emerging
segment of enlightened advice literature, increasingly designed particularly for parents and other lay readers involved in the care of
children, advised its readership to render the bark palatable for children by administering it in almond milk, raspberry syrup or, like the

---

[50] Buchan, *Domestic Medicine*, 264. A Peruvian recipe collection contained two recipes
joining the bark together with lemon syrup. See Anon., 'El Medico verdadero,' 446–47.
See also Dundas, 'Collections of medical and cookery receipts in English,' 22.

[51] 'Tasación hecha por Manuel Seminario de la botica que Juan Lucas Camacho legó al
Hospital de la Caridad. Se incluye inventario de los bienes existentes en dicha botica,
indicando precios,' *Archivo Histórico / Instituto Riva Agüero PUCP*, Colección
Maldonado, A-I-42, Lima, 1759-09-13.

[52] Günergun and Etker, 'From Quinaquina to "Quinine Law,"' 45. On rose flavouring in
Ottoman cuisine, see Constance L. Kirker and Mary Newman, *Edible Flowers. A Global
History* (London: Reaktion Books, 2016).

[53] Claude-Adrien Helvétius, *Manière de donner le quinquina aux pauvres* (Versailles:
Imprimerie François Muguet, 1686), 2.

[54] For French recipes relying on 'capillaire syrup', see, for instance, 'Collection of medical
receipts, with a few household and veterinary receipts: in French,' 35; 'Remèdes
éprouvés contre les Fiévres tierces, double tierces, & autres Fiévres intermittentes,'
*Journal Oeconomique, ou Memoires, Notes et Avis sur l'Agriculture, les Arts, le Commerce,
& tout ce qui peut avoir rapport à la santé, ainsi qu'à la conservation & à l'augmentation des
biens des Familles, & c.* (1764). On that syrup's popularity in the South of France, see
John L. Russell, 'The Botanical and Horticultural Literature of the Olden Times, with
Remarks on the Species and Sorts,' *The Magazine of Horticulture, Botany and all Useful
Discoveries and Improvements in Rural Affairs* 24 (1858), 174.

popular Swedish physician-author Nils Rosén von Rosenstein (1706–1773), 'watery chocolate' (in schwacher Chocolade).[55] Sugar and honey were still often attributed medicinal properties around 1800 – the ability to cleanse and dilute humours and to 'impede putrefaction'[56] – and valued for the keeping qualities they imparted.[57] Their principal purpose, however, and the chief reason for their persistent presence in preparations of the bark across the Atlantic World, was by all accounts that of rendering these bitter-tasting medicines agreeable to the sufferers' own taste.

Another common solvent and carrier to 'conceal the bitter taste of cinchona'[58] and to render the dry bark ingestible and appealing to sufferers was alcohol in the form of beverages, fundamental to consumer culture and pharmaceutical practice across the Atlantic World at the time.[59] Depending on a sufferer's economic possibilities, culinary partiality or fancy, cinchona was variously, and without much ado, 'shook up in Red Port Wine',[60] taken 'in a little warm beer',[61] or infused in strong distilled liquors – 'Holland gin'[62] or brandy[63] – for varying lengths of

[55] Nils Rosen de Rosenstein, *Anweisung zur Kenntniss und Cur der Kinderkrankheiten* (Göttingen: J.C. Dieterich, 1768), 148; 299. For syrup- and bark-based recipes for children, see also Buchan, *Domestic Medicine*, 114.

[56] Joseph Masdevall, *Relacion de las epidemias de calenturas pútridas y malignas que en estos últimos años se han padecido en el Principado de Cataluña y principalmente de la que se descubrió el año pasado de 1783 en la ciudad de Lérida, Llano de Urgel y otros muchos corregimientos y partidos, con el método feliz, pronto y seguro de curar semejantes enfermedades* (Madrid: Imprenta Real, 1797), 16–17. See also Ken Albala, 'Medicinal Uses of Sugar,' in *The Oxford Companion to Sugar and Sweets*, ed. Darra Goldstein (Oxford: Oxford University Press, 2005).

[57] Jane O'Hara-May, 'Foods or Medicines? A Study in the Relationship between Foodstuffs and Materia Medica from the Sixteenth to the Nineteenth Century,' *Transactions of the British Society for the History of Pharmacy* 1, no. 2 (1971), 70.

[58] Pinto de Azeredo, *Ensaios sobre algumas enfermidades d'Angola*, 65.

[59] On the use of alcohol and wine in medicine, see Jonathan Reinarz and Rebecca Wynter, 'The Spirit of Medicine: The Use of Alcohol in Nineteenth-Century Medical Practice,' in *Drink in the Eighteenth and Nineteenth Centuries*, ed. Barbara Schmidt-Haberkamp and Susanne Schmid (Routledge: Pickering & Chatto, 2014), 130. On alcohol in the Atlantic World, locally varied yet ubiquitous, see, for instance, Frederick H. Smith, *Caribbean Rum: A Social and Economic History* (Gainesville: University Press of Florida, 2005).

[60] See, for instance, Samuel Curtis, *A Valuable Collection of Recipes, Medical and Miscellaneous. Useful in Families, and Valuable to Every Description of Persons* (Amherst, N.H.: Elijah Mansur, 1819), 23. See also Clark, Observations on Diseases which prevail in Long Voyages to Hot Countries, 185.

[61] Maria Dunham, 'Maria Dunham Her Book,' *Wellcome Library*, Archives and manuscripts, Closed stores WMS 4, MS.8301, n.p., 1781. Medical authors also advised beer as a convenient 'vehicle' for the bark. Irving, *Experiments on the Red and Quill Peruvian Bark*, 173.

[62] Buchan, *Domestic Medicine*, 183.

[63] See, for instance, 'Collection of medical receipts by several hands: in Italian.' The official Portuguese Pharmacopeia also contained a 'compound tincture of cinchona' that

time. Most commonly, however, cinchona was administered in wine. To a wide range of practitioners – Ottoman doctors,[64] Portuguese hospital orderlies,[65] Spanish court physicians,[66] Russian apothecaries[67] and householders in New Spain[68] – 'a good wine', as the Loja healer Fernando de la Vega put it,[69] either red or white, was the vehicle of choice for the bark by the late 1700s and early 1800s. Infusions or macerations of the bark in wine had been among the earliest methods for administering cinchona in the mid-1600s – recommended in the 1649 'Roman Pamphlet' (*Schedula Romana*), and by Robert Talbor's (1642–1681) famous 'English remedy' (*remède anglois*)[70] – a circumstance that may in part account for their great prevalence and currency in the late 1700s and early 1800s. Wine and distilled spirits as well as beer presumably served, however, like sugars, a variety of medical, culinary and chemical purposes in the making of plant-based remedies, not least as a preservative to halt the deterioration of plant ingredients, ensuring infusions a

contained ground yellow cinchona, macerated in spirit of wine, for six days. Tavares, *Pharmacopeia Geral para o reino, e dominios de Portugal*, 213. See also the Russian pharmacopeia, which contained a recipe for 'Peruvian bark tincture' (*Tinctura Corticis Peruviani*): Peruvian bark infused in French brandy (*Spiritus Vini Gallici*). *Pharmacopoea Rossica* (St Petersburg: Academiae Scientiarum, 1778), 130. See also Ramón Calbo y López, 'Experimentos [sobre las] propiedades de la Quina o cascarilla nuevamente descubierta,' *Archivo General de Indias*, Indiferente 1557, n.p., 1816–17, 1028.

[64] Ottoman physicians particularly recommended the administration of cinchona powder in wine, though pills were prepared for those who had a sensitive stomach or abstained from wine. Günergun and Etker, 'From Quinaquina to "Quinine Law,"' 47. See also Aydüz and Yildirim, 'Bursalı Ali Münşî ve Tuhfe-i Aliyye,' 98.

[65] The Portuguese religious João de Jesus Maria's 1760 manuscript recipe collection contained a recipe for cinchona-infused white wine. Jesus Maria, 'Colleção medica de receitas,' 92–93.

[66] See, for instance, Joseph Alsinet, *Nuevas utilidades de la quina, demostradas, confirmadas y añadidas por el Doctor Don Josef Alsinet, Medico de la Familia de su Magestad, y Jubilado del Real Sitio de Aranjuez. Se manifiesta el modo cómo cada uno en su casa podrá quitar el amargor á la Quina, sin perjuicio de su virtud febrifuga* (Madrid: Don Miguél Escribano, 1774), 165–66.

[67] Wylie, *Pharmacopoeia castrensis Ruthena*, 42. On the presence of cinchona and other American drugs in Russia outside the court in the seventeenth century, see Clare Griffin, 'Russia and the Medical Drug Trade in the Seventeenth Century', *Social History of Medicine* 31, no. 1 (2018), 16–17.

[68] Esteyneffer, *Florilegio medicinal*, 297. On that home guide's persistent popularity, see George M. Foster, 'On the Origin of Humoral Medicine in Latin America,' *Medical Anthropology Quarterly* 1, no. 4 (2013), 363–65.

[69] Cited in Estrella, 'Ciencia ilustrada y saber popular,' 56.

[70] Crawford, 'An Empire's Extract,' 218–20. Talbor's remedy, which successfully cured Charles II of England in 1679, caused an important break in public awareness of the bark. Klein and Pieters, 'The Hidden History of a Famous Drug,' 413. On the contents of the Schedula Romana, see Jarcho, *Quinine's Predecessor*, 12–14; 262–67; Maehle, *Drugs on Trial*, 226.

reasonably long shelf life.[71] Practitioners also clearly recognized their ability to render medicines more appealing to particular groups of sufferers. As Michael O'Ryan (d. 1794), chief physician at the Lyon Hospital (Grand Hôtel-Dieu de Lyon), expressed himself in the context of his endeavour to institute preventative cinchona consumption in the army, 'the military men of this country' might be reluctant to take a 'preparation of the Peruvian bark in water' on a regular basis. They would not refuse to take 'a warm bitter dram', however, 'made with some sort of spirituous liquor'.[72] Or, as John Pringle (1707–1782), a British military physician and temporarily the Physician-General of the army, phrased it in his *Observations on the Diseases of the Army*, 'the most effectual way to make a soldier continue the Bark, [was] to mix it with equal parts of brandy and water'.[73] It was their capacity to appeal and tie themselves to different medical and drinking cultures and localities, it would seem, that made spirituous preparations of the bark both varied and ubiquitous.

Other than recipes for vehicles that rendered the bark ingestible and palatable, there was a series of more elaborate procedures, generally destined both to render the bark 'more pleasant to the taste' and to enhance its effect, and assist it in the 'prevention and cure of diseases'.[74] Popular complements to that end were citrus fruits – 'Sevil Orange, or Lemon',[75] most commonly – ingredients considered, like the bark, to be antiseptics capable of counteracting putrefaction at a time when fevers were frequently conceived as a putrid change in the blood.[76] By the late 1700s and early 1800s, Mexican hospital orderlies, English housewives, Andean healers, Ottoman physicians, Dutch practitioners and Portuguese friars alike compounded 'bittersweet' (*agridulce*) 'febrifugal lemonades' and decoctions that contained cinchona and lemon or orange juice,

---

[71] I take the idea that some preparations were kept in readiness and stored, and that there were variations in a medicine's shelf life, from Elaine Leong, 'Making Medicines in the Early Modern Household,' *Bulletin of the History of Medicine* 82, no. 1 (2008), 158. Elaine Leong has conducted groundbreaking research on various aspects of the history of recipes in England. *Recipes and Everyday Knowledge. Medicine, Science and the Household in Early Modern England* (Chicago: University of Chicago Press). On the tenuous division between aliments and medicaments, see O'Hara-May, 'Foods or Medicines?'

[72] O'Ryan, *A Letter on the Yellow Peruvian Bark*, 23.

[73] Pringle, *Observations on the diseases of the Army*, 217.

[74] Buchan, *Domestic Medicine*, 177; 241; 64–65. See also Skeete, *Experiments and Observations on Quilled and Red Peruvian Bark*, 149.

[75] Huxham, *An Essay of Fevers, and Their Various Kinds*, 122.

[76] See Chakrabarti, *Medicine and Empire*, 45. According to John Pringle, 'alcaline salts', snakeroot and cinchona were also 'antiseptic'. Pringle, *Observations on the diseases of the Army*, iii–xiv.

usually sweetened with various kinds of sugars.[77] Advice manuals like Buchan's, popular the world over – with dozens of English-language editions as well as Dutch, Italian, German, French, Spanish, Swedish, Portuguese, Russian and Japanese translations –[78] likewise recommended administering the bark in ready-made orange water, 'bitter orange preserve' (*eingemachte Pomeranzenschaalen*)[79] or lemon syrup.[80] Infusions, or macerations of the bark in white or red wine, 'acidulated with Juice of Sevil Orange, or Lemon', were common, too.[81] Some sufferers relied

---

[77] Recipe number 223 in the Mexican collection 'Medical and Surgical Secrets (*Secretos medicos, y chirurgicos*)', recommended 'six drachms of ground cinchona (*Poluos de Kina Kina*), an ounce of wormwood salt (*sal de ajenjos*), half a pound of orange juice, and one and a half scruples of white sugar' 'for tertian fevers, after the purge'. 'Secretos medicos, y chirurgicos.' See also Brother Rosuela's orange, cinchona and sugar-based 'bittersweet' (*agridulce*) remedy against tertian fevers. Rosuela, 'Pertenece este al uso de Frai Joseph Rosuela,' Recipe 194. For a Portuguese recipe for a 'febrifugal lemonade', see Jesus Maria, 'Colleção medica de receitas,' 101. A recipe for the 'ague' pasted into 'Miss Myddleton's Book' recommended 'one Dram of Bark', '½ D[ram] of Venice Treacle', '3 Large spoonfulls of Lemon Juice', and '3 D[rams] Mountain Vine'. 'Receipts copied from Miss Myddleton's Book.' The Andean healer Fernando de la Vega recommended 'cinchona salt', in orange juice, in 'all kinds of fevers'. Estrella, 'Ciencia ilustrada y saber popular.' For an Ottoman recipe of cinchona, water and lemon, see Aydüz and Yildirim, 'Bursalı Ali Münşî ve Tuhfe-i Aliyye,' 96. For a Lima decoction of bitter orange peel, sugar and cinchona, see Anon., 'El Medico verdadero,' 446. For a Dutch recipe relying on lemon juice, cinchona and sugar, see Terne, *Verhandelingen*, 735.

[78] Between the 1769 publication of its first edition in Edinburgh and the appearance of its last English-language version just over a century later in Philadelphia, Buchan's *Domestic Medicine* appeared in at least 142 English-language editions. Rosenberg, *Explaining Epidemics*, 33. For a comprehensive list of the book's various editions, and references to its Italian, French, German and Swedish translations, see 'BUCHAN, William, 1729–1805,' in *An Annotated Catalogue of the Edward C. Atwater Collection of American Popular Medicine and Health Reform: A–L*, ed. Christopher Hoolihan (Rochester, NY: University of Rochester Press, 2001), 132. On Spanish translations, see Perdiguero, 'The Popularization of Medicine during the Spanish Enlightenment,' 171. On the book's circulation in Spanish America, see Palmer, *Doctors*, 21. On its circulation in Portugal and Portuguese America, see Jean Luiz Neves Abreu, 'A Colônia enferma e a saúde dos povos: a medicina das "luzes" e as informações sobre as enfermidades da América portuguesa,' *História, Ciências, Saúde-Manguinhos* 14, no. 3 (2007), 766. The 1817 Japanese translation was based on part 2 of the 1780 second printing of the Dutch translation of 'Domestic Medicine'. Grant Kohn Goodman, *Japan: The Dutch Experience* (London: Bloomsbury, 1986), 142. On the book's translation into Russian from a French edition, see Robert L. Nichols, 'Orthodoxy and Russia's Enlightenment, 1762–1825,' in *Russian Orthodoxy under the Old Regime*, ed. Robert L. Nichols and Theofanis George Stavrou (Minneapolis: University of Minnesota Press, 1978), 70.

[79] Rosen de Rosenstein, *Anweisung zur Kenntniss und Cur der Kinderkrankheiten*, 149.

[80] Buchan, *Domestic Medicine*, 264. A Peruvian recipe collection contained two recipes combining the bark with lemon syrup. See Anon., 'El Medico verdadero,' 446–47. English gentlewomen likewise employed lemon syrup for their cinchona-based preparations. Dundas, 'Collections of medical and cookery receipts in English,' 22.

[81] Huxham, *An Essay of Fevers, and Their Various Kinds*, 122. See also 'A booke of divers receipts,' *Wellcome Library*, Archives and manuscripts, MS.1322/54, n.p., c. 1660–1750,

solely on citrus fruit when beset by illnesses attributed to putrefaction,[82] drinking a 'whole Lemon' squeezed into wine[83] or a 'Gill of the Best French Brandy',[84] or 'eating a boiled lemon with the rind'.[85] However, most practitioners and sufferers would have agreed with authors like Buchan that medicines that were the result of citrus fruits and the bark 'joined together' were more effectual 'than either of them separately'.[86] The bark was also thought to be rendered agreeable and, particularly in obstinate agues, 'much more efficacious when assisted' by 'warm and aromatic bitters', such as 'the Virginian snakeroot, canella alba, orange peel' or ginger.[87] 'Virginian snakeroot' from North America, also an antiseptic, thought to be 'good in Fevers' and a range of other illnesses – 'hysteric Complaints', or 'Worms' –[88] was perhaps the most popular addition to the bark in ailments as diverse as epilepsy, ague or nervous fevers.[89] Preparations of snakeroot, orange peel and the bark, infused in a base liquor, were standard formulae in recipe books and medical guides

---

53. A late eighteenth-century recipe book begun by a Mrs Finger and continued by Anna Maria Reeves contained a recipe 'to cure an ague' that contained 'the best Bark in Powder, one Drachm, venice treacle half a Drachm', mixed 'with the juice of half a good Lemon & 2 or 3 spoonsfull of white wine'. Finger and Reeves, 'Collection of medical, cookery, and household receipts,' 4; 10.

[82] António Nunes Ribeiro Sanches, *Tratado da conservaçam da saude dos povos. Obra util, e igualmente necessaria aos Magistrados, Capitaens Generaes, Capitaens de Mar, e Guerra, Prelados, Abbadessas, Medicos, e Pays de familias* (Lisboa: Na Officina de Joseph Filippe, 1757), 298. See also Huxham, *An Essay of Fevers, and Their Various Kinds*, 122. This is the first English edition; the book was first published in Latin in 1739.

[83] A French domestic recipe collection recommended 'against tertian fevers' wine with the juice of 'half a lemon'. Collection of medical receipts, with a few household and veterinary receipts: in French, 10. 'Collection of medical receipts, with a few household and veterinary receipts: in French,' 10. A Mexican translation and paraphrase of a treatise by the Jesuit missionary Filippo Salvadore Gilii (1721–1789) recommended a concoction of orange juice, sugar and wine against 'tertian and quartan fevers'. 'Remedios singulares usados por los Misioneros de tierra firme, q[u]e se allan en el Apendice I° de la H[istori]a del Orinoco del Abate Felipe Salvador Gilij. tomo 3° en Roma año 1782,' *Wellcome Library*, Archives and Manuscripts, Closed stores WMS/ Amer.8, n.p., 1782.

[84] 'Collection of medical receipts, with a few cookery receipts: by several hands,' *Wellcome Library*, Archives and manuscripts, Closed stores WMS 3, MS.4057, n.p., n.d., 10.

[85] Buchan, *Domestic Medicine*, 177.   [86] Ibid.

[87] Buchan, *Domestic Medicine*, 113; 288; Skeete, *Experiments and Observations on Quilled and Red Peruvian Bark*, 152. On the 'salutary effects' of 'aromatic bitters' like 'Peruvian bark' or 'orange-peel', see also William Blair, *The Soldier's Friend: or, the Means of Preserving the Health of Military Men; Addressed to the Officers of the British Army* (London: Mr. Longman et al., 1798), 59.

[88] John Hill, *A History of the Materia Medica: Containing Descriptions of all the Substances used in Medicine; their Origin, their Characters when in Perfection, the Signs of their Decay, their Chymical Analysis, and an Account of their Virtues, and of the several Preparations from them now used in the Shops* (London: T. Longman et al., 1751), 612.

[89] Buchan cited a recipe for an 'anti-epileptic' that combined 'Jesuits bark in powder three ounces, Virginian Snake-root powdered one ounce, as much syrup of paeony or cloves as

by the late 1700s and early 1800s.[90] 'Wild cinnamon' (*Canella alba*) and other cinnamon varieties were also commonly added to cinchona both on account of their ability to render preparations palatable and to alter a consumer's bodily state in ways perceived to be beneficial.[91] Gentlewomen in the Scottish Lowlands,[92] Winterthur surgeons,[93] Lisbon apothecaries,[94] Königsberg physicians[95] and Lima hospital orderlies[96] alike devised or collected recipes that combined cinchona with cinnamon – or 'spirituous cinnamon water' or 'cinnamon spirit' – sometimes together with snakeroot or orange peel. So did Chinese physicians, who had adopted a recipe blending cinnamon and cinchona from 'the barbarians at Macao'.[97] Ginger was, like snakeroot or cinnamon, thought to 'render the Bark agreeable' and 'at the same time to increase its efficacy', and was made with it into decoctions,[98] electuaries[99] or 'Peruvian Bark Gingerbread', of the Scottish surgeon Robert Robertson's

is sufficient to form it into a soft electuary'. Buchan, *Domestic Medicine*, 506. For a Portuguese 'antifebrile decoction' containing cinchona, spirit and Virginian snakeroot, see Manuel Joaquim Henriques de Paiva, *Farmacopéa Lisbonense, ou Collecçao dos simplices, preparaçoes e composiçoes mais efficazes e de major uso* (Lisboa: Officina de Filippe da Silva e Azevedo, 1785). For a French febrifuge, administered in Saint-Domingue, see Pierre Pomme, 'Mémoire et observations cliniques sur l'abus du Quinquina,' in *Supplément au traité 'Les affectations vaporeuses des deux sexes', ou Maladies nerveuses* (Paris: Chez Cussac, 1804). See also Dancer, *The Medical Assistant; or Jamaica Practice of Physic*, 72.

90 See, for instance, the recipe for a cinchona-based tincture in 'Collection of medical receipts by several hands: in Italian.' The recipe was also contained in the official Portuguese pharmacopeia, see Tavares, *Pharmacopeia Geral para o reino, e dominios de Portugal*, 213. See also Buchan, *Domestic Medicine*, 174.

91 Cook and Walker, 'Circulation of Medicine in the Early Modern Atlantic World,' 2. See also Paula De Vos, 'The Science of Spices: Empiricism and Economic Botany in the Early Spanish Empire,' *Journal of World History* 17, no. 4 (2006), 409.

92 Dundas, 'Collections of medical and cookery receipts in English,' 22.

93 Paul Volmar's collection contained a recipe for a febrifuge that combined cinchona, cinnamon, absinthe, sugar and '*Flores Salmiac*'. Paul Volmar. 'Gehört Paulus Volmar. Med. Pract.,' *Archiv für Medizingeschichte*, Universität Zürich, Rezeptbücher, MS J 4, Winterthur, 1772.

94 The 1794 official 'Portuguese Pharmacopoeia' contained a recipe for an 'antifebrile decoction' (*cozimento antifebril*) that contained 'ground yellow cinchona', Virginia snakeroot, water and cinnamon spirit. See Tavares, *Pharmacopeia Geral para o reino, e dominios de Portugal*, 144.

95 Johann Jacob Woyts, *Abhandlung aller innerlichen und äußerlichen Kranckheiten, in zwei Theilen, in welchen jedwede Kranckheit deutlich beschrieben, und zur Kur die bewährtesten Arzney-Mittel aus denen Schrifften derer berühmtesten Aerzte und die Hand gegeben werden* (Leipzig: Friedrich Lanckisches Erben, 1753), 714.

96 Anon., 'El Medico verdadero,' 446–47.

97 The recipe is taken from the Pen-ts'ao kang mu shih-I, compiled in 1765 by Chao Hsüeh-min (1719–1805). Cited in Unschuld, *Medicine in China*, 166.

98 See, for instance, Buchan, *Domestic Medicine*, 176.

99 Curvo Semedo's 'fourteenth secret' was an 'electuary that healed quartan, and tertian intermittents', consisting of sagapeno root, 'legitimate cinchona', ginger, 'diagridio' and

(1742–1829) invention.[100] By the late 1700s and early 1800s, a varied, but limited set of febrifuges was commonly added to the bark across the Atlantic World to assist it in the cure.

Many formulae, rather than concealing the bark's taste, would have augmented the resultant medicines' bitterness. Peruvian bark recipes devised or employed by Luanda physicians, Italian householders and Lima hospital orderlies alike frequently joined cinchona, sometimes together with the various 'warm aromatics',[101] to one or several other bitter-tasting plant ingredients, on the assumption that these would likewise enhance their medicines' effectualness. Gentian root,[102] bitter-wood – also referred to as quassia[103] – dried wormwood, or lesser centaury,[104] a thistle-like flowering plant, were among the plants commonly added to the bark. In many instances, as chemical ingredients were increasingly combined with conventional materia medica,[105] bitter-tasting plant ingredients were administered alongside cinchona and aromatics, in the processed form of 'alkaline salts' made from calcination: centaury salt,[106] Cardo Santo salt[107] or salt of

honey. Suarez de Ribera, *Ilustracion, y publicacion de los diez y siete secretos del Doctor Juan Curvo Semmedo*, 192–95.

[100] Robert Robertson, *Directions for Preparing and Administering Peruvian Bark Gingerbread as a Preventive and Cure of Tertian and Remitting Fever, Extracted from Dr. Robertson's Synopsis Morborum* (London: D.N. Shury, 1812), 3–5.

[101] The 1785 'Lisbon Pharmacopoeia', for instance, held a recipe for cinchona-based wine that consisted of cinchona bark, gentian root, bitter orange peel and white wine. Henriques de Paiva, *Farmacopéa Lisbonense*, 227. See also Dundas, 'Collections of medical and cookery receipts in English,' 8; Mary White and Mary Downing Cartwright. 'Receipts [sic] Physick & Chirugery,' *Wellcome Library*, Archives and manuscripts, Closed stores WMS 4, MS.8300, n.p., 1715–1719. See also Theobald, *Every Man His Own Physician*, 2; 'Receipts copied from Miss Myddleton's Book,' 31.

[102] See, for instance, Hueber, 'Handbüchlein angefangen den 6ten Mey 1727 in Coullioure.' See also 'Collection of medical receipts by several hands: in Italian.' Curvo Semedo's 'Lusitanian water' was thought to contain cinchona, gentian root, sen leaves, tartar, centaury, zedoary root, ammoniac salt and wormwood salt, among other ingredients. Suarez de Ribera, *Ilustracion, y publicacion de los diez y siete secretos del Doctor Juan Curvo Semmedo*, 126. See also Pinto de Azeredo, *Ensaios sobre algumas enfermidades d'Angola*, 65.

[103] Thomson, *A treatise on the diseases of negroes*, 14.

[104] See, for instance 'Collection of medical receipts by several hands: in Italian.' Portuguese recipes also often combined ground cinchona, minor centaury and wine. Jesus Maria, 'Colleção medica de receitas,' 41–42.

[105] De Vos, 'From Herbs to Alchemy,' 136–37; 50–52. On the use of alkaline salts in intermittent fevers, separately or together with the 'bark', see Lind, *An Essay on Diseases incidental to Europeans in hot Climates*, 230; Buchan, *Domestic Medicine*, 177.

[106] For a French example, see 'Livre des remedes de Madame la presidante de Maliverny,' 109. For a Portuguese recipe collection, see Jesus Maria, 'Colleção medica de receitas,' 102. For an example from the Swiss Confederacy, see Molinari, 'Ricete,' 235.

[107] See recipe number 15, in Rosuela, 'Pertenece este al uso de Frai Joseph Rosuela.'

wormwood,[108] this last also occasionally, as in Masdevall's report, referred to as absinthe salt.[109] As with snakeroot or cinnamon, the bitter plant extracts and salts commonly administered alongside cinchona were medicines and febrifuges in their own right. Wormwood salt was a 'very famous Febrifuge and Stomachic', as was centaury, both age-old antipyretics recommended in the Hippocratic Corpus.[110] Other bitters gained, or regained, popularity in the context of endeavours to find surrogates for cinchona outside the Spanish American territories. Quassia,[111] for instance, or gentian root, combined with astringent gall nuts, oak bark and willow bark, had been tried with some success in the treatment of intermittent and other fevers.[112] The authors of Swiss, Jamaican and Swedish vernacular medical literature were agreed that home-grown bitters like centaury, quassia or wormwood were often quite as useful as cinchona,[113] and some advised their readership to substitute the bark, if it was too expensive or could not 'be supplied in sufficient quantity', with 'any bitter tincture', since it would 'in some measure answer the same purpose'.[114] Wherever possible, however, medical practitioners relied on compound medicines that combined the bark with these other bitters.

[108] For examples from popular advice manuals, see Buchan, *Domestic Medicine*, 177; Theobald, *Every Man His Own Physician*, 2. See also Suarez de Ribera, *Ilustracion, y publicacion de los diez y siete secretos del Doctor Juan Curvo Semmedo*, 126. For an English-language recipe book, see Dundas, 'Collections of medical and cookery receipts in English,' 22. See also Amaro, *Introdução da medicina ocidental em Macau e as receitas de segredo da Botica do Colégio de São Paulo*, 20. See also a Mexican cinchona, wormwood salt, orange juice and sugar-based recipe 'for tertian fevers, after the purge'. Recipe number 223 in 'Secretos medicos, y chirurgicos.' For examples from British households, see White and Downing Cartwright, 'Receipts [sic] Physick & Chirugery,' 9; Dundas, 'Collections of medical and cookery receipts in English.'

[109] A French remedy 'against the fevers' combined cinchona, ammoniac salt, 'absinthe salt', vitriolated tartar and betony syrup. 'Collection of medical receipts, with a few household and veterinary receipts: in French,' 2. See also 'Collection of medical receipts, preceded by some domestic receipts, receipts for liqueurs,' *Wellcome Library*, Archives and manuscripts, Closed stores WMS 4, MS.4082, n.p., c. 1750–1775, 180. For an early example from the Iberian context, see Suarez de Ribera, *Ilustracion, y publicacion de los diez y siete secretos del Doctor Juan Curvo Semmedo*, 126. See also 'Livre des remedes de Madame la presidante de Maliverny,' 109. For an example from the Swiss Confederacy, see Hueber, 'Handbüchlein angefangen den 6ten Mey 1727 in Coullioure,' 238; Molinari, 'Ricete.'

[110] Hill, *A History of the Materia Medica*, 352. On the use of 'absinthe' and centaury as febrifuges in the Hippocratic Corpus, see Jean-Claude Dousset, *Histoire des médicaments des origines à nos jours* (Paris: Payot, 1985), 43.

[111] Murray, *Vorrath von einfachen, zubereiteten und gemischten Heilmitteln*, 1, 1120.

[112] Maehle, *Drugs on Trial*, 277. Gentian was hailed as 'the cinchona of Europe' for a while. Boumediene, *La colonisation du savoir*, 246.

[113] Tissot, *Anleitung für das Landvolk*, 266; 642. See also Murray, *Vorrath von einfachen, zubereiteten und gemischten Heilmitteln*, 1, 1120. Thomson, *A treatise on the diseases of negroes*, 150.

[114] Blane, *A Short Account of the Most Effectual Means of Preserving the Health of Seamen*, 42.

They treated the slaves ailing from fevers on Jamaican plantations with 'a few doses of bark and bitterwood' added together,[115] 'sick and convalescent' sailors of the Royal Navy with an ale of cinchona, 'snakeroot, gentian, [and] quassia, [...] in sweet wort, and fermented with a little barm',[116] and customers at Macao's St Paul College pharmacy with the 'Finest Febrifugal Water' (*Agoa Febrefuga da Botica do Coll' de Macao. Optima*) – cinchona, wormwood salt, contrayerba, chicory and endive water, sweetened with confection of Alkermes[117] – because they expected these combinations to potentiate the febrifugal effect, or, as Buchan put it, the various ingredients to assist one another in the cure.[118] Intensely bittersweet formulations of the bark, like Masdevall's absinthe salt and syrup-based Antipyretic Opiate, would have been a familiar and shared experience for men and women from the West Indies to the Maghreb.

Peripatetic recipes for bittersweet aromatic liquors, opiate, and febrifugal lemonades were by no means the only conceivable preparations of cinchona in the late 1700s and early 1800s. Formulations of a more limited circulation and popularity combined cinchona with chemical ingredients like 'magnesia'[119] or ammoniac salt,[120] with aromatics like sandalwood[121] or nutmeg,[122] with panaceas like

[115] Thomson, *A treatise on the diseases of negroes*, 14.

[116] Robert Robertson, *Directions for Administering Peruvian Bark, in a Fermenting State, in Fever and Other Diseases in Which Peruvian Bark is Proper; and More Especially in Such Cases as the Usual Formulae of the Bark are Rejected by the Stomach, or Nauseated by the Sick; with some Experiments to Ferment the Peruvian Bark with Different Sweets* (London: Marche and Teape, 1799), 10.

[117] Amaro, *Introdução da medicina ocidental em Macau e as receitas de segredo da Botica do Colégio de São Paulo*, 21–22. The remedy was named after its inventor, João Curvo Semedo (1635–1719). Curvo Semedo, *Polyanthea Medicinal*, 570.

[118] Buchan, *Domestic Medicine*, 177.

[119] 'Collection of medical receipts, in Italian. Perhaps originally part of a Pharmacist's Prescription book. With indented alphabet,' *Wellcome Library*, Archives and manuscripts, Closed stores WMS 4, MS.4106, n.p., n.d; Skeete, *Experiments and Observations on Quilled and Red Peruvian Bark*, 160.

[120] 'Collection of medical receipts by several hands: in Italian.'; Hueber, 'Handbüchlein angefangen den 6ten Mey 1727 in Coullioure.' See also 'Collection of medical receipts, preceded by some domestic receipts, receipts for liqueurs,' 180. For an example from the Mexican context, see Rosuela, 'Pertenece este al uso de Frai Joseph Rosuela.' See also Lewis and Rotheram, *The Edinburgh new dispensatory*, 484. John Pringle made the bark 'for common practice' 'into an electuary, in which, to an ounce of the powder [of the Bark], a drachm of crude Sal Ammoniacum was sometimes added'. Pringle, *Observations on the diseases of the Army*, 213. See also Molinari, 'Ricete,' 238. Ali Münşî administered cinchona and ammonia in 'holy water' to a fever patient. Aydüz and Yildirim, 'Bursalı Ali Münşî ve Tuhfe-i Aliyye.'

[121] An Italian recipe for a 'Tincture of Cinchona by Masdevall (*Tintura di China del Masdeval*)' combined cinchona with sandalwood, distilled liquor and sugar. 'Collection of medical receipts by several hands: in Italian.'

[122] The anonymous author of the 'booke of divers receipts' had been given a recipe 'for an ague [...] the 13 Apl. 1739 by Mr John Barnard' that contained a 'dram of Bark, ½ of

theriac[123] and with narcotics like opium[124] for 'the most wonderful relief it produced'.[125] Other preparations blended the bark with herbs and fruits far more reluctant about processes of worldwide convergence and integration – Mesoamerican sossocoyoli,[126] or Abyssinian Woogi-noos.[127] Again, other febrifuges would have dispensed with the bark entirely – using anything from cobwebs[128] to 'medicinal clay' (*rothe terra sigillata*),[129] in its stead – because of opposition,[130] or obliviousness, to it.

Venice Treacle, ½ Nutmeg grated, Juice of Cest of Lemon', all mixed 'together in a glass of red wine'. 'A booke of divers receipts,' 53.

[123] 'Collection of medical receipts by several hands: in Italian.' See also Amaro, *Introdução da medicina ocidental em Macau e as receitas de segredo da Botica do Colégio de São Paulo*, 45. Theriac, a complex compound medicine based on 'anything between 50 and 70 ingredients', including viper's flesh, cinnamon and cardamom, as well as opium, from the time of its first formulation in Greek and Byzantine medical texts, continued to be held in high esteem as a universal panacea in early modern Persian, Urdu and European-language pharmacopoeias. Attewell, 'Interweaving Substance Trajectories,' 9–11.

[124] The recipe collection of Daniele Molinari, a Swiss bureaucrat, for instance, contained a recipe for a remedy against 'fevers' that joined together 'ground Peruvian bark', 'volatile ammoniac salt', and 'opium (*laud. opiat*)'. Molinari, 'Ricete,' 238. Samuel Curtis, the author of the 'Valuable Collection of Recipes, Medical and Miscellaneous', a pharmaceutical handbook, would have his readership cure 'fever and ague' by administering opium '1 or 2 hours before you expect the cold fit', followed by 'the best Red Bark, shook up in Red Port Wine'. Curtis, *A Valuable Collection of Recipes, Medical and Miscellaneous*, 23.

[125] Clark, *Observations on Diseases which prevail in Long Voyages to Hot Countries*, 176–78.

[126] Esteyneffer, *Florilegio medicinal*, 297.

[127] Bruce, *Travels to discover the source of the Nile*, 5, 71.

[128] Some Scots ingested cobwebs and spiders during intermittent fevers. Risse, *New Medical Challenges during the Scottish Enlightenment*, 178. A mid-eighteenth-century Portuguese recipe collection also praised the use of cobwebs in 'tertian' and 'quartan' fevers, presumably since it was largely a verbatim copy of chapters from the early fifteenth-century medical work *Menor daño de Medicina* by Alfonso Chirino (c. 1365–1429) and the thirteenth-century treatise 'Libro de medicina llamado Tesoro de pobres'. See chapter 60, 'P[ara] remediar la Calentura tertiana', and chapter 61, 'P[ara] sanar e remediar la Calentura quartana', in 'Receitas medicinais para diversas enfermidades,' *Biblioteca Nacional de Portugal*, Manuscritos Reservados, COD. 6177// 3, n.p., –1759, f. 42. Both chapters are culled from Juan XXI (Pedro Julião / Pedro Juliano), *Libro de medicina llamado Tesoro de pobres: en que se hallàrán remedios muy aprouados para la sanidad de diuersas enfermedades* (Sevilla: Nicolas Rodriguez, 1655), 54–56.

[129] See, for instance, Anna Katharina Haydtin's 'fever powder for the tertian and quartan fever' (3. oder 4. Tägliche Fieber). 'Medicin Buch von der verstorbenen Frauen Anna Katharina Haydtin,' *Wellcome Library*, Archives and manuscripts, MS.2797, n.p., 1756, 93.

[130] In some contexts, acceptance of the bark was suspended owing to resistance from influential 'schools' of physicians, like that of Georg Ernst Stahl (1659–1734) at Halle. According to Stahl, Peruvian bark, by its astringency, trapped the materia peccans in the body and thus oppressed nature's efforts to eject it – a version of Galenists' old objection that the bark did not eliminate the morbific, febrile matter from the humours. Maehle, *Drugs on Trial*, 274; 84. See also Geyer-Kordesch, 'Fevers and Other Fundamentals,' 118.

The structurally similar cinchona formulae, continually replicating the combination of sweetener, solvent and kindred febrifuge – common from Lisbon to Macao, from Paris to Constantinople and from Edinburgh to Lima – speak, therefore, not to the exclusion of the possibility of other, perhaps more inwardly directed preparations, nor to any sort of unrestrained erosion of difference, or universal sameness, in the realm of medicine. They speak merely, but quite distinctly, to the reality of epistemic exchange, in the realm of therapeutic practice and bodily care, along the veins of Atlantic trade, proselytizing and imperialism by the late 1700s and early 1800s, and to the vulnerability of men, women and children the world over 'to cultural metamorphosis', and their potential for internalizing novelties of taste, and medical experience.[131] Recipes for very particular preparations of the bark traversed societies, empires and oceans by the late 1700s and early 1800s because of both of these veins and these vulnerabilities, and of the material and therapeutic properties that cinchona brought to bear in practices of administration and consumption.

Proprietary Medicines

While self-dosing, family care and lay medicine remained the most common ways in which patients received and acquired drugs throughout the late eighteenth century and the beginning of the nineteenth, the face of drug production and lay healing also changed considerably over the period under consideration. Alongside the continued importance of remedies prepared by sufferers and practitioners, the eighteenth century witnessed the growing popularity of ready-made, proprietary and eponymous medicines, advertised and bought over the counter from an expanding urban network of fixed shops.[132] A sizeable share of cinchona would have been distributed in that form.

The English equivalent of Masdevall's Antipyretic Opiate, for instance, was Huxham's Tincture of Bark, a compound remedy common

---

[131] Norton, 'Tasting Empire,' 691.

[132] On the rise of commercialized remedies in England from the late 1700s, see David Boyd Haycock and Patrick Wallis, *Quackery and Commerce in Seventeenth-Century London: The Proprietary Medicine Business of Anthony Daffy*, Medical History Supplement (London: Wellcome Trust Centre for the History of Medicine at UCL, 2005); Hill Curth, 'Introduction: Perspectives,' 5. On the rise of patent medicines in Portugal, see Figueiredo, 'A "Água de Inglaterra" em Portugal,' 16; Sousa Dias, *A água de Inglaterra*, 13–15. On proprietary medicines in North America from the early eighteenth century, see James Harvey Young, *The Toadstool Millionaires: A Social History of Patent Medicines in America before Federal Regulation* (Princeton, N.J.: Princeton University Press, 1972 (1961)).

to many apothecary books, pharmacopoeias and health advice manuals. Devised by Dr John Huxham (d. 1768) and first published in his 1739 *Essay on Fevers*, the preparation, which Huxham claimed to have 'used for many Years with Success, not only in intermittent and slow nervous Fevers, but also in the putrid, pestilential, and petechial, especially in the Decline', contained

two ounces of best Peruvian bark in powder (*R. Cort. Peruv. Opt. pulv.*), one and a half ounces of Spanish orange peel (*Flaved. Aurant. Hispal.*), three drams of Virginia Snake root (*Rad. Serpent. Virgin.*), four scruples of English saffron (*Croc. Anglic.*), two scruples of cochineal (*coccinel*) [...] infused for several days (three or four) in twenty ounces of French Brandy (*Sp. Vini Gallici*) and strained thereafter.[133]

If a patient were 'costive, or hath a tense and tumid Abdomen', a practitioner could 'premise a Dose of rhubarb, Manna or the like'. The practitioner could also accompany it, if the fever continued, with 'supporting Drinks and Diet', such as 'a generous red Wine', 'especially, when acidulated with Juice of Sevil Orange, or Lemon, [...] impregnated with some Aromatics, as Cinnamon, Sevil Orange Rind, red Roses, or the like', or by adding 'a few Drops of Elixir Vitrioli'.[134] While some advice manuals, recipe collections and pharmacopoeias copied the recipe nearly verbatim – Francis Spilsbury's 1773 *The Friendly Physician*,[135] the 1796 *Edinburgh new dispensatory*,[136] an Italian 'collection of medical receipts'[137] and Samuel Curtis's 1819 *Valuable Collection of Recipes, Medical and Miscellaneous* which catered to the necessities of a North American readership[138] – others contained amended or reduced variations of the preparation. The 1794 official 'Portuguese Pharmacopoeia', for instance, contained a recipe for a 'composed tincture of cinchona or antiseptic essence of Huxham' (*TINTURA DE QUINA COMPOSTA ou Essencia antiseptica d'Huxham*) that contained, like Huxham's original formulation, two ounces of Peruvian bark, one and a half ounces of orange peel and three drams of Virginia snakeroot infused in twenty

[133] Huxham, *An Essay of Fevers, and Their Various Kinds*, 122. [134] Ibid.
[135] Francis Spilsbury, *The friendly physician: A new treatise: containing rules, schemes, and particular instructions, how to select and furnish small chests with the most approved necessary medicines; to which are added many excellent receipts for particular disorders* (London: Sold by Mr. Wilkie et al., 1773), 23.
[136] Lewis and Rotheram, *The Edinburgh new dispensatory*, 483–84.
[137] An Italian recipe collection contained a recipe for 'Huxham's Antiseptic, or Febrifugal, Elixir' (*Elixir antisetico, o febrifugo dell Huxam*), consisting of Peruvian bark, orange peel, saffron, cochineal and alcohol. 'Collection of medical receipts by several hands: in Italian.'
[138] Curtis, *A Valuable Collection of Recipes, Medical and Miscellaneous*, 14.

ounces of spirit, but that dispensed with cochineal and saffron.[139] So did William Buchan's *Domestic Medicine*, which encompassed a cinchona, snakeroot and orange peel-based antidote against the ague, to be infused 'for five or six days in a bottle of brandy'.[140] Huxham's Tincture of Bark also circulated in and through various translations of Huxham's *Essay on Fevers* – editions were published in Amsterdam, Munich, Paris and Venice – which often introduced slight variations owing to the culinary and cultural background consumer groups brought to bear in their consumption practices. Translators and editors substituted English for Viennese saffron, Spanish oranges for Portuguese lemons, and French brandy for 'any good wine'.[141]

Other than in the form of compound tinctures named after their inventor, prepared and commonly kept in apothecaries' shops, like Huxham's Tincture of Bark or Masdevall's Opiate, cinchona was also available in the form of commercialized, patent remedies. A classic throughout the Kingdom of Portugal and its overseas dominions, particularly Angola and Brazil, was bottled English Water (FIGURE 3.1).[142] English Water was generally considered 'a sovereign remedy' in 'periodical or intermitting nervous fevers', but no less efficacious in any 'period ailment' with regular seizures, such as ophthalmias, 'periodical epilepsy', or typhoid; nervous and hectic fevers if they remitted; and, since its virtue was 'corroborant or tonic', in a wide range of 'asthenic diseases' proceeding from 'weakness, looseness or atony of the solids', such as catarrhs and coughs, as well as haemorrhages.[143] Initially a secret concoction

[139] Tavares, *Pharmacopeia Geral para o reino, e dominios de Portugal*, 213.
[140] Buchan, *Domestic Medicine*, 174.
[141] Compare, for instance, the versions of the recipe contained in the English original and the Munich edition of the German translation, and the French translation. John Huxham, D. Johann Huxham, *Mitglieds der Königlichen Gesellschaft in London, Abhandlung von Fiebern, welche von der Beschaffenheit des Geblütes herrühren* (München / Augsburg: Merz u. Mayer, 1756), 177; Huxham, *An Essay of Fevers, and Their Various Kinds*, 122. In the French translation, which juxtaposes the French and Latin recipe, the Latin '*Flaved. Aurant. Hispalens*' is – presumably purposefully – (mis)translated as '*de l'écorce de citron de Portugal*'. John Huxham, *Essai sur les différentes espèces de fièvres, avec des dissertations sur les fievres lentes, nerveuses, putrides, pestilentielles & pourprées* (Paris: D'Houry, Impr. Lib. de Mgr. le Duc D'Orléans, 1776), 149. On the Italian and Latin editions, see William Schupbach, 'The Fame and Notoriety of Dr. John Huxham,' *Medical History* 25, no. 25 (1981), 415.
[142] On the consumption of English Water in Portugal and its dominions, see Figueiredo, 'A "Água de Inglaterra" em Portugal.'; Sousa Dias, *A água de Inglaterra*.
[143] Manuel Joaquim Henriques de Paiva, *Memoria sobre a excellencia, virtudes, e uso medicinal da verdadeira agua de Inglaterra da invenção do Dr. Jacob de Castro Sarmento, Membro do Real Collegio de Medicos de Londres, e Socio da Sociedade Real. & c. Actualmente preparada por José Joaquim de Castro, na sua Real Fabrica, por Decretos de Sua Alteza Real o Principe Regente* (Lisboa: Impressão Regia, 1816), 2–42.

Figure 3.1 Bottle of English Water (*Água de Inglaterra*), c. 1800.
© *Museu da Farmacia*, Portugal.

commercialized from the 1730s onwards by Jacob de Castro Sarmento
(1692–1762) and his descendants, the recipe for English Water was
made available to the Portuguese public by the end of the eighteenth
century, when an emerging ethic of making cures available for the
welfare of mankind made the keeping of medical secrets immoral in
Portugal, as elsewhere in Europe,[144] through charity books and official
pharmacopoeias, issued by the authority of the royal government. As
with Huxham's Tincture of Bark and Masdevall's Antipyretic
Opiate, adaptations of English Water were popular in and beyond the
confines of the Portuguese Empire. Lisbon hospital orderlies,[145] Siena

---

[144] Schiebinger, 'Prospecting for Drugs,' 120. On the late medieval and early modern
tradition of 'secret' remedies, see Leong and Rankin, 'Introduction: Secrets and
Knowledge,' 12. On the decline of the literary tradition of the 'books of secrets', see
Eamon, *Science and the Secrets of Nature*, 357–58. In England, too, those compound
medicines approved as 'patent medicines' had to reveal their composition. James
Harvey Young, 'Patent Medicines and the Self-Help Syndrome,' in *Medicine without
Doctors. Home Health Care in American History*, ed. Guenter B. Risse, Ronald
L. Numbers and Judith Walzer Leavitt (New York: Science History Publications,
1977), 96–97.

[145] João de Jesus Maria's manuscript recipe compilation already claimed to divulge the
'legitimate recipe' for English Water – a ground cinchona, cornflower (*centaura menor*)
and wine-based decoction – by 1760. Jesus Maria, 'Colleção medica de receitas,'
41–42.

charlatans[146] and Spanish physician-authors alike adapted, reused and devised perfected varieties of English Water that supposedly exceeded it in efficacy.[147]

By the late 1700s and early 1800s, a considerable share of shop-bought cinchona-based medicines were marketed as extracts[148] or concentrated tinctures, that is, preparations advertised not by virtue of the ingredients they joined to the bark but of their saturation with its 'efficacious particles' and their being 'free from the useless'.[149] Extraction, a process whereby the medical virtue of a plant product was removed and concentrated into a resinous substance using chemical techniques, was a common pharmaceutical procedure, and cinchona extracts were fabricated from the late 1600s.[150] By the late 1700s and early 1800s, ready-made, water-soluble extracts – some produced, sold and distributed directly from manufacturers in the Spanish American harvest areas[151] – were available on the medical market, from Göttingen[152] to Manila,[153] from Saint Petersburg[154] to Constantinople.[155] Remedies that contained 'the effectiveness of the best barks packed together (*zusammengedrängt*)'[156] were uniformly advertised, and merchandised, by virtue of their palatability and digestibility, their portability – owing to their reduced volume – and their stability in transport. Catering to an

---

[146] The Siena charlatan Giovanni Greci sold 'health drops' (*gocciole di salute*), which contained, in addition to 'balsam of Mecca', sugar and powdered coffee, also English Water. David Gentilcore, *Medical Charlatanism in Early Modern Italy* (Oxford: Oxford University Press, 2006), 200.

[147] Suarez de Ribera, *Ilustracion, y publicacion de los diez y siete secretos del Doctor Juan Curvo Semmedo*, 125.

[148] George Brown, 'The concentrated tincture of yellow Peruvian bark: prepared by George Brown, chemist, at no. 79, St. Paul's Church Yard: a certain cure for the intermittents, and very useful in all complaints which require large doses of the bark,' *Wellcome Library*, Med. ephemera, Drug Advertising: Pre-1850: Box 1, EPH381G. 19, London, 1805. Johan Andreas Murray's Vorrath likewise referred to a 'fever bark extract'. Murray, *Vorrath von einfachen, zubereiteten und gemischten Heilmitteln*, 1, 1222.

[149] Crawford, 'An Empire's Extract,' 224.

[150] Ibid, 219–20; Boumediene, *La colonisation du savoir*, 244.

[151] Crawford, 'An Empire's Extract,' 224.

[152] Murray, *Vorrath von einfachen, zubereiteten und gemischten Heilmitteln*, 1, 1225.

[153] The majority of cinchona remittances shipped from New Spain to Asia via the Manila galleons were derivatives – processed remedies such as powders, extracts or ointments. Pacheco Olivera, 'Análisis del intercambio de plantas entre México y Asia de los siglos XVI al XIX,' 126–27.

[154] See the formula for a 'cinchona bark extract' (*Extractum Chinae Corticis*), in the 1778 'Russian Pharmacopoeia,' *Pharmacopoea Rossica*, 88.

[155] Historians of Ottoman medicine have suggested that cinchona arrived in Turkey in all its available preparations. Günergun and Etker, 'From Quinaquina to "Quinine Law,"' 49.

[156] Murray, *Vorrath von einfachen, zubereiteten und gemischten Heilmitteln*, 1, 1225.

expanding peripatetic sector of the population – to colonists, officials and merchants – extracts like George Brown's Concentrated Tincture of Yellow Peruvian Bark, a 'certain cure for intermittents', were said to retain their virtues for 'any length of time, in any climate', where unprocessed bark was liable to decay and loss 'of its efficacy on being long kept'.[157] Extracts, or concentrated tinctures, were also popular owing to their presumed greater effectiveness. Even in small doses they were supposed to be 'more powerful than the largest [dose] of the powders' and, thus, 'very useful in all complaints which require large doses of the bark'.[158] Particularly in the face of widespread, wilful fraud, but also natural variation in the bark's colour, taste and texture, it was also generally hoped that extracts of cinchona that contained 'its active principles in a concentrated form' would assist the bark's epistemic and medical standardization and stabilization.[159]

Whereas brand names suggest originality, inimitability and novelty, cinchona-based patent medicines bore a close resemblance to some of the more popular combinations of ingredients commonly contained in laypersons' medical guides, family notebooks of medical recipes and the writings of army surgeons and hospital orderlies. As the author of the 1785 'Lisbon Pharmacopoeia', Manuel Joaquin Henriques de Paiva (1752–1829) lamented, the 'avaricious, ambitious' inventors of English Water sought to persuade the public that their waters' effectiveness lay in the 'extraordinary and to this day unknown liquid' in which cinchona was dissolved. In fact, however, 'men of judgement' were well aware all febrifugal waters sold under the name of English Water were 'compound wines of the bark, more or less' (*vinhos de quina, mais, ou menos compostos*), containing cinchona, gentian root and orange peel, macerated in brandy and white wine.[160] Eponymous medicines named after their inventor, like Huxham's Tincture of Bark and Masdevall's Antipyretic Opiate, also frequently resembled classic, popular formulae. Recipes joining together cinchona, citrus peel and snakeroot and infusing them in a base liquor were a standard formula in recipe books and medical

[157] Brown, 'The concentrated tincture of yellow Peruvian bark.'   [158] Ibid.
[159] Chakrabarti, 'Empire and Alternatives,' 89.
[160] Henriques de Paiva, *Farmacopéa Lisbonense*, 228. For similar complaints against 'secret remedies' in general, and the manufacturers of English Water in particular, see Francisco Tavares, José Correa Pieanço and José Francisco Oliveira, 'Sendo S. Magestade Servida Determinar por Avizo expedido pela Secretaria d'Estado dos Negocios do Reyno em data de seis d'Abril d'este corrente anno que a Junta do Proto-Medicato naõ comprehenda a André Lopes de Castro no processo e execuçao annunciados no Edital de 15 de Março,' *Arquivo Nacional da Torre do Tombo*, Ministério do Reino / Negócios diversos do Físico-Mor, Maço 469/ Caixa 585, Lisbon, 1799-08-12.

guides by the late 1700s and early 1800s. Combinations of cinchona with wormwood salt and syrups, preceded by antimonial drugs that functioned as purgatives, were also common. Whether some of the most popular combinations of ingredients metamorphosed into patent, eponymous and proprietary medicines towards the late 1700s and early 1800s, or whether home-made remedies and medicines prepared by apothecaries replicated these renowned medicines is neither in every case ascertainable, nor is it relevant, for both likely evolved in conjunction with each other. Patent, eponymous and proprietary medicines responded not so much to alterations in the period's medical system as to its commercialization. They were, therefore, frequently a change in presentation rather than in substance. They contributed, however, to establishing particular arrangements and preparations of the bark in the medical canon and to making them ever more popular with men and women across the Atlantic World.

★★★

By the late 1700s and early 1800s, not only was cinchona conveyed across large distances along the veins of trade, proselytizing and imperialism. So, too, were therapeutic practices, routines of medication and bodily care attendant to its administration. Courtiers in the sultanate of Morocco, householders in the Scottish Lowlands and the beneficiaries of Mexican religious pious charity – men, women and children who otherwise had precious little in common – would have been familiar with the comfort offered by a bitter dram, 'made with some sort of spirituous liquor', or the sense of relief after the 'proper evacuations' preceding it.[161] Discourses and practices about how to concoct preparations of the bark that were palatable and efficacious had, so to speak, woven themselves into the fabric of everyday medicinal practice in Andean villages, North American plantations, West African hospitals and Cantonese cabinets alike, and into the lives of men and women who, in that regard at least, belonged to one and the same 'community of ideas and healing practice'.[162] Methods for arranging and administering the bark had coalesced into a series of identifiable formulae and recipes by the late 1700s and early 1800s – extracts, bittersweet febrifugal lemonades and aromatic compound wines of the bark – that would have been familiar

---

[161] O'Ryan, *A Letter on the Yellow Peruvian Bark*, 23.
[162] This is taken from a passage by Charles E. Rosenberg, who wrote about 'the shared knowledge and assumptions' that bound professional physicians like William Buchan and lay people together 'in a community of ideas and healing practice'. Rosenberg, *Explaining Epidemics*, 32.

and recognizable to men and women in Tangier, Lisbon and New Orleans, in Graubünden, Luanda or Lima. They derived, in turn, from sustained proximity and the commercial, ecological and cultural integration of the Atlantic World, much of which shared by then a fondness for sweetness, a liking for alcohol, a taste for lemons or cinnamon and, not least, an appreciation of the salutary virtue of the Peruvian bark. Patterns and possibilities of medical consumption were not, to be sure, boundless or placeless, and they did, in some measure, remain contingent. The distribution and passage of preparations of the bark was supported and confined by the reach of trading companies and the sphere of influence of the Iberian, British or French empires. Extracts, proprietary medicines and formulae were commonly devised in and bore the mark of the societies most concerned in the bark's consumption, trade or advocacy – in Portugal, New Granada, England or Spain. Most importantly, preparations of the bark underwent alterations in transfer, with men and women, because of their readings, palate or creed, replacing English with Viennese saffron, wine with vinegar and Spanish oranges with Portuguese lemons. Indeed, it was often due to the fact that people could tinker with recipes, to subtly adapt, appropriate and situate them, that rendered preparations of the bark appealing for them. It was their capacity to tie these formulae to different medical cultures and localities that made preparations of the bark both varied and ubiquitous.

4   Febrile Situations

To those causes, which introduce fevers into the fleet, may be added another source of mortality, which prevails among our sailors in the West Indies, that the surgeons of the navy are not supplied with the most essential medicine for their cure, at least in proper quantity: I mean the Peruvian bark: nor can they afford to purchase it in that part of the world.
— John Hunter, *Observations on the Diseases of the Army*, 1788.

Fevers, as John Clark (1744–1805), a former East India Company surgeon, put it in his 1773 *Observations on Diseases in Long Voyages to Hot Countries*, though they had 'been divided into many GENERA', and though various appellations had been 'given to them both [...], derived from the time of their duration, from some remarkable predominant symptom, from the state of the fluids, and from various other circumstances', were yet alike not only 'in their essential symptoms', but also in their cure. That, according to Clark, depended 'entirely upon [...] the Peruvian bark, in as large doses as the patient's stomach will bear'. In the insalubrious parts of Newcastle and Bengal alike, Clark wrote, he had laid bloodletting aside, used opium but 'as an auxiliary to the bark', and 'vomits and purges' just as a preparation for the bark's administration.[1] For, like practitioners in so many other parts of the world – and he cited colleagues in Britain, Africa, South Asia and the Caribbean – he had found 'the early and liberal use of the bark' to be the proper and by far the most successful method for the 'prevention and cure' of fevers.[2] Not only were fevers everywhere alike in their symptoms and cure, so they were 'in the causes which produce[d] them'. All fevers were 'the offspring' of the 'heat and moisture' of hot countries, of 'marsh exhalations', or 'of confined air loaded with human effluvia'. They invariably were the effect of 'season; situation; and climate'.[3]

---

[1] Clark, *Observations on Diseases which prevail in Long Voyages to Hot Countries*, 170–71; 176–78.
[2] Ibid., vii.   [3] Ibid., 150; 62; 67.

Fevers were probably 'the most common disease or group of diseases' medical practitioners encountered, and sufferers experienced, in the late 1700s and early 1800s. They were not only the period's most general but presumably also its most fatal ailment.[4] Prevalent across the Atlantic World and beyond, fevers might differ in their severity – in tropical climates, they were generally considered to be 'greatly more violent in their attack, quicker in their progress, and more fatal in their termination, than [...] in Europe'[5] – but most contemporaries would have agreed with Clark on the subject of unfamiliar fevers' basic intelligibility. They, too, would have reasoned that fevers did not vary either in their nature, their cure – the bark was considered 'the ordinary method' of treatment in the 'common ague'[6] as well as in intermittent, 'slow nervous' or 'bilious remittent fevers'[7] – or, indeed, in their cause. Invariably, fevers were thought to proceed from the interaction between an insalubrious environment and the sensibility and constitution of a distinct, yet porous, body.[8]

This chapter charts a series of insalubrious, febrile environs linked to high levels of cinchona consumption: the world's low-lying marshes, the sickly air of close, crowded spaces – the narrow confines of ships, army camps and rapidly growing urban spaces – and the hot and humid climates of the tropics that a growing number of settlers, sailors and

[4] Dancer, *The Medical Assistant; or Jamaica Practice of Physic*, 65. Hispanic sources stressed the prevalence, and importance, of 'fevers', in similar terms. See, for instance, Pereyra, *Tratado completo de Calenturas*. For a discussion of the prevalence of fevers, see also William F. Bynum and Vivian Nutton, introduction to *Theories of Fever from Antiquity to the Enlightenment*, ed. William F. Bynum and Vivian Nutton (London: Wellcome Institute for the History of Medicine, 1981), vii. On the prevalence of 'fevers' in parts of New Spain and coastal Peru, see Fields, *Pestilence and Headcolds*, 19–22; Gänger, 'In Their Own Hands.' In relation to the cultural imaginary, see, for instance, Candace Ward, *Desire and Disorder. Fevers, Fictions, and Feeling in English Georgian Culture* (Lewisburg, Pa.: Bucknell University Press, 2007).

[5] Hunter, *Observations on the Diseases of the Army*, 15. According to Dancer, fevers were 'the disorders that carry off the greater part of mankind in all climates, more especially in the hot ones'. Dancer, *The Medical Assistant; or Jamaica Practice of Physic*, 65. Portuguese authors assumed a similar identity between the causes and nature of the 'putrid fevers' of Portugal, America and Angola. Nunes Ribeiro Sanches, *Tratado da conservaçam da saude dos povos*, 59.

[6] Baker, 'Observations on the Late Intermittent Fevers to which is added a Short History of the Peruvian Bark,' 661; 79.

[7] Dancer, *The Medical Assistant; or Jamaica Practice of Physic*, 87–88; 89–92. See also Lind, *An Essay on Diseases incidental to Europeans in hot Climates*, 41; 53; 286. On the 'slow nervous fever', see Huxham, *An Essay of Fevers, and Their Various Kinds*, 124.

[8] Ward, *Desire and Disorder*, 153. For more examples, see James Ewell, *The planter's and mariner's medical companion: treating, according to the most successful practice* (Philadelphia: John Bioren, 1807), 41–43; M. F. B. Ramel, *De l'Influence des marais et des étangs sur la santé de l'homme, ou Mémoire couronné par la ci-dev. Société Rle de médecine de Paris* (Marseille: J. Mossy, 1801), xiii–xv.

military personnel encountered in a period of global imperialism, warfare and settlement. The chapter argues that, by the late 1700s and early 1800s, bark knowledge was common across various Atlantic societies not only in the form of imaginative stories or culinary practices, as the previous chapters have shown, but also in that of diagnostics, of expertise in indications for the bark and of a topographic literacy of sorts that associated even widely different environments – Newcastle or Calcutta – with the same, familiar kind of febrile threat.[9] The historiography of cinchona, though it has commonly acknowledged the bark's widespread use in fevers – especially in those retrospectively diagnosed as malaria – has paid scant attention to how knowledge and experience of that use were part of the dominant cultural repertoire of men and women from various societies and all ranks around 1800: familiar to, and popular with, a Luanda merchant, a Hamburg physician and a Lima householder alike.[10] Indeed, it was presumably in its intense preoccupation with the effects of the bark's consumption – particularly the alleged role of quinine in the high tide of imperialism[11] – that the literature has neglected the

---

[9] On French military and civil medicine and medical topography, see Brockliss and Jones, *The Medical World of Early Modern France*, 751–53; Jean-Michel Derex, 'Géographie sociale et physique du paludisme et des fièvres intermittentes en France du XVIIIe au XXe siècles,' *Histoire, économie & société* 27 (2008). According to historians, the British military and naval medical services were 'the vanguard' of that 'neo-Hippocratic medicine'. Harrison, 'Disease and Medicine in the Armies of British India, 1750–1830,' 93. On neo-Hippocratic medicine and medical topography in the eighteenth-century German-speaking medical world, see Tanja Zwingelberg, 'Medizinische Topographien und stadthygienische Entwicklungen von 1750–1850, dargestellt an den Städten Berlin und Hamburg,' in *Natur und Gesellschaft. Perspektiven der interdisziplinären Umweltgeschichte*, ed. Manfred Jakubowski-Tiessen and Jana Sprenger (Universitätsverlag Göttingen: Universitätsverlag Göttingen, 2014). On climatic determinism in British India, see Mark Harrison, *Climates & Constitutions: Health, Race, Environment, and British Imperialism in India, 1600–1850* (New Delhi: Oxford University Press, 2002 (c1999)). On Spanish America, see Jean-Pierre Clément, 'El nacimiento de la higiene urbana en la America Española del siglo XVIII,' *Revista de Indias* 43, no. 171 (1983). On eighteenth-century Portugal and Brazil, see Neves Abreu, *Nos Domínios do Corpo*, 88–95. On North America, see Bolton Valencius, *The Health of the Country*, 4. On New Spain, see Fields, *Pestilence and Headcolds*, 115–24. Islamic societies shared similar concepts in the eighteenth century; see Hormoz Ebrahimnejad, 'Medicine in Islam and Islamic Medicine,' in *The Oxford Handbook of the History of Medicine*, ed. Mark Jackson (Oxford: Oxford University Press, 2011), 169–89.

[10] Earlier historians of the bark have focused mostly on the bark's acculturation in academic medicine or professional therapeutic practice or on earlier and later periods. Boumediene, *La colonisation du savoir*; Klein and Pieters, 'The Hidden History of a Famous Drug'; Jarcho, *Quinine's Predecessor*.

[11] For a discussion of the literature, see 'An Appraisal of the Historiography' in the Introduction.

grounds and foundations of that consumption. Drawing on a wide range of sources, from royal orders to travel writings from across the Atlantic World, as well as environmental histories of imperialism, settlement and warfare,[12] this chapter exposes and explores the prevalence of familiarity with indications for the bark among professional medical practitioners and self-dosing sufferers alike. Men and women across the Atlantic World and its various entrepôts who inhabited or moved temporarily into insalubrious environs shared a belief that their ability to preserve or restore bodily well-being was contingent on a litany of precautions and cares. Cinchona bark, this chapter contends, had become a fundamental element of that register by the late 1700s and early 1800s.

Marshes and Wetlands

By the late 1700s and early 1800s, medication, or self-medication, with 'impudent' (*dreist*)[13] quantities of the bark was among the most common of a series of measures – ranging from the drainage of clay soils to regimen, that is, 'a rational diet', moderation or frequent baths[14] – which the inhabitants of saline and riverine marshlands in different parts of the world adopted to lessen the burden of morbidity and mortality imposed upon them by their environs. Indeed, few places were considered as febrile and harmful to health in the eighteenth century as low-lying marshland districts. Physicians, medical topographers and laymen alike were convinced of a link between stagnant, salt- or freshwater sources, 'insalubrious air', and the wearying ailments contemporaries referred to as 'marsh fever', the ague, and tertian and quartan fevers– or, collect-ively, intermittent fevers[15] – ailments so designated because they entailed febrile accessions that recurred at regular intervals, on the third and fourth day, respectively, with every accession beginning suddenly with a chill, accompanied by headache, nausea and 'bilious vomiting', followed by a period of 'burning' fever, difficulty breathing, limb pain and

---

[12] For environmental histories of imperialism, settlement and warfare that make references to uses of the bark, see, for instance, McNeill, *Mosquito Empires*; Susy Sánchez, 'Clima, hambre y enfermedad en Lima durante la guerra independentista (1817–1826),' in *La independencia del Perú. De los Borbones a Bolívar*, ed. Scarlett O'Phelan Godoy (Lima: PUCP Instituto Riva-Agüero, 2001).

[13] Rambach, *Versuch einer physisch-medizinischen Beschreibung von Hamburg*, 310–12.

[14] For an example of regimen advice from Lima, see Cosme Bueno, *El conocimiento de los tiempos; efemeride del año de 1788* (Lima: Imprenta Real, 1787), 2; Cosme Bueno, *El conocimiento de los tiempos, Efemeride del Año de 1796* (Lima: Imprenta Real, 1795), 3. On regimen advice in 'intermittent fevers' in France, see Derex, 'Géographie sociale et physique du paludisme,' 44.

[15] Dobson, *Contours of Death and Disease*, 15.

delirium, and ending in profuse sweating.[16] By the early nineteenth century, medical practitioners had long tried the bark's propriety and effects in a series of other ailments, but it was only in intermittent fevers, as Jacob Bigelow (1787–1879) phrased it in 1822, that its efficacy and value was to them undoubted and unchanged.[17] Indeed, cinchona's 'earliest and most durable reputation [had been] acquired in the treatment of intermittent fever'.[18] Presumably a part of pre-conquest Andean pharmacopoeia as a remedy for the 'rigors and chills' (*rigor et frigor*) caused by exposure to cold water, the bark had first entered Jesuit and, subsequently, Hispanic and Italian medical practice around the mid-seventeenth century to alleviate the 'rigors and chills' that came with the ague.[19]

The use of the bark was common throughout the North Sea Basin by the late 1700s and early 1800s, where marsh fevers were then endemic. '*Tredjedagsfrossa*' (tertian fevers), the ague, or '*Wechselfieber*' (intermittent fevers) had long imposed a heavy burden of suffering, poverty and death on men and women in the brackish coastal zones of Flanders, the Baltic and East Friesland – where they were the direct or indirect cause of every fourth to fifth death – and in the saline and riverine marshlands of Denmark and Scandinavia.[20] The Enlightenment's demand that medical knowledge be made available and be communicated to a wider lay public conspired with the steadfast consensus on the bark's efficacy in academic

---

[16] For descriptions of the classic sickness episode, see, for instance, Dávalos, 'Specimen Academicum de morbis nonnullis Limae grassantibus,' 13–14; Buchan, *Domestic Medicine*, 157.

[17] Jacob Bigelow, *A treatise on the materia medica: intended as a sequel to the Pharmacopoeia of the United States* (Boston: Ewer, 1822), 131.

[18] Ibid.

[19] According to Gaspar Caldera de Heredia (1591–ca. 1669), whose *De Pulvere Febrifugo Occidentalis Indiae* was first published in Antwerp in 1663, the Jesuits learned of the bark's virtues from 'Indians', who took it, dissolved in hot water, to alleviate the 'cold and chills' (*rigor et frigor*) caused by exposure to cold water. Jarcho, *Quinine's Predecessor*, 4–8.

[20] On various aspects of 'intermittent fevers', and attendant bark consumption, in northern Europe, see Otto S. Knotterus, 'Malaria around the North Sea: A Survey,' in *Climate Development and History of the North Atlantic Realm*, ed. Gerold Wefer (Berlin: Springer, 2002); Bruce-Chwatt and Zulueta, *Rise and Fall*; Derex, 'Géographie sociale et physique du paludisme'; Huldén, Huldén and Heliövaara, 'Endemic malaria.' On the relationship between fevers or morbidity and poverty, see Dobson, *Contours of Death and Disease*; Armando Alberola Romá and David Bernabé Gil, 'Tercianas y calenturas en tierras meridionales valencianas: una aproximación a la realidad médica y social del siglo XVIII,' *Revista de Historia Moderna* 17 (1998–1999). 'Intermittent fevers' gradually diminished over the course of the eighteenth century in Denmark, mainly with a pronounced increase in the number of animal herds. Thorkild Kjærgaard, *The Danish Revolution, 1500–1800: An Ecohistorical Interpretation* (Cambridge: Cambridge University Press, 2006), 184. On cinchona consumption in the North Sea Basin, see also Chapter 1.

medical circles to contribute to that medicine's growing popularity in these parts around 1800. Especially its inclusion in some of the most best-selling enlightened medical advice manuals – William Buchan's 1769 *Domestic Medicine* in particular – that hailed it as an infallible remedy in intermittent fevers helped establish its reputation in the popular medical culture of several societies rimming the North Sea Basin.[21] So did hundreds of less conspicuous and peripatetic handbooks, almanacs and periodicals that recommended its use in agues to poor Finnish parishioners,[22] Swedish nurses[23] and English householders, many of whom evidently revised and adjusted their therapeutic arsenal on those grounds.[24] The mediation of medical knowledge about the bark in conversations among laymen and in bedside encounters between sufferers and university-trained physicians, apothecaries or charitable hospital orderlies was significant, too.[25] Family notebooks of medical recipes from England, where the ague had long been prevalent along the coasts and estuaries of the south-east, particularly in the salty marshlands of Kent and Essex and in the fenlands of Lincolnshire, Cambridgeshire and Norfolk, often speak to how elite and middle-class householders adopted their physicians' confidence in the bark.[26] Owing also to increasingly systematic medical poor relief and the provision of remedies by irregular practitioners, in England the upper and middle strata of society as well as village folk in the marshlands had come to use the bark 'in every form' by the early 1800s, consuming a variety of cinchona tinctures, infusions and other 'ague formulations' (FIGURE 4.1).[27] Up to 1.2

---

21 Buchan recommended the use of the 'Peruvian bark' in intermittent fevers. Buchan, *Domestic Medicine*, 111. Tissot spoke of the bark as an 'infallible remedy'. Tissot, *Anleitung für das Landvolk*, 288; 94; see also Samuel Auguste André Tissot, *Avis au peuple sur sa santé* (Lausanne: J. Zimmerli, 1761), 284.
22 Huldén, 'The First Finnish Malariologist,' 5.
23 Rosen de Rosenstein, *Anweisung zur Kenntniss und Cur der Kinderkrankheiten*, 148; 299.
24 The notebook of medical recipes kept by 'Miss Myddleton', for instance, a housewife in the – by all accounts – very 'insalubrious' and febrile English county of Essex, shows that she relied on advice literature in her endeavours to cure or ward off fevers, copying bark recipes from manuals and periodicals. 'Receipts copied from Miss Myddleton's Book.'
25 For the systematization of 'popular medicalization', see Stolberg, 'Medical Popularization,' 96.
26 Anna Maria Reeves, a housewife from Berkshire, in England, evidently followed the directives of a certain Dr Steward in preparing her cinchona-based remedy. See the entry 'Dr Steward's Recipe for an intermitting Fever,' in Finger and Reeves, 'Collection of medical, cookery, and household receipts,' f. 4.
27 Risse, *New Medical Challenges during the Scottish Enlightenment*, 78; 173; Mary J. Dobson, 'Mortality Gradients and Disease Exchanges: Comparisons from Old England and Colonial America,' *Social History of Medicine* 2, no. 3 (1989); Dobson, *Contours of Death and Disease*, 358. On the accessibility of drugs in England's rural areas, see also King, 'Accessing Drugs in the Eighteenth-Century Regions.'

Figure 4.1 Lambeth Delftware pill jar, decorated and labelled EX:
CORT; PERUV:D (hard extract of cinchona bark), c. 1710–1740
© The Science Museum / Science and Society Picture Library.

million doses of the bark were available in England annually, at a time
when the counties where the ague caused the most havoc – Essex, Kent
and Sussex – taken together only had a population of 340,000 inhabitants
by the 1720s and some 695,000 by 1801.[28] Likewise, in the Hanseatic
city of Hamburg, where '*Wechselfieber*' (intermittent fevers), attributed to
'the aqueous exhalations of our marshlands, usually flooded in the
spring', were prevalent in the spring and summer, physicians administered 'brazen', 'impudent' doses of the bark both to paying customers
and paupers who benefitted from medical relief programmes.[29] Indeed,
the value of Hamburg's imports of Peruvian bark was greater than that of
any other medicinal substance in the period.[30] The ague was a wearisome,

---

[28] On cinchona consumption, see Wallis, 'Exotic Drugs and English Medicine,' 34;
Dobson, *Contours of Death and Disease*, 70. On the population history of England, see
Wrigley and Schofield, *The Population History of England*.
[29] Rambach, *Versuch einer physisch-medizinischen Beschreibung von Hamburg*, 310–12. On
medical relief, see Chapter 2.
[30] Pfister and Fertig, 'Coffee, Mind and Body,' 229.

onerous disorder for 'the natives',[31] but it proved 'particularly severe,
and sometimes fatal' to strangers. 'Newly arrived soldiers' sent to guard
the coast of France,[32] recently arrived schoolmasters, vicars and young
wives in the saline marshlands of south-east England,[33] or seasonal
workers in Friesland's coastal marshes[34] were generally believed to be
more prone to the intermittent fevers of 'marshy situations' and advised
to supply themselves 'with a quantity of good tincture of Peruvian bark',
and to 'take a tea-spoonful or two twice a day, in a glass of water or
wine'.[35] For men and women in the eighteenth and nineteenth centuries,
'the human form was steeped in its physical citizenship'[36] and displace-
ment – be it emigration, the movement of troops or the journeys of
seamen – would unavoidably register in and affect a wanderer's constitution. Those who did not belong in a given terrain were well aware they
could expect to suffer for their temerity.[37]

Some areas of the Mediterranean experienced a recrudescence of
intermittent fevers and high levels of cinchona consumption in the
affected areas in the late eighteenth and early nineteenth centuries.[38] In
Portugal, where tertian and quartan fevers were endemic along rice fields
and river valleys,[39] and in the Kingdom of France, where they haunted

---

[31] Among the inhabitants of saline and riverine marshlands, though the 'agues' were feared
for their perseverance, they were commonly considered a comparatively benign and
generally nonfatal ailment. For a Peruvian source, see, for instance, Dávalos,
'Specimen Academicum de morbis nonnullis Limae grassantibus,' 15. On 'benign
tertian fevers' in Spain, see Pereyra, *Tratado completo de Calenturas*, 249. For a
discussion of similar assertions in relation to 'intermittent fevers' in France, see Derex,
'Géographie sociale et physique du paludisme,' 41.

[32] François-Emmanuel Fodéré, *Recherches expérimentales, faites à l'hôpital civil et militaire de
Martigues, sur la nature des fièvres à périodes et sur la valeur des différens remèdes substitués au
Quinquina* (Marseille: Jean Mossy, 1810), 36.

[33] Dobson, *Contours of Death and Disease*, 294–301.

[34] Knotterus, 'Malaria around the North Sea: A Survey.'

[35] Blair, *The Soldier's Friend*, 96.   [36] Bolton Valencius, *The Health of the Country*, 34.

[37] Ibid.; Rod Edmond, 'Island Transactions. Encounter and Disease in the South Pacific,'
in *The Global Eighteenth Century*, ed. Felicity A. Nussbaum (Baltimore: Johns Hopkins
University Press, 2003), 251.

[38] Derex, 'Géographie sociale et physique du paludisme.' See also Alberola Romá and
Bernabé Gil, 'Tercianas y calenturas en tierras meridionales valencianas,' 101–02.

[39] On 'intermittent fevers' and cinchona consumption in Portugal, see the Portuguese
edition of Buchan's *Domestic Medicine*: William Buchan, *Medicina domestica, ou Tratado
de prevenir, e curar as enfermidades com o regimento, e medicamentos simplices , escripto em
inglez pelo D.r Guilherme Buchan, Socio do Collegio dos Medicos de Edimburgo, Traduzido em
Portuguez, com varias notas, e observações concernentes ao Clima de Portugal, e do Brazil, com
o receituario correspondente, e hum Appendice sobre os Hospitaes Navaes, Cura, e Dieta dos
Enfermos dos mesmos Hospitaes, por Manoel Joaquim Henriques de Paiva*, 4 vols. (Lisboa:
Typografia Morazziana, 1787), 27. Late eighteenth-century documents mention
epidemic outbreaks during the French invasion of Portugal in 1807 in the context of
the Napoleonic Wars. Sousa Dias, *A água de Inglaterra*, 12; 36.

the south-western and northern coastal provinces of Guyenne and Gascony, Brittany and Normandy, the Alsace and the Vendée and Loire river valleys,[40] cinchona was a staple, and its indications in agues were discussed and familiar among noblemen,[41] mendicant friars[42] and common householders[43] alike. The bark was a necessity also to the Knights of St John on the Maltese archipelago, frequently haunted by epidemics of tertian fevers.[44] It was likewise popular with sufferers on the Italian Peninsula, among whom the Pontine marshes, the coastal plains, Sardinia and Sicily had long been notorious for their insalubrity.[45] In peninsular Spain, where tertian fevers were prevalent along the coastline in the south, around the calm waters of the rice fields of Valencia and along some of the Douro, Tagus, Guadiana, Guadalquivir and Ebro river valleys in the summer, cinchona was the mainstay of public health. It was the foundation of the welfare and prosperity of families, a preserver of labourers' and farmworkers' productivity and the protector of entire armies of soldiers in these parts, as Manuel Hernandez de Gregorio, the Spanish King's Apothecary, phrased it in an 1804 memorandum.[46] Not only did *cabildos*, hospitals and other establishments 'hoard that precious article' from mid-March, before assaults of intermittent fevers were to be expected. So did many 'private citizens' (*particulares*), who,

[40] Derex, 'Géographie sociale et physique du paludisme,' 44–46. See also Bruce-Chwatt and Zulueta, *Rise and Fall*, 69. On fevers and cinchona consumption in Lyon and the nearby marshlands of Dombes, Bresse and Dauphiné, see O'Ryan, *A Letter on the Yellow Peruvian Bark*.

[41] Sousa Dias, *A água de Inglaterra*, 31–32.

[42] Jesus Maria, 'Colleção medica de receitas,' fs. 43; 94–95; 101.

[43] See, for instance, 'Livre des remedes de Madame la presidante de Maliverny,' f. 95; 'Receitas medicinais para diversas enfermidades,' f. 67.

[44] The Maltese archipelago, ruled by the Knights of St John after 1530, suffered epidemics of tertian fever in Rabat and Mdina in 1707–1709, 1754, 1755, 1775 and 1801, imputed to stagnant waters in the nearby valley. Boxes of cinchona commissioned by the Grand Master of Malta arrived in Cadiz in that period. Miguel de Muzquiz, 'Carta a Julián de Arriaga,' *Archivo General de Indias*, Indiferente 1554.

[45] On the history of malaria in Rome, see Robert Sallares, *Malaria and Rome: A History of Malaria in Ancient Italy* (Oxford: Oxford University Press, 2002). 'Intermittent fevers' were especially rife in the south of Italy, in Sicily and Sardinia. Giuseppe Dodero, 'Ippocratismo, malaria e medicina didascalica in Sardegna,' in *Pietro Antonio Leo. Di alcuni pregiudizii sulla così detta Sarda intemperie*, ed. Alessandro Riva and Giuseppe Dodero (Cagliari: CUEC, 2005).

[46] Hernandez de Gregorio, 'Memoria,' ff. 1019; 29; 34. On cinchona consumption in areas afflicted with 'fevers', see Alberola Romá and Bernabé Gil, 'Tercianas y calenturas en tierras meridionales valencianas,' 111; Juan Riera Palmero, *Fiebres y paludismo en la España Ilustrada. Félix Ibáñez y la epidemia de La Alcarria, 1784–1792* (Valladolid: Universidad de Valladolid, Secretariado de Publicaciones, 1984); Riera Palmero, 'La Medicina en la España del siglo XVIII.'; Riera Palmero, 'Quina y malaria en la España del siglo XVIII.' Other areas joined the traditional hotbeds in the eighteenth century. Riera Palmero, 'La Medicina en la España del siglo XVIII,' 16.

aware of intermittent fevers' seasonality and the bark's effectiveness against them, likewise routinely stored the bark in advance of the ague season to safeguard their own health and that of their kin.[47] The Crown also gave away sizeable amounts of cinchona to charity.[48] During the 1780s, when epidemics of intermittent fevers haunted the Iberian Peninsula, the Spanish Crown dispatched cinchona for hundreds of 'poor sufferers' in Toledo,[49] San Fernando,[50] Castilla la Vieja[51] and Segovia and the nearby village of Santa Maria la Real de Nieva,[52] pressing colonial officials for ever more fresh supplies of the bark.[53] Indeed, it is possible that it was their and other paupers' sustained exposure to cinchona through relations of free-of-charge medical attention that accounted for the bark's inclusion in their medical repertoire and their pleas for it on future occasions. When Guadalajara, a city with a large population of poor workers and their families, struggled to accommodate patients and provide the required medications in the 1786 epidemic of fevers, the city's inhabitants petitioned that the Crown, the city council and wealthier residents give relief to 'so much misery' in the form of

---

[47] 'Aprovado con la advertencia de que si se pudiese despachar toda la quina en rama seria mas sencilla,' *Archivo General de Indias*, 1557, Madrid, 1807-01-28 / 1807-02-03.

[48] See Chapter 2.

[49] 'El Cura Párroco, el Alcalde, Regidores, y Procurador Síndico del Lugar de Vargas del Arzobispado de Toledo Hacen presente à V.M. que la mayor parte de su vecindario son pobres jornaleros,' *Archivo General de Indias*, Legajo 961, 1786-08-09.

[50] Antonio Valdez, 'El Presbítero Don Antonio María Chacón, Administrador del Real Hospital de San Fernando del Sitio de San Ildefonso, ha hecho presente que con motivo de haberse aumentado considerablemente el vecindario de aquel pueblo, ha crecido también el numero de los pobres enfermos,' *Archivo del Palacio Real*, Copias de Ordenes comunicadas por el Ministro de Hacienda al Señor Sumiller de Corps. Real Botica, Reinados Carlos III / Legajo 197, 3, San Lorenzo, 1787-11-21.

[51] José (Conde de Floridablanca) Moñino y Redondo, 'En carta confidencial que me ha escrito el capital general de Castilla la Vieja me dice que continúan las crueles tercianas de los Pueblos de aquella provincia,' *Archivo del Palacio Real*, Copias de Ordenes comunicadas por el Ministro de Hacienda al Señor Sumiller de Corps. Real Botica, Reinados Carlos III / Legajo 197, 3, San Lorenzo, 1788-11-01.

[52] José (Conde de Floridablanca) Moñino y Redondo, 'El Procurador Sindico general del lugar de la Moraleja [...] ha representado [...] que teniendo cerca de quatrocientos tercianarios, para cuya curación carecen de los auxilios necesarios, está expuesto todo el Pueblo a un contagio perlífero, si no se le socorre con una competente porción de Quina para su pronto remedio,' *Archivo del Palacio Real*, Copias de Ordenes comunicadas por el Ministro de Hacienda al Señor Sumiller de Corps. Real Botica, Reinados Carlos III / Legajo 197, 3, San Ildefonso, 1788-09-13; José (Conde de Floridablanca) Moñino y Redondo, 'El Rey ha mandado que á la Villa de Santa María la Real de Nieva se de una porción de Quina para mas de 300 enfermos pobres que tiene,' *Archivo del Palacio Real*, Copias de Ordenes comunicadas por el Ministro de Hacienda al Señor Sumiller de Corps. Real Botica, Reinados Carlos III / Legajo 197, 3, San Ildefonso, 1788-09-17.

[53] 'El Sub.do de la Real Hacienda de Lima contesta a la Real Orden de remisión de Quina,' *Archivo General de Indias*, Indiferente 1554, Lima, 1787.

Figure 4.2 Albarello drug jar from Spain used to store cinchona bark, 1731–1770; calisay[a] was one popular cinchona variety.
© The Science Museum / Science and Society Picture Library.

hospitals beds, food and the bark.[54] The Crown also bestowed cinchona upon monasteries in the affected areas. The convent of Capuchin nuns in Huesca, Aragón, repeatedly applied to the Crown for cinchona, since it was, as the convent's abbess lamented, 'situated in a place utterly susceptible to tertian fevers', 'such that the majority of the community [was] suffering from that illness most of the time and thus consume[d] large portions of cinchona'[55] (FIGURE 4.2).

Just as in the Mediterranean and the North Sea Basin, elsewhere, too, knowledge of indications for the bark trailed the contours of the period's low-lying marshes rather accurately. The bark was popular in the marshy plains along the course of the Danube and its tributaries,[56] in the soggy

[54] Riera Palmero, *Fiebres y paludismo*, 34–35; 40.
[55] Sor María Margarita, 'La Abadesa y Religiosas del Convento de capuchinas de la ciudad de Huesca en el Reyno de Aragon con el debido respetto a los Rs. Ps. de V.M. exponen,' *Archivo del Palacio Real*, Reinados Carlos III / Legajo 197, 3, Huesca, 1759–1788.
[56] Bruce-Chwatt and Zulueta, *Rise and Fall*, 46; 61–62. On fevers, and cinchona consumption in Transylvania during the late 1700s, see also Lochbrunner, *Der Chinarindenversuch*, 36.

terrain of the Lower Mississippi Valley in Louisiana[57] and in the Pearl
River wetlands around Canton, a landscape of embankments and
canals,[58] where Portuguese sailors and missionaries had spread the word
about its admirable effectiveness in intermittent fevers – as well as the
consequences of excessive indulgence in wine.[59] The inhabitants of the
coastal plains of the Peruvian Viceroyalty – living, as they did, mostly
along the valleys of rivers that bring rainwater down from the Andes –
likewise suffered grievously from tertian and quartan fevers, and generally trusted in the bark for their cure, on the recommendations of health
advice manuals, the yearly almanac or family recipes.[60] Pharmaceutical
inventories and the estimates of contemporary physicians suggest an
annual consumption of between 0.4 and 1.4 tons of cinchona in the
Viceroyalty of Peru – between 25,000 and 86,000 doses – and the bulk
of that was apparently consumed by the roughly 50,000 inhabitants of the
viceregal capital.[61] Cinchona consumption was low throughout the Viceroyalty – even at the heart of the harvest areas – as the Spanish official
Miguel de Santisteban wrote in a 1751 report, with the only 'exception of
Lima and some places along the coast'.[62] While the origin of expertise
and experience in indications for the bark, transmitted among most
laymen – Finnish parishioners, French ladies and Spanish paupers alike –
would invariably have been learned medicine,[63] the one place where this
was not necessarily the case was Andean South America. Particularly in
the Viceroyalties of Peru and New Granada, where the principal harvest
areas were located, some, including creole sectors of society, would have
experienced the bark in intermittent fevers at the hands of Andean

[57] On the insalubrity of Louisiana, see Marion Stange, 'Governing the Swamp: Health and
Environment in Eighteenth-Century Nouvelle-Orleans,' *French Colonial History* 11
(2010). On cinchona consumption, see Chapters 2 and 3.
[58] On 'malaria' in southern China, see Erhard Rosner, 'Gewöhnung an die Malaria in
chinesischen Quellen des 18. Jahrhunderts,' *Sudhoffs Archiv* 68, no. 1 (1984). On
cinchona consumption, see Chapters 2 and 3.
[59] That quote is taken from Chao Hsüeh-min's (1719–1805) 'Addenda and corrigienda to
the Pen-ts'ao kang-mu (Pen-ts'ao kang mu shih-i)', cited in Unschuld, *Medicine in
China*, 166.
[60] Gänger, 'In Their Own Hands.'
[61] On the population of Lima, and the department, see Pilar Pérez Cantó, 'La población de
Lima en el siglo XVIII,' *Boletín Americanista* 32 (1982), 396; Paul Gootenberg,
'Population and Ethnicity in Early Republican Peru: Some Revisions,' *Latin American
Research Review* 26, no. 3 (1991), 113–15.
[62] Miguel de Santisteban, 'Copia de Carta,' *Archivo General de Indias*, Indiferente 1554,
Santa Fé, 1753-06-04, f. 797. As outlined in Chapter 2, the entirety of the Viceroyalty of
Peru consumed between 0.4 and 1.4 tons of cinchona.
[63] For a similar diagnosis in relation to early modern Europe, see Stolberg, *Homo
patiens*, 112.

healers,[64] who played an important part in Spanish American healthcare more broadly – not least because an Atlantic Enlightenment, in its empiricist dictates and nostalgic critique of civilization, both fashioned and fetishized the simple, illiterate and humble knower as healer and 'botanist by instinct'.[65] Whether Andeans had inherited their knowledge of the bark from their ancestors eludes us, of course, but it is possible that they owed it at least not primarily or exclusively to Hispanic medicine.

Cities, Ships and Camps

In the late eighteenth and early nineteenth centuries, while fevers were invariably thought to be caused by foul, 'putrefying', 'morbid' air, marshes and swamps were by no means the only conceivable source of such 'exhalations'. 'Morbid effluvia or emanations' communicating distempers to those in health could also issue either from 'corrupted animal or vegetable substances', or, indeed, from fellow humans, particularly persons 'labouring under fever'.[66] The sickly air of close, crowded spaces was considered as much as a place's natural situation – its climate, altitude or terrain – to be able to generate disease.[67] The narrow confines of ships, army camps and rapidly growing urban spaces were, accordingly, some of the period's other most important sites for cinchona consumption.

Late eighteenth- and early nineteenth-century physicians and surgeons – from John Pringle (1707–1782) to Pedro María González (1764–1838) – dwelt extensively on the putrid fevers that occurred aboard ship and the necessity of having antiseptics like cinchona, so 'marvellously useful' (*maravillosamente útil*) in their prevention and cure, always provided on board.[68] Indeed, ship's surgeons employed by the

---

[64] On the bark collector and curandero Fernando de la Vega see, for instance, Estrella, 'Ciencia ilustrada y saber popular,' 56.

[65] See Chapter 1.

[66] Clark, *Observations on Diseases which prevail in Long Voyages to Hot Countries*, 150.

[67] On the English context, see Dobson, *Contours of Death and Disease*, 16. On French environmentalism and urban environments, see Brockliss and Jones, *The Medical World of Early Modern France*, 753. On Spanish and Spanish American anxieties about urban health, see, Adam Warren, *Medicine and Politics in Colonial Peru: Population Growth and the Bourbon Reforms* (Pittsburgh, Pa.: University of Pittsburgh Press, 2010); Clément, 'El nacimiento de la higiene urbana.'

[68] John Pringle recommended cinchona, alongside other 'antiseptics', as well as cleanliness, discipline and exercise, as a cure to counteract putrefaction. John Pringle, *Observations on the Nature and Cure of Hospital and Jayl-Fevers in a Letter to Doctor Mead, Physician to his Majesty* (London: A. Millar and D. Wilson, 1750), 45; Pringle, *Observations on the diseases of the Army*, appendix xiii. On cinchona's 'marvellous' usefulness in 'putrid fevers' aboard ships, see Gonzalez, *Tratado de las enfermedades de*

Dutch East India Company and the Royal Navy alike at the time were routinely provided with medicine chests containing potions and herbs, with their curative properties noted on prescribed lists. Next to the administration of opium, ipecac, powdered jalap and mercury, British and Dutch ships' crews would have been routinely exposed to that of the Peruvian bark.[69] So would sailors on board slave ships off the African coast – which usually held, next to opium or gentian root, large amounts of cinchona[70] – Portuguese expeditionary ships[71] or the frigates of the United States Navy.[72] Indeed, by the early 1800s, many ship's surgeons and their crews alike had learned to think of the bark as a necessity and to complain bitterly if it was not made available to them 'in proper quantity'.[73] The use of cinchona was also common and familiar in the narrow confines of army camps. Outbreaks of fatal epidemics of fevers in the camps, barracks and hospitals of stationed armies – for the 'unwholesome effluvia from so many bodies always occasion[ed] sickness'[74] – were commonly treated with 'evacuations', and the bark.[75] So were the

---

la gente de mar, 203. On Pedro María González (1764–1838), see Manuel Martínez Cerro, 'Don Pedro María González, navegante y erudito. Aclaratoria solicitud de licencia,' *Apuntes* 2, no. 4 (2004). John Clark was likewise persuaded of the necessity of providing 'the bark (...) for the Royal Navy', for the 'prevention and cure of diseases' and 'the health of seamen and soldiers'. Clark, *Observations on Diseases which prevail in Long Voyages to Hot Countries*, x.

[69] Iris Bruijn, *Ship's Surgeons of the Dutch East India Company: Commerce and the Progress of Medicine in the Eighteenth Century* (Amsterdam: Amsterdam University Press, 2009), 350.

[70] Harrison, *Medicine in an Age of Commerce and Empire*, 133.

[71] The medicine chests on the ships of the first Portuguese naturalist expedition to Brazil, led by Alexandre Rodrigues Ferreira (1756–1815), contained bottled English Water. Beltrão Marques, *Natureza em Boiões*, 240.

[72] The North American frigate *New York*, sent to North Africa by the Thomas Jefferson administration in 1801 to confront Algiers, Tunis and Tripoli, carried more than 100 remedies, including 'China cortex ruber' and 'China cortex flavus'. Rutten, *Dutch Transatlantic Medicine Trade*, 38.

[73] Hunter, *Observations on the Diseases of the Army*, 138.

[74] Gerard Swieten, Freiherr van, *The diseases incident to armies: with the method of cure* (Philadelphia: R. Bell, 1776), preface. The book was first published in Vienna in 1759, in French (*Description abrégée des maladies qui régnent le plus communément dans les armées, avec la méthode de les traiter*).

[75] See, for instance, Pringle, *Observations on the diseases of the Army*, 207; José Celestino Mutis, 'Borrador y copia de un oficio de José Celestino Mutis al arzobispo virrey Antonio Caballero y Góngora comunicándole que ha redactado un "Plan de curación para las enfermedades que padecen las tropas del rey establecidas en el Darién" que consisten en diversas clases de calenturas y fiebres. Le notifica que ha enviado un cargamento de quina al hospital de estas tropas. Adjunto, borrador manuscrito de Mutis y copia del citado plan de curación, ambos incompletos,' *Archivo del Real Jardín Botánico*, Real Expedición Botánica del Nuevo Reino de Granada (1783–1816), José Celestino Mutis, Documentación oficial, Oficios de José Celestino Mutis, RJB03/0002/0002/0069, Mariquita (Colombia), 1786-05-18.

fevers that afflicted localities on account of the camps' presence. When French troops caused 'numerous and grave epidemic[s]' in Madrid,[76] on Tenerife[77] or in Aragón and Cataluña,[78] Spanish physicians invariably administered cinchona-based medicines.

Perhaps the most obvious settings susceptible to crowd diseases and heavy levels of bark consumption, were the cities of the Atlantic World, which were growing at an unprecedented rate and in a manner 'uneven, usually unplanned, and sometimes unwanted' between 1750 and 1850, as a consequence of industrialization, reform policies and faster population growth.[79] Given their easier access to remedies,[80] their greater likelihood of illness – with many a city's atmosphere 'fatally tainted' by the harmful vapours arising from open sewers, burial grounds or overcrowded barracks[81] – and their greater exposure to both professional medical treatment and health advice, townsfolk and city dwellers drugged themselves with the bark at a scale out of all proportion. In viceregal Cartagena de Indias, a city on the Caribbean coast where 'confined air' conspired with a 'hot climate' (*temperamento ardiente*) and the neighbouring sea, the resident population was evidently given to

[76] Tomas de Salazar employed cinchona in 1782 and 1783, when the French army caused 'a numerous and grave epidemic' in Madrid, and allegedly 'lost but few patients'. Salazar, *Tratado del uso de la quina*, 22.

[77] Francisco de Saavedra and Felipe Carrillo, 'Informe sobre una porción de Quina pedida al Rey por el Sindico Procurador gral. de la Isla de Tenerife,' *Archivo del Palacio Real*, Caja 22283 / Expediente 7, San Lorenzo, 1797-12-06 / Canarias, 1797-10-13.

[78] Riera Palmero, 'La Medicina en la España del siglo XVIII,' 16. See also Masdevall's own account: Masdevall, *Relacion de las epidemias de calenturas pútridas y malignas*.

[79] Between 1750 and 1910, the percentage of the European population living in communities numbering 5,000 or more inhabitants more than tripled, and between 1800 and 1900 the percentage living in cities of at least 100,000 inhabitants more than quadrupled. British overseas expansion depended heavily on urbanization, and by the eighteenth century market towns and seaports dotted the landscape from Maine to Georgia. Andrew Lees and Lynn Hollen Lees, *Cities and the Making of Modern Europe* (Cambridge: Cambridge University Press 2007), 1; 18–21. On urbanization in Spanish America, see Richard Morse, 'The Urban Development of Colonial Spanish America,' in *The Cambridge History of Latin America*, ed. Leslie Bethell (Cambridge: Cambridge University Press, 1984), 99.

[80] See Chapter 2.

[81] On Spanish and Spanish American anxieties about urban health, see, Warren, *Medicine and Politics in Colonial Peru*; Clément, 'El nacimiento de la higiene urbana.' On the English context, see Dobson, *Contours of Death and Disease*, 16. On French environmentalism and urban environments, see Brockliss and Jones, *The Medical World of Early Modern France*, 753. See also Lees and Hollen Lees, *Cities and the Making of Modern Europe*, 59–65. On Prussian cities, see Ute Frevert, *Krankheit als politisches Problem 1770–1880. Soziale Unterschichten in Preußen zwischen medizinischer Polizei und staatlicher Sozialversicherung*, Kritische Studien zur Geschichtswissenschaft (Göttingen: Vandenhoeck & Ruprecht, 1984), 25–26. On Rome's 'fatally tainted' atmosphere, see Richard Wrigley, *Roman Fever. Influence, Infection and the Image of Rome, 1700–1870* (New Haven: Yale University Press, 2013), 61.

regular, and presumably preventive, consumption of the bark.[82] In 1787, at a time when that city had but around 20,000 inhabitants,[83] these resorted to some 600 *libras* – 276 kilograms – of the bark, an amount that, calculating with an average dose of 16 grams, would have sufficed for around 17,000 doses.[84] In other populous, ill-situated cities, like Lima, New Orleans, Marseille and Constantinople, medication and self-medication with the bark ranged at similar levels.[85] Rome, one of the earliest cities of the Old World to embrace the bark in fevers[86] and a notoriously unhealthy, fever-laden spot,[87] imported, according to José Nicolás de Azara (1730–1804), the Spanish ambassador in that city, by 1785 some 4,600 kilograms – that is, 287,500 doses – of bark per annum legally, in addition to the 'very great portion of it' that was 'smuggled' (*que và de fraude*), at a time when Rome had a population of roughly 163,000 inhabitants.[88] According to the ambassador, it was not easy to ascertain the real scale of consumption, but it had 'to be immense, owing to the practice these [Roman] doctors have of administering such quantity of cinchona that if it was told here no one would believe it'.[89]

'Hot Climates'

The understanding that fevers in 'hot climates' were more severe and that particularly Europeans were more vulnerable to them – that moving out of one's accustomed environment, and exchanging it 'for a distant

[82] On Cartagena de Indias and its 'particular disrepute regarding health conditions', see, Gómez, *The Experiential Caribbean*, 42–45.

[83] According to the 1815 census, the city had some 20,000 to 22,000 inhabitants. Maruja Redondo Gómez, *Cartagena de Indias: cinco siglos de evolución urbanística* (Bogotá: Fundación Universidad de Bogotá Jorge Tadeo Lozano, 2004), 64.

[84] The city of Cartagena de Indias had consumed 600 *libras* of cinchona annually by 1787. Salvador Vázquez, 'Las quinas del norte de Nueva Granada,' 406. On Cartagena's 'ardent temperament', see Mutis, 'Borrador del oficio de José Celestino Mutis al virrey Pedro Mendinueta y Muzquis.'

[85] On Constantinople, see Günergun and Etker, 'From Quinaquina to "Quinine Law,"' 50. When Napoléon Bonaparte (1769–1821, r. as consul 1799–1804, as emperor 1804–1814/15) had 150 *quintales* of powdered bark – some 6,900 kilograms – distributed in 42 French cities between 1808 and 1812, Marseille received a share of 460 kilograms, some 28,750 doses. Bruce-Chwatt and Zulueta, *Rise and Fall*, 71. On New Orleans, see Stange, 'Governing the Swamp.' On bark consumption in Lima, see Gänger, 'In Their Own Hands.'

[86] On the introduction and early uses of the bark in Rome from the 1630s, see Anagnostou, *Missionspharmazie*, 298; 304; 21.

[87] Wrigley, *Roman Fever*, 61. See also Sallares, *Malaria and Rome*.

[88] Paolo Malanima, 'Italian Cities 1300–1800. A Quantitative Approach,' *Rivista di storia Economica* XIV, no. 2 (1998), 113.

[89] Azara, 'Carta a Pedro de Lorena'; 'Informe del Ministro de V.M. en Roma,' *Archivo General de Simancas*, Legajo 961/2, Roma, 1785.

Climate',[90] would invariably come at a cost – was a fact firmly impressed on the British, French, Portuguese, Dutch and Spanish imaginations alike around 1800.[91] Already from the middle of the fifteenth century, the luxuriant, prodigious nature of the tropics had become associated with febrile illnesses.[92] How to counter it was a particularly pressing preoccupation in the decades ranging from 1760 to 1830, an age of Atlantic Revolutions, global imperialism and the Great Divergence – when Britain, Napoleonic France and Russia first gained military, political and economic superiority over Qing China, Mughal India and the Ottoman Empire – that took unprecedented numbers of soldiers, colonial officials, merchants and colonists to remote, alien and, all too often, insalubrious environs.[93] It was widespread anxieties about bodily health that account for a large share of the vast quantities of cinchona consumed in colonialism, warfare and settlement in the world's warm climates during the late 1700s and early 1800s.[94]

Cinchona consumption among merchants, colonists and officials in the British, Portuguese and Dutch coastal enclaves of Africa's Guinea region, Mozambique and the Cape colony was general, high and systematic. The merchant community in Portuguese Luanda, Western Africa's largest slaving port, was accustomed to regularly being administered – at the hands of physicians or in the hospital – and to self-administering cinchona-based medicines to cure or ward off the fevers that killed the

---

[90] Lind, *An Essay on Diseases incidental to Europeans in hot Climates*, 2.

[91] See, for instance, Ward, *Desire and Disorder*, 150–51. From the 1730s, men and women in the tropics were acutely aware that newcomers from cooler climates stood a far greater chance of falling ill and of dying than did locals. Margaret Humphreys, *Malaria. Poverty, Race, and Public Health in the United States* (Baltimore: Johns Hopkins University Press, 2001), 23; McNeill, *Mosquito Empires*, 65.

[92] Portuguese encounters with fevers in West Africa during the fifteenth century engendered uncertainty about the Hippocratic framework, which had associated verdant nature with health. According to Hugh Cagle, the invention of the 'tropics' began in the mid-fifteenth century and culminated in the late seventeenth century. Hugh Cagle, 'Beyond the Senegal: Inventing the Tropics in the Late Middle Ages,' *Journal of Medieval Iberian Studies* 7, no. 2 (2015); Cagle, *Assembling the Tropics*, 9. On the history of 'tropicality' as an environmental imaginary, see also the classic work of David Arnold in *The Problem of Nature: Environment and Culture in Historical Perspective* (Oxford: Blackwell, 1996), chapter 8, 'Inventing Tropicality,' 141–68.

[93] On 'The First Age of Global Imperialism', Christopher A. Bayly, 'The First Age of Global Imperialism, c. 1760–1830,' *The Journal of Imperial and Commonwealth History* XXVI, no. 2 (1998). On the 'Great Divergence', see Kenneth Pomeranz, *The Great Divergence: China, Europe, and the Making of the Modern World* (Princeton, N.J.: Princeton University Press, 2000); Jean-Laurent Rosenthal and R. Bin Wong, *Before and Beyond Divergence: The Politics of Economic Change in China and Europe* (Cambridge, Mass.: Harvard University Press, 2011).

[94] On how the creation of the Portuguese empire also entailed the increasing prominence of febrifuges, see Cagle, *Assembling the Tropics*, 307.

majority of their new arrivals, as José Pinto de Azeredo (d. 1810), a Brazilian doctor appointed surgeon-general of Portuguese Angola in 1790, put it, 'within weeks or months'.[95] So did the factors of the British Royal African Company. The company exported 'at its own expense large quantities of cinchona' in order to distribute it among its factories along the Guinea coast where fevers 'destroyed many of [their] men every year'.[96] As a matter of fact, according to the company's late eighteenth-century records, about half of the Europeans sent to West Africa died within one year, while only one in ten ever returned to England, and their deaths were almost invariably attributed to fevers.[97] British, French, Dutch and Spanish merchants, settlers and colonial officials in the West Indies were likewise 'attended with great mortality',[98] and, acutely aware of the dangers the environs posed to their health, confided their 'whole reliance [...] to the bark – that great sheet-anchor of West India practice', as George Pinckard phrased it in his 1806 *Notes on the West Indies*, 'prescribed in ample quantity, and in various forms'.[99] Physicians in Dutch Berbice – to become a British plantation colony after 1814, and British Guiana in 1831 – prescribed 'emetics and the bark [...] in all cases of fever',[100] and so did French practitioners in Port-au-Prince,[101] the Spanish and Spanish Americans in the Panamá, Santa Marta and Maracaibo provinces and in the Captaincies General of Yucatán and Cuba,[102] and British plantation doctors on the island of Jamaica.[103] As James Clark, a physician with 'twenty-five years constant practice in the West Indies', put it, very few died of

---

[95] On cinchona consumption in Luanda, see Pinto de Azeredo, *Ensaios sobre algumas enfermidades d'Angola*, 64–66. On the consumption of 'Agoa de Inglaterra' in Angola, see Figueiredo, 'A "Água de Inglaterra" em Portugal,' 123–24.

[96] Salazar, *Tratado del uso de la quina*, 86.

[97] Kenneth F. Kiple, *The Caribbean Slave. A Biological History* (Cambridge: Cambridge University Press, 1984), 13. See also Linda A. Newson and Susie Minchin, *From Capture to Sale. The Portuguese Slave Trade to Spanish South America in the Early Seventeenth Century* (Leiden: Brill, 2007), 74–75.

[98] Hunter, *Observations on the Diseases of the Army*, 12.

[99] George Pinckard, *Notes on the West Indies written during the expedition under the command of the late General Sir Ralph Abercromby: including observations on the island of Barbadoes, and the settlements captured by the British troops, upon the coast of Guiana; likewise remarks relating to the Creoles and slaves of the western colonies, and the Indians of South America: with occasional hints, regarding the seasoning, or yellow fever of hot climates*, 3 vols., vol. 3 (London: Longman, Hurst, Rees, and Orme, 1806), 58.

[100] Ibid., 21.

[101] Pomme, 'Mémoire et observations cliniques,' 148. See also Sheridan, *Doctors and Slaves*, 303.

[102] Jijon y León, 'Recomendaciones,' 139. On cinchona consumption in Cartagena, Havana and Panamá, see also Salvador Vázquez, 'Las quinas del norte de Nueva Granada,' 406.

[103] Thomson, *A treatise on the diseases of negroes*, 166.

remittent and intermittent fevers if they were given the Peruvian bark 'as early as possible, in substance'.[104] Colonists on Saint Domingue, or Jamaica, also treated slaves ailing from fevers with 'a few doses of bark',[105] but physicians generally testified to the low incidence of intermittent, malignant and yellow fever among men and women of African descent.[106] While they were not believed to be 'entirely exempted from them', they supposedly 'suffer[ed] infinitely less than Europeans', as the army physician John Hunter (1754–1809) put it.[107] The British expansion in Bengal, where fevers were, together with fluxes – that is, diarrhoea or dysentery – the most fatal disorder for Europeans,[108] also relied upon a regular and large-scale supply of cinchona. During the 1770s and 1780s, Peruvian bark administered in wine was one of the most popular remedies among the British in India for the treatment of intermittent and remittent fevers.[109] The settler population on the Spanish Philippines, where fevers, particularly tertian and quartan fevers, were prevalent, resorted to the bark[110] and so did the merchants and soldiers of the VOC in Batavia – today's Jakarta, on the Island of Java – 'ill-situated on a low coastal plain with swamps'. The 'unhealthiness of Batavia' killed more than half of the new arrivals after 1733 within a year and weakened survivors for many years, and by the 1760s expenditure for cinchona among Europeans was, according to contemporaries, lavish.[111]

---

[104] Clark, *A Treatise on the Yellow Fever*, 89.

[105] Thomson, *A treatise on the diseases of negroes*, 14; 19; 27; Gauché, 'Description d'un Quinquina indigène á St. Domingue,' 2–3.

[106] Sheridan, *Doctors and Slaves*, 186.

[107] Hunter, *Observations on the Diseases of the Army*, 24. See also Ward, *Desire and Disorder*, 156–57.

[108] Hospital records from Bengal, for instance, show that the most common causes of admission to the military hospital at Fort William in Calcutta in 1767 were fevers and fluxes, in that order: 283 cases of 'fevers', 117 of 'fluxes', 57 of 'pains', and the remainder fewer than 10 cases each of 'ulcers', 'obstruction', 'dropsy'. Harrison, 'Disease and Medicine in the Armies of British India, 1750–1830,' 91–92. See also Arabinda Samanta, *Malarial Fever in Colonial Bengal 1820–1939. Social History of an Epidemic* (Kolkata: Firma KLM Private Limited, 2002). On rates and causes of death in Monsoon Asia, particularly on the western side of the Bay of Bengal, see also Curtin, *Death by Migration*, 30.

[109] Particularly from the 1810s until the 1830s, cinchona fell into disuse in British India and purgatives came into vogue instead, particularly mercury. The use of mercury in the treatment of fever became paramount in British India and the West Indies, though that practice was rare in Britain itself, where practitioners still tended to recommend the bark and stimulants in fevers. Harrison, 'Disease and Medicine in the Armies of British India, 1750–1830,' 96–101.

[110] Linda A. Newson, *Conquest and Pestilence in the Early Spanish Philippines* (Honolulu: University of Hawaii Press, 2009), 14. On cinchona consumption, see Chapter 2.

[111] Average mortality among seamen in Asia before 1733 was three times as high as that of male adults in Europe; after 1733, when Batavia became 'unhealthy', it was five times

Settlers along the world's socio-ecological frontiers around 1800 – in South Africa or the Americas – also relied on cinchona to shield them in lands with fever-laden air that seeped into newcomers' every pore.[112] With the westward movement of settlers in post-revolutionary North America, especially along the tidewater settlements of the Atlantic coast, the expansion of the British farming frontier to the east from Cape Province and internal colonization processes in Atlantic Brazil – where surface deposits of gold and diamonds brought Portuguese immigrants and African mine workers to the interior – agues, malignant fevers and yellow fever frequently left settlers vulnerable, prostrate or dead.[113] In Atlantic Brazil, New Orleans and New England, accordingly, cinchona was a standard item in apothecary shops, plantation medicine chests and the provisions of itinerant barber-surgeons.[114] The bark was liberally administered to colonists in the frontier regions' few hospitals, but primarily, it seems, self-administered by them – a crucial component of the arsenal they had at their disposal as they sought to gain their bearing in foreign terrain that held both promise of wealth and independence, and foreboding with regard to frontiersmen's physical selves.[115] Vernacular self-help manuals like José Antônio Mendes's (fl. 1750s–1770s) 'Guidance for Miners' (*Governo de Mineiros*), a popular medical handbook that offered counsel to the inhabitants of the Brazilian gold mine

---

as high. For soldiers, Asia was even more dangerous: mortality rates were ten times as high as in Europe in the period between 1730 and 1775. Peter Harmen van der Brug, 'Malaria en malaise. De VOC in Batavia in de achttiende eeuw' (PhD dissertation, Rijksuniversiteit te Leiden, 1994), 27–28. See also Peter Harmen van der Brug, 'Malaria in Batavia in the 18th Century,' *Tropical Medicine and International Health* 2, no. 9 (1997), 896. James Lind mentions the example of cinchona consumption on the English ship of war *Panther*, which 'touched at Batavia' in the years 1762 and 1764 and suffered greatly 'by the malignant and fatal diseases of that climate'. Lind, *An Essay on Diseases incidental to Europeans in hot Climates*, 85–87.

[112] Bolton Valencius, *The Health of the Country*, 12. On 'fever-laden air', see Stange, 'Governing the Swamp,' 2.

[113] On disease along the North American Frontier, see Dobson, 'Mortality Gradients and Disease Exchanges'; Bolton Valencius, *The Health of the Country*; John Duffy, 'The Impact of Malaria on the South,' in *Disease and Distinctiveness in the American South*, ed. Todd L. Savitt and James Harvey Young (Knoxville: University of Tennessee Press, 1988); Webb, *Humanity's Burden*, 85–91. On South Africa, see James L. A. Webb, *The Long Struggle against Malaria in Tropical Africa* (New York: Cambridge University Press, 2014), 37. On Atlantic Brazil, see Neves Abreu, *Nos Domínios do Corpo*.

[114] On North Americans' access to the bark, see Gevitz, '"Pray Let the Medicine be Good",' 15; Fett, *Working Cures*, 68; Humphreys, *Malaria*, 39; Duffy, 'The Impact of Malaria on the South,' 32. For references to cinchona consumption in Brazil, see Walker, 'The Medicines Trade in the Portuguese Atlantic World,' 102; Beltrão Marques, *Natureza em Boiões*, 240; Júnia Ferreira Furtado, 'Barbeiros, cirurgiões e médicos na Minas colonial,' *Revista do Arquivo Público Mineiro* 41 (2005).

[115] Bolton Valencius, *The Health of the Country*, 15.

districts of Minas Gerais[116] or John C. Gunn's (d. 1863) *Domestic Medicine, or, Poor Man's Friend,* which guided household doctoring in North America's antebellum borderlands,[117] advised the use of cinchona in 'agues' (*sezões*) and bilious fever, and also as a general tonic, to 'increase the tone of the muscular fibres, and thereby strengthen the whole body',[118] in a 'by-your-own-bootstraps approach to healing that dovetailed' with a frontier rhetoric of self-reliance,[119] of the industrious, resilient, able-bodied pioneer.

Cinchona became a standard element of military medical supplies at least from the age of Atlantic Revolutions, concomitant uprisings in British North America (1765–1783), France (1776–1799), Saint-Domingue (1791–1804) and Spanish America (1808–1830) that took large armies of soldiers to 'marshy situations', 'the neighbourhood of the sea', and particularly hot, tropical climates 'fraught with pestilential vapours' and 'unwholesome exhalations' understood to cause putrid, intermittent and other fevers.[120] At a time when the increasing scale and frequency of warfare, as well as a series of developments in technology that made it costlier to train men for battle, provided an important stimulus to the conservation of manpower, the supply and provision of medicines, alongside attention to regimen – clothing and nutrition, cleanliness, temperance and exercise – gained relevance more generally as part of military strategy in the armed forces of Russia, the Netherlands, France and England, as well as the Iberian empires.[121] The troops of Spain's royal army, for instance, were regularly supplied with tons of

---

[116] José Antonio Mendes, *Governo de Mineiros mui necessário para os que vivem distantes de professores, seis, oito, dez e mais léguas, padecendo por esta causa os seus domésticos e escravos, queixas que pela dilação dos remédios se fazem incuráveis, e as mais das vezes mortais* (Lisboa: Officina de Antônio Rodrigues Galhardo, 1770). Cited in Maria Cristina Cortez Wissenbach, 'Gomes Ferreira e os símplices da terra: experiências sociais dos cirurgiões no Brasil–Colônia,' in *Luís Gomes Ferreira, Erário mineral,* ed. Júnia Ferreira Furtado (Rio de Janeiro: Editora FIOCRUZ, 2002), 144.

[117] Rosenberg, *Explaining Epidemics,* 57–73.

[118] John C. Gunn, *Gunn's Domestic Medicine: Or Poor Man's Friend, in the Hours of Affliction, Pain and Sickness: This Book Points Out, in Plain Language, Free from Doctor's Terms, the Diseases of Men, Women, and Children, and the Latest and Most Approved Means Used in Their Cure and is Expressly Written for the Use of Families in the Western and Southern States* (Springfield: John M. Gallagher, 1835), 575; 623. The book was originally published in Knoxville, in 1830.

[119] Bolton Valencius, *The Health of the Country,* 54.

[120] See, for instance, Blair, *The Soldier's Friend,* 54–59; John B. Davis, *A scientific and popular view of the fever of Walcheren, and its consequences, as they appeared in the British troops returned from the late expedition; with an account of the morbid anatomy of the body, and the efficacy of drastic purges and mercury in the treatment of this disease* (London: Samuel Tipper, 1810), ii–iii.

[121] Mark Harrison, 'Medicine and the Management of Modern Warfare,' *History of Science* 34, no. 106 (1996), 382–86; Harold J. Cook, 'Practical Medicine and the British Armed

cinchona, at least from the reign of Charles III onwards, to care for the
sick during the Great Siege of Gibraltar (1779–1783)[122] or for the
defence of Mallorca (1799) – with amounts sufficient for over 10,000
treatments, as many as the army had soldiers.[123] The preparation and
allocation of cinchona-infused wines for Portuguese army and navy
hospitals from Lisbon[124] to Bahia[125] were likewise standard practice,
and, after 1799, regulated and monitored by Portugal's royal *protomedicato*.[126] Cinchona, the 'indispensable fever bark' (*die unentbehrliche Fieberrinde*), was also frequently included in the provisions of Prussian army
surgeons,[127] the pharmaceutical supplies of Ottoman military
hospitals[128] and the Habsburg armed forces. For, as Gerard van Swieten
(1700–1772), personal physician and advisor to Empress Maria Theresa,
put it in his essay on the *Diseases Incident to Armies*, first published in
Vienna in 1759, in a variety of sicknesses that commonly attended a
military life – the obstinate 'spring fevers' and 'autumnal intermittents',
the quartan and 'continued fevers' or gangrene – the bark had to be given
immediately.[129] Cinchona was likewise a standard component in the
medicine chests allotted to British and continental regimental surgeons
alike during the American War of Independence. The Peruvian bark, as
the army surgeon Robert Jackson put it, 'properly administered', had
'seldom failed of cutting short' the course of 'the mild and regular

Forces after the "Glorious Revolutions,"' *Medical History* 34 (1990), 8; Harrison,
Medicine in an Age of Commerce and Empire, 16–18.
[122] Over 400 *libras* were administered in Spanish military hospitals in Algeciras, San
Roque, Medina Sidonia, Tarifa, Marbella and Estepona during military action in
Gibraltar. Andrés Turrión and Terreros Gómez, 'Organización administrativa,' 422.
[123] An army of 10,000 men received 1,150 *libras*, some 529 kilograms, of cinchona, enough
for 10,580 treatments. Ibid., 424. On pharmaceutical provisions for the military in the
Spanish army, see also Andrés Turrión, 'Quina de la Real Hacienda para el ejército
español en el siglo XVIII,' 421.
[124] Lisbon's military hospital S. João de Deus kept stocks of English Water. 'Enformes que
deram os Medicos d'esta Corte,' f. 59. English Water and cinchona were also
administered in Lisbon's Hospital da Marinha. See the report by Doctor Antonio
Soares de Macedo Lobo, Lisbon, April 17, 1799. Ibid., 19.
[125] Beltrão Marques, *Natureza em Boiões*, 137.
[126] The royal protomedicato was an official examination board created in 1782, responsible
primarily for the licensing of physicians, and the inspecting of pharmacies. 'ALVARÁ
(CÓPIA), do príncipe regente (D. João), ordenando que a Junta do Protomedicato
ficasse responsável pela preparação das 'águas de Inglaterra' (quina) para consumo das
Armadas Reais e nos domínios ultramarinos,' *Arquivo Histórico Ultramarino*, CU 076 –
REINO RESGATE 20121023 / Cx. 30-A, Pasta 16, Queluz, 1799-06-24. Also cited in
Beltrão Marques, *Natureza em Boiões*, 239.
[127] Johann Christian Anton Theden, *Unterricht für die Unterwundärzte bey Armeen, besonders
bey dem königlich-preußischen Artilleriecorps* (Berlin: Friedrich Nicolai, 1782), 168;
90–91.
[128] Günergun and Etker, 'From Quinaquina to "Quinine Law,"' 49.
[129] Swieten, *The diseases incident to armies*, 50; 54–55; 94.

intermittent' in the late American war, 'even in the southern and more unhealthy provinces of that extensive country'.[130] In the wars of Spanish American Independence, too, cinchona was, together with emetic tartar, storax ointment – a liquidambar-based balsam – and 'Saturn extract' – liquid lead acetate – a staple in army physicians' medicine chests.[131] Additional supplies of cinchona – often thousands of doses – were dispatched or acquired quickly in the event of epidemics in wartime: in the Walcheren Campaign (1809), the British Siege of Havana (1762) and the epidemic of 'marsh fevers, intestinal fevers, and dysentery' that haunted José San Martín's (1778–1850) troops upon disembarking in coastal Huaura.[132]

Cinchona was not only administered curatively. In line with a broader shift in medicine from reactionary treatment to preventive medicine,[133] the bark also appears to have been administered prophylactically by the

---

[130] Jackson, *An Outline of the History & Cure of Fever, Endemic and Contagious*, 276. On medical supplies during the American War of Independence, see Paul E. Kopperman, 'The British Army in North America and the West Indies, 1755–83: A Medical Perspective,' in *British Military and Naval Medicine*, ed. Geoffrey L. Hudson (Amsterdam: Editions Rodopi B.V., 2007), 74. On the prevalence of 'remittents and intermittents' during the revolutionary wars, particularly in the south, see Humphreys, *Malaria*, 28–29; Duffy, 'The Impact of Malaria on the South,' 36–37.

[131] José Fernando de (Marqués de la Concordia) Abascal y Sousa, Virrey del Perú, 'Expediente que la Real Hacienda envía al Marqués de La Concordia, virrey del Perú incluyendo relación de las medicinas proporcionadas por el farmacéutico Luis de Montes, para el auxilio de los escuadrones de guerra, quien solicita el pago por este servicio,' *Biblioteca Nacional del Perú*, Sección Manuscritos – XPB/D27, Lima, 1816. The 51 different medicines that were sent to supply the medicine case kept by the physician of the Squadron of Hussars in May 1816 encompassed 5 *libras* of cinchona (*quina entera*). Luis de Montes, 'Relación de las Medicinas, y utensilios que he entregado al físico del Equadron de Usares, Dn. Geronimo Maria del Aguila, de las Recirculas del Profesor de Farmacia Dn. José Solano, y Razón a continuación de las existentes, en mi poder en estado de exclusión,' *Biblioteca Nacional del Perú*, Expediente que la Real Hacienda envía al Marqués de La Concordia, virrey del Perú incluyendo relación de las medicinas proporcionadas por el farmacéutico Luis de Montes, para el auxilio de los escuadrones de guerra, quien solicita el pago por este servicio, Sección Manuscritos – 1816 – XPB/D27, Lima, 1816. On cinchona provisions for the Spanish American troops, see also Sánchez, 'Clima, hambre y enfermedad en Lima,' 244–45; Arias-Schreiber Pezet Jorge, *Los médicos en la independencia del Perú* (Lima: Editorial Universitaria, 1971), 50–53.

[132] In Huaura, 3,000 of the 5,000 soldiers who had disembarked in Pisco fell ill and pharmacists sent medicines worth 4,000 pesos, including 100 *libras* of cinchona, enough for some 3,000 doses. Sánchez, 'Clima, hambre y enfermedad en Lima,' 244–45. Similarly, during the English siege of Havana, in the context of the Seven Years' War, Vice-Admiral George Pocock spent £1,500 on cinchona bark in January 1763, in the context of an epidemic of yellow fever. McNeill, *Mosquito Empires*, 185. On Walcheren, see Martin R. Howard, 'Walcheren 1809: A Medical Catastrophe,' *British Medical Journal* 319, no. 7225 (1999).

[133] On preventive medicine in the military context, see Erin Spinney, 'British Military Medicine during the Long Eighteenth Century: A Relationship between Preventive

late eighteenth and early nineteenth centuries. In some regiments of the French, Habsburg, British and Spanish armies, cinchona was administered to soldiers to shield them 'against the miasmata' of intermittent fevers and 'other putrid diseases', which were, as contemporaries were acutely aware, invariably so 'far more destructive' to them 'than the enemy's sword'.[134] Already during the Austro-Turkish War (1716–1718), cinchona was said to have preserved those regiments of the Habsburg army stationed near Belgrade – where 'an epidemic of fatal fevers, typical of that country, [...] and a pernicious dysentery [...] did more damage among the troupes than the Turks ever did' – that drank it 'twice or thrice a day' 'dissolved in liquor'.[135] A similar measure purportedly had the most 'happy effect on the French forces' during the Revolutionary Wars.[136] When Spain's imperial troops were sent off to protect Navarra, an area where tertian fevers were endemic, in 1793 during the Revolutionary Wars, its medical provisions comprised ten times as many doses of cinchona as the army had soldiers.[137] Officers, soldiers and other persons who could not 'be expected to have the same knowledge' as physicians were encouraged to self-medicate, and to medicate their subordinates, prophylactically with the bark. Self-help manuals like William Blair's 1798 *The Soldier's Friend*, William Buchan's

and Reactionary Medicine, Supply, and Empire' (master's thesis, University of New Brunswick, 2011), 10; Chakrabarti, *Medicine and Empire*, 43. On preventive medicine, see chapters 18 and 19 in Harry Wain, *A History of Preventive Medicine* (Springfield: Charles C. Thomas, 1970); James C. Riley, *The Eighteenth-Century Campaign to Avoid Disease* (Basingstoke & London: Macmillan, 1989).

[134] O'Ryan, *A Letter on the Yellow Peruvian Bark*, 23. See also Blair, *The Soldier's Friend*, 7–8. For a similar observation by a Portuguese author, see, Nunes Ribeiro Sanches, *Tratado da conservaçam da saude dos povos*, 203.

[135] Salazar, *Tratado del uso de la quina*, 84–85.

[136] 'Los felices efectos que ha hecho la Quina en los exercitos franceses,' *Archivo del Palacio Real*, Papeles del Almacén de la Quina, Caja 22283 / Expediente 7, Madrid, 1797–1798. On earlier cinchona provisions for the French army, see Perez, 'Les médicins du roi et le quinquina aux XVIIe et XVIIIe siècles,' 186.

[137] The two most numerous armies – some 8,000 men taken together – which were to protect Navarra and Guipuzcoa and to invade France through Rosellón were supplied with around 3,397 *libras* – some 1.6 tons – in 1793, the following year with some 9,857 *libras* – around 4.5 tons – and 7,888 *libras* – 3.6 tons – in 1795, amounts that could have yielded 37,200, 104,651 and 83,720 doses, respectively, ten times the number of soldiers. Andrés Turrión, 'Quina de la Real Hacienda para el ejército español en el siglo XVIII,' 421. On Navarra as a site of endemic 'intermittent fevers', see Manuel Gil y Alveniz, *Colección de Memorias Médicas: Contiene una memoria premiada por la Real Academia de Medicina práctica de Barcelona, sobre las epidemias generales de España en los años de 1803 y 4. La práctica de la vacunación en cinco memorias presentadas y aprobadas por la Real Junta Superior Gubernativa de Medicina. Y un tratado sobre la pronta, fácil, segura y económica curación de las calenturas intermitentes por medio de la quina* (Madrid: Iearra, 1820), 206–07.

Domestic Medicine and Gilbert Blane's (1794–1834) 1780 *Short Account of the Most Effectual Means of Preserving the Health of Seamen* uniformly advised sailors and soldiers in 'situations [...] inimical to the health of troops' to take a daily dose of 'half a wine glass [...] of the tincture of Peruvian bark',[138] as an antidote to 'prevent fevers, and other fatal diseases'.[139] Indeed, soldiers and sailors, thus accustomed to and familiar with the bark's regular administration, were often credited with spreading the word about the Peruvian bark at home and abroad.[140] In the Ottoman Empire, for instance, sailors and other travellers were said to have taken 'the cinchona plant to other cities and lands far beyond Constantinople', as the physician Ali Münşî of Bursa (d. 1733) put it, where it then quickly won recognition as a febrifuge.[141]

The same counsel that self-help manuals extended to naval and military men was also given to civilians. Medical advice books designed for the benefit of emigrants in the world's hot climates encouraged their readership to take for 'prevention, a wine glass of an infusion of the bark and orange peel in water', or 'some bark, garlic and rhubarb' infused in brandy, 'or what will prove more effectual, table spoonful of a strong tincture of the bark, in spirits, diluted occasionally with water', every day, morning and evening.[142] Apparently many civilians in thrall to the moist and humid climates of Europe's colonial, commercial and evangelizing entrepôts adopted the habit of taking cinchona preventively, from advice literature or, presumably, one another. Several self-dosing residents and travellers on the island of Martinique, in Ottoman Tripoli, Abyssinian Gondar and the Surinam bush claimed never to have fallen sick from tropical fevers since they had begun taking a daily dose of the bark,[143]

138 Buchan, *Domestic Medicine*, 136. See also Blane, *A Short Account of the Most Effectual Means of Preserving the Health of Seamen*, 33, 42; Blair, *The Soldier's Friend*, 59.
139 Buchan, *Domestic Medicine*, 136.
140 On how 'military men' and sailors in particular were generally 'well placed to gather and exchange information', see Linda Colley, 'Writing Constitutions and Writing World History,' in *The Prospect of Global History*, ed. James Belich, et al. (New York: Oxford University Press, 2016), 166. On the spread of Enlightenment idea[l]s among sailors, see also Manning and Cogliano, 'Introduction. The Enlightenment and the Atlantic.'
141 Aydüz and Yildirim, 'Bursalı Ali Münşî ve Tuhfe-i Aliyye,' 92.
142 Lind, *An Essay on Diseases incidental to Europeans in hot Climates*, 41; 53; 140; 286. On preventive cinchona consumption in antebellum North America, see Bolton Valencius, *The Health of the Country*, 28.
143 On French Martinique, see Harrison, *Medicine in an Age of Commerce and Empire*, 133. On the Surinam bush, see McNeill, *Mosquito Empires*, 74. On Tripoli, see Ramel, *De l'Influence des marais et des étangs sur la santé de l'homme*, 76. James Bruce, in his 'Travels to Discover the Source of the Nile', self-administered 'small dozes of ipecacuanha under the bark', when assaulted with 'dysenteries'. Bruce, *Travels to discover the source of the Nile*, 5, 70.

while colonists in South Carolina were 'always chewing' it 'as a preventative [...] during the fever months'.[144] Historians have long dated the beginnings of cinchona – or rather, quinine – prophylaxis among Europeans in the tropics to the mid-nineteenth century,[145] but the bark had by then long been a fundamental element in a litany of precautions and cares – regimen, relocation to healthier climes, as well as the construction of levees and drainage ditches, forest clearance and crop planting – adopted by settlers, officials and soldiers to prevent contagion and to manage seasoning and acclimation.[146] Men and women who went to live for an extended period of time, or settle permanently, in South Asia, North America, West Africa or the Caribbean were well aware that they inhabited places where their health depended on the 'skill of the physician', as well as their own, to preserve and maintain bodily well-being.[147] To them, cinchona had long become one imperative means to that end.

★★★

The sharp gradients in contours of insalubrity around marshland districts, congested urban areas and warm climates, with which the fortunes of men and women were still so closely aligned in the late 1700s and early 1800s,[148] governed, and grounded, registers of cinchona consumption the world over. Cinchona consumption thus speaks to the world's medical unevenness in the late 1700s and early 1800s – it being interspersed with insalubrious, febrile, environments – but also, and rather eloquently, to its levelness and consistency. It bears eloquent testimony to how men and women from the fenlands of Cambridgeshire to the river valleys of Lima shared ideas about their environs, and about how to preserve and restore bodily well-being within them. By the late 1700s and early 1800s, the remedies, skills and conceptions that belonged to the cure of fevers, the period's most general, and most fatal ailment, had become second nature to marsh folk, townspeople and the many soldiers,

---

[144] The words are those of a 'visitor' to South Carolina, commenting on the prevalence of cinchona consumption among the resident population, 'attacked almost every year by intermittent fevers'. Cited in Peter Wood, *Black Majority* (New York: Knopf, 1974), 76.

[145] Webb, *The Long Struggle against Malaria in Tropical Africa*, 17; 23–25; Chakrabarti, *Medicine and Empire*, 9.

[146] Clark, *A Treatise on the Yellow Fever*, 81; 89. On preventive medicine in British India between 1750 and 1830, see Harrison, 'Disease and Medicine in the Armies of British India, 1750–1830.' On the building of levees and drainage ditches, forest clearance and sanitary measures, to improve health conditions, see Stange, 'Governing the Swamp.'

[147] Bolton Valencius, *The Health of the Country*, 32.   [148] Ibid., 4.

colonists and merchants – Americans, Portuguese, Turks or Dutch –
who went to trade, fight or settle in insalubrious, foreign terrain where
they did not belong. The Peruvian bark was a fundamental element of the
litany of precautions and cares that men and women across the Atlantic
World had at their disposal to preserve or restore bodily well-being in
febrile, insalubrious environs.

5   Harvests of Change

The places, and sites, where the trees grow that yield cinchona bark are not estates, or plantations [...]. The [trees] [...] have grown wild from time immemorial on the humid slopes and [...] hills of the Cordillera Real, interspersed amid countless other tree species, and occupy large swaths of the great jungles and forests. That is where the harvesters will go, some by conjecture, and at a venture as they say, others guided by some report. There are large areas where they will not find one tree of that species, in others they will be so scattered, and distant one from the other, that it is of little use, and not worth their while, to work in these parts. And so they roam through the forest until they reach a place where there is an abundance of these trees all together, or where they are not so dispersed [...] and where they can harvest a considerable quantity of *arrobas*.'
– Clemente Sánchez de Orellana y Riofrío, *Dictamen*, 1776.

In April 1782, the governor of Loja, Pedro Xavier de Valdivieso y Torre, instigated an interrogation of the inmates of the local jailhouse. Most of the prisoners kept under lock and key there were not perpetrators of violence or common thieves, as one would perhaps expect, but had been jailed as debtors of cinchona bark. As one of the prisoners, Mariano Villa, registered in the report as a thirty-year-old 'Indian from the village of Vilcabamba', related under oath, he found himself in jail by order of his lordship the governor because he had been paid in advance for

a *quintal* of cinchona [...] but had only been able to deliver three *arrobas* and thus still owed one *arroba*, which he had been unable to procure, in spite of his assiduous endeavours to do so; he had searched all over the forest where [the bark] was cut, but the hills were exhausted, there was no more of that specific (*espesifico*) to be found. He knew that all of those who were imprisoned with him owed said cinchona, and had been unable to deliver it on account of its notorious shortage.[1]

Mariano Villa's statement was remarkably consistent with those of his fellow inmates, most of them young Indian men from the Loja area.

---

[1] Testimony of Mariano Villa, dated as of April 27, 1782, in: 'Sobre la conservacion de Montes de Cascarilla,' fs. 18–19.

All of them had been contracted as cinchona bark harvesters by merchants or Crown officials, as Clemente Sánchez de Orellana y Riofrío (1707–1782) described it in his 1776 dictum, to venture into the mountains in search of the trees, which grew wild in scattered batches, high up 'on the humid slopes and [...] hills of the Cordillera Real'.[2] Like Villa, his companions had been paid in advance for a specified amount of bark they were expected to bring back from the mountains, but had found only a fraction of the quantities they had received advance payment for.[3] They uniformly related that they had spent weeks or months roaming through the mountains in search of cinchona trees, but in vain. The forests had been 'lumbered and destroyed entirely', they said, and 'there was no way of getting [cinchona] neither nearby nor at a distance'[4].

This chapter is concerned with the changes that the bark's wide recognition and extensive use among geographically disperse and socially diverse societies wrought in cinchona's natural habitat, in the Viceroyalties of New Granada and Peru. It details how the bark's extensive harvest and recruitment to the various corollaries of its processing, passage and commerce affected the human and physical ecology adjacent to its paths. Drawing on extant literature on the Spanish imperial economy, the bark trade and extractive system, the chapter opens with a section detailing how the harvest areas' linkage to an Atlantic market altered these regions' economies and shifted patterns of migration and settlement. In the second part, extending its boundaries beyond human institutions, the chapter studies how the harvest transformed the landscape and ecology of the central and northern Andes. Though the cinchona forests' 'notorious shortage' has long been a common trope in the literature, the chapter contributes new detail to our understanding of the scale and dynamics of forest loss driven by the cinchona trees' extraction. It breaks fresh ground especially in examining the spectre of the cinchona forests' shortage – their being 'lumbered and destroyed entirely', as Mariano Villa and his companions put it – against the backdrop of the emergence of the epistemic possibility of species

---

[2] Clemente Sánchez de Orellana y Riofrío (Marqués de Villa de Orellana), 'Dictamen sobre la conveniencia, ó perjuicio, que estancandosse la Quina pueda resultar a la causa Publica, y comercio de estas Provincias,' *Archivo General de Indias*, Indiferente 1554, n. p., 1776, 947–48; 51. On how cinchona trees grew in 'patches (*manchones*)', see also 'El Presidente Guzman acompaña una representación del Corregidor de Loxa, y del Botanico, encargado del acopio de la Quina (Tomás Ruiz Gómez de Quevedo y Vicente Olmedo y Rodríguez) en que tratan prolijamente del Estado de su Comision,' *Archivo General de Indias*, Indiferente 1556, Quito, 1794-11-21, fs. 120–21.

[3] For more cases of imprisonment, see Moya, *Auge y Crisis de la Cascarilla*, 98.

[4] Testimony of Melchor Bera, dated as of April 27, 1782, in 'Sobre la conservacion de Montes de Cascarilla.'

loss, anthropogenic rarity and extinction around 1800. The chapter returns, in the last part, to the bark cutters themselves. It seeks to understand how the extensive demand for and trade in the bark affected those most immediately concerned in cinchona's ingathering. Parting from the system of advance payment, debt contracts and the threat of imprisonment and loss of freedom that unfolds through the prison records, the chapter adds new detail to our understanding of the harvesters' predicaments and the various forms of corvée – through tribute, servitude or the *reparto* – that came to replace free labour during the late 1700s and early 1800s. The chapter reminds readers, at parting, how plant trade, therapeutic exchange and epistemic brokerage are not extricable from time and space: from landscapes of possession, commerce and demographics, from the distribution and abundance of vegetation, nor from the livelihood, health and fate of its producers. Patterns of consumption invariably begin, as historians have argued, 'with changes to the material world, to physical nature' and society.[5]

The Growth Regions

During the late 1700s and early 1800s, the bulk of cinchona for the world market came from harvest areas in the Quito *Audiencia*, then part of the Viceroyalty of New Granada: between 1750 and 1775, mainly from the aforementioned Loja *corregimiento*, and between 1775 and 1787, from the Cuenca *corregimiento*, joined from around 1785 in some measure by growth regions in the Jaén de Bracamoros *gobierno*, and later, the Riobamba *corregimiento*.[6] From the 1770s, harvest sites in the Viceroyalty of Peru, principally in the Huánuco and Piura provinces,[7] and in the Santa Fé Audiencia – Santa Fé, the Fusagasugá Valley, Ocaña and Santa Marta – in the north of the Viceroyalty of New Granada, temporarily

---

[5] Klingle, 'Spaces of Consumption in Environmental History,' 94. Environmental historians have long argued that particularly modern imperial consumer societies and their extractive and agrarian economies had a profound impact on natural environments. William Beinart and Lotte Hughes, introduction to *Environment and Empire*, ed. William Beinart and Lotte Hughes (Oxford: Oxford University Press, 2007), 1.

[6] Moya, *Auge y Crisis de la Cascarilla*, 16. On Loja, see Petitjean and Saint-Geours, 'La ecomomía de la cascarilla'. On cinchona from Cuenca, see also Jacques Poloni-Simard, *El Mosaico Indígena. Movilidad, estratificación social y mestizaje en el corregimiento de Cuenca (Ecuador) del siglo XVI al XVIII* (Quito: ABYA-YALA, 2006), 397.

[7] On cinchona from Peruvian harvest areas, see in particular the work of Jaime Jaramillo Arango and Miguel Jaramillo Baanante. Jaime Jaramillo Arango, 'Comercio y ciclos económicos regionales a fines del período colonial. Piura, 1770–1830,' in *El Perú en el siglo XVIII. La Era Borbónica*, ed. Scarlett O'Phelan Godoy (Lima: Pontificia Universidad Católica del Perú / Instituto Riva-Agüero, 1999); Jaramillo Baanante, 'El comercio de la cascarilla,' 72. Cinchona trees in Huánuco were apparently discovered in 1776. See Crawford, *The Andean Wonder Drug*, 133–34.

gained popularity, until misgivings about their virtue, identity and efficacy discredited these areas' barks.[8]

Particularly in the Loja and Cuenca *corregimientos*, the oldest and
longest-lasting harvest sites, the trade in cinchona had, by the late
1700s and early 1800s, become a principal sector of the local economy.
In fact, as don Francisco Palacio y Vallejo (b. 1735), mayor (*alcalde ord
(inar)io*) of Loja, and don Felipe Joaquin Xaramillo y Celi (fl. 1770s), its
procurador general, put it in 1779,

> [...] ever since the middle of the preceding century [...], and at present (*en la actual
> presente estacion*), from our own experience, and the knowledge of others
> (*conosimiento que nos asiste*), this city has not had, nor has it, any commerce worthy
> of such a name other than that in the bark, or cinchona (*el de la cascarilla, o Quina*),
> that was discovered now a hundred and thirty years ago, more or less [...].[9]

Similarly, by 1791, the Cuenca government asserted that there was 'no
other commerce' (*otro comercio*) in that *corregimiento*. There was hardly an
individual not 'engaged in this occupation' – the men as well as the
women, as officials noted – and no one who was not, in one way or
another, 'relieved by means of that commerce'.[10] Cuenca and Loja
notables were presumably a little excessive in their assertion – the bark
trade was not quite the areas' 'only [...] commerce'[11] – but there can be
little doubt that cinchona was the principal sector of the Loja economy
during the 1700s[12] and the most lucrative branch of the Cuenca economy.[13] At least during the boom years, the late 1700s and early 1800s,

---

[8] On the (re-)discovery of cinchona forests near Popayán, see Moya, *Auge y Crisis de la
Cascarilla*, 44. On the controversy over barks from the Santa Fé *Audiencia*, remittances of
which were discouraged after 1789, see also Nieto Olarte, *Remedios para el imperio*,
197–206; Salvador Vázquez, 'Mutis y las quinas del norte de Nueva Granada'; Pilar
Gardeta Sabater, *Sebastián José López Ruiz (1741–1832)* (Málaga: Universidad de
Málaga, 1996); Salvador Vázquez, 'Las quinas del norte de Nueva Granada.'
[9] 'Representation' of Francisco Palacio y Vallejo and Felipe Joaquin Xaramillo y Celi,
dated as of July 23, 1779, in 'Expediente sobre el corte de cascarilla en los Montes de
Loxa,' f. 1.
[10] Report of the Cuenca Governorship 1791 (*Informe de la Gobernacion de Cuenca 1791*),
cited in Moya, *Auge y Crisis de la Cascarilla*, 179.
[11] This quote is taken from a report dated as of September 9, 1779, by Matías de Salazar, a
Spanish official. Transcribed in Alfonso Anda Aguirre, *La Quina o Cascarilla en Loja*
(Quito: Universidad Técnica Particular de Loja, 2003), 58. On other sectors of the Loja
economy, see Silvia Palomeque, 'Loja en el mercado interno colonial,' *Revista
Latinoamericana de Historia Económica y Social* 2 (1983).
[12] Kenneth J. Andrien, *The Kingdom of Quito, 1690–1830: The State and Regional
Development* (Cambridge: Cambridge University Press, 2002), 96.
[13] According to Silvia Vega Ugalde, the bark trade was the most lucrative economic activity
in the area. Silvia Vega Ugalde, 'Cuenca en los movimientos independentistas,' *Revista
del Archivo Nacional de Historia*, 6 (1986), 15–16.

cinchona was the entire Quito south sierra's most profitable, most important and, with the exception of some minor gold and silver deposits, its single export product for a world market.[14] By the late 1700s, the bark amounted to 28 per cent of the registered exports that left the area[15] and it accounted for 56 per cent of the total value of goods – which included merchandise as diverse as leather, soap and wax – shipped through the port of Paita, in the Piura region, southward to Callao, or northward to Guayaquil and Panamá.[16] As a matter of fact, historians have argued that the decline of Paita as an export port during the 1780s was related to the rerouting of cinchona exports through the port of Guayaquil, on account of the growth regions from Cuenca southwards to Jaén de Bracamoros being declared exhausted in 1783.[17] Unlike earlier commodities – silver or spices[18] – the bark trade was not at the beginning, or formative, to major trade links, but it did have a role in the permanence or damage, and the blossoming or decline, of commercial routes and entrepôts (FIGURE 5.1).

From the city of Cuenca southwards to Jaén de Bracamoros, the bark trade brought on economic booms, that is, periods of increased growth and prosperity.[19] The Cuenca *corregimiento* became the most dynamic region of the Quito *Audiencia*, as a testament to which it was elevated to the status of a *gobernación* in 1771 and became the seat of a bishopric

[14] Ibid. On the bark's economic relevance as an export product, see also Andrien, *The Kingdom of Quito*, 50.

[15] This refers to the area from Paita to Cuenca in the north and Chachapoyas to the east. Jaramillo Baanante, 'El comercio de la cascarilla,' 88.

[16] Given that the bark was packed in wooden boxes too heavy for mules, it was but rarely taken across the overland route – the Carrera de Lima – which went from Piura southward to Lima through the valleys of Lambayeque and Trujillo. Jaramillo Arango, 'Comercio y ciclos económicos regionales a fines del período colonial. Piura, 1770–1830,' 55.

[17] Ibid. See also Jaramillo Baanante, 'El comercio de la cascarilla,' 74. Highland products made up between a quarter and half of the Guayaquil exports during the 1770s and 1780s and cinchona constituted a large part of these (72 per cent between 1784 and 1788). Carlos C. Contreras, *El sector exportador de una economía colonial. La costa del Ecuador: 1760–1830* (Quito: FLACSO, 1990), 43.

[18] Whereas the spice trade is closely associated with European expansion, the origin of world trade and the establishment of the Manila Galleon trade route are now commonly linked to the Asian demand for American silver. Flynn and Giráldez, 'Born with a "Silver Spoon."'

[19] According to the historian Luz del Alba Moya, it was primarily cinchona and only secondarily textiles that brought about the Cuenca economy's boom – because the value of cinchona exports through Guayaquil was far greater than that of textile exports, and because monetarization occurred through cinchona exports. Also, the region went into crisis after the Cuenca cinchona forests were declared exhausted in 1784 and the commerce in cinchonas from Santa Fé liberalized in 1790. On the south sierra boom on account of cinchona, see Moya, *Auge y Crisis de la Cascarilla*, 178–79; 24; 235; Petitjean and Saint-Geours, 'La ecomomía de la cascarilla,' 38.

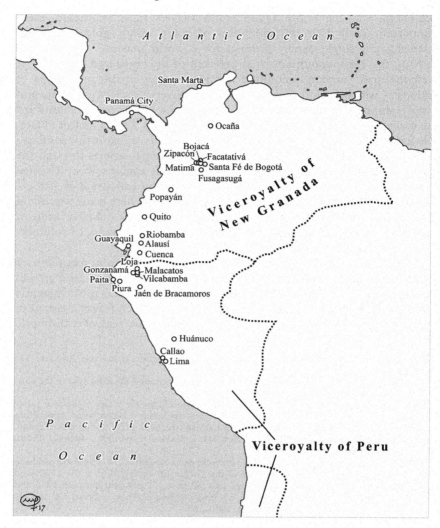

Figure 5.1 Some of the principal harvest areas and transfer sites for cinchona in the Viceroyalties of New Granada and Peru in the late 1700s and early 1800s.

in 1779.[20] As economic historians of the central and northern Andes have argued for some time now, cinchona exports also brought about the monetarization of the region. Given that Cuenca, Loja and Piura

---

[20] Moya, *Auge y Crisis de la Cascarilla*, 235.

merchants sold their produce in Lima, Panamá, Portobello and even Cádiz, where cash was in use,[21] and given that by the late 1700s the Crown advanced money rather than goods, and some merchants gave money as well as merchandise to the bark cutters,[22] for many in the central and northern Andes, the exploitation of cinchona was an unprecedented source of monetary income.[23] About 70 per cent of the total production costs in the bark trade was incurred in the harvest areas themselves, and men and women from these parts, with their integration into wider commercial and monetary circuits, acquired the ability to purchase imported products.[24] By the 1770s, in the Loja *corregimiento* and the Jaén de Bracamoros *gobierno*, the harvest and sale of bark had become one of very few opportunities not just to settle debts but to procure necessary naturalia, in particular garments and 'other kinds of European produce' (*los generos de Europa*).[25] In Cuenca, too, 'all manner of foodstuffs and clothes' (*todo género de Viveres, y Ropas de Castilla*) entered the area as a consequence of the bark trade.[26] Cinchona merchants were commonly not just cinchona export traders, but involved in a series of commercial activities – from muleteering to cotton cultivation, sugar exportation or mining – and often reinvested their monetary earnings, dynamizing other sectors of the economy.[27] Even where merchants and harvesters did not receive cash for the bark, its harvest 'fomented and sustained commerce', since cinchona was in many places equivalent to a currency.[28] In the port of Piura, cinchona was traded against slaves,[29] and even in the capital city of Quito and the port of Callao, where cash was available, transactions were frequently conducted exchanging cinchona for merchandise.[30] As Manuel Hernandez de Gregorio, the King's

---

[21] Ibid., 178.
[22] See Petitjean and Saint-Geours, 'La ecomomía de la cascarilla,' 34–36.
[23] Moya, *Auge y Crisis de la Cascarilla*, 175.
[24] Jaramillo Baanante, 'El comercio de la cascarilla,' 89.
[25] 'Expediente y cartas de José García de Leon y Pizarro,' fs. 232–34; Juan Pablo de Blad, 'Carta a Salvador Rizo,' *Archivo del Real Jardín Botánico*, Real Expedición Botánica del Nuevo Reino de Granada (1783–1816). José Celestino Mutis, Correspondencia, Correspondencia a Salvador Rizo, RJB03/0001/0003/0149, Neiva (Colombia), 1789-10-27.
[26] Report of the Cuenca Governorship 1791 (*Informe de la Gobernacion de Cuenca 1791*), cited in Moya, *Auge y Crisis de la Cascarilla*, 179.
[27] Jaramillo Baanante, 'El comercio de la cascarilla,' 89–90.
[28] 'Expediente y cartas de José García de Leon y Pizarro,' fs. 232–34; Blad, 'Carta a Salvador Rizo.'
[29] Martin Minchom, 'The Making of a White Province: Demographic Movement and Ethnic Transformation in the South of the Audiencia de Quito (1670–1830),' *Bulletin de l'Institut français d'études andines* XII, no. 3–4 (1983), 36.
[30] Moya, *Auge y Crisis de la Cascarilla*, 179; Petitjean and Saint-Geours, 'La ecomomía de la cascarilla,' 35–37.

Apothecary, phrased it in an 1804 memorandum, cinchona commerce had transformed the harvest areas, 'employing its wretched inhabitants, dressing them with clothes from Europe and the country, advancing agriculture and industry, [and] nourishing the population'.[31]

An important corollary of the growth regions' increased prosperity was demographic – the in-migration of hands and, occasionally, their kin, as involvement in cinchona trade became part of the portfolio of opportunities available to Andeans. For much of the seventeenth century, the economy of the Quito *Audiencia* had relied principally on the textile sector – it had been a secondary regional market supplying the silver mining zones in the Viceroyalty of Peru with woollens. The Quito economy's place in the colonial order eroded with the late seventeenth century, owing principally to the Crown's decision to allow the introduction of cheap European cloth, which undercut Quiteño woollens.[32] In contrast to the declining north-central zone, the agricultural economy of the south sierra experienced steady growth from the 1690s, with many estates, particularly in the east and south-east, specializing in organizing the harvesting of cinchona.[33] As textile manufacturing declined, many Andeans migrated south, where vacant land, lower tribute assessments and the promise of labour in cinchona harvest and upstream sectors – carpentry to manufacture the wooden chests, tailoring to cut to size the cowhides and muleteering to transport the bark to the ports[34] – worked as 'a magnet for the poor and dispossessed from the northern provinces'.[35] While the population of the south-central highlands of the Quito *Audiencia* experienced an overall loss of about 15 per cent in the second half of the eighteenth century, that of Loja grew of the order of 20 to 40 per cent between 1780 and the onset of the struggles for independence[36] – a population growth historians have attributed to inter-regional in-migration attracted by, among other factors, cinchona

[31] Hernandez de Gregorio, 'Memoria,' 1034.
[32] Andrien, *The Kingdom of Quito*, 16. According to contemporary observers, cinchona commerce aided the region as the *obrajes* went into decline. Moya, *Auge y Crisis de la Cascarilla*, 179.
[33] Andrien, *The Kingdom of Quito*, 80; 95.
[34] Moya, *Auge y Crisis de la Cascarilla*, 140–48. See also Petitjean and Saint-Geours, 'La ecomomía de la cascarilla,' 28.
[35] Andrien, *The Kingdom of Quito*, 33; 94. See also Jaramillo Baanante, 'El comercio de la cascarilla,' 89.
[36] By the 1770s, the Loja province had some 23,000 inhabitants: a majority indigenous population (53.9 per cent), as well as significant populations of creoles and mestizos (23.6 per cent) and free blacks (22.6 per cent). Martin Minchom, 'Demographic Change in Eighteenth-Century Ecuador,' in *Equateur 1986*, ed. D. Delaunay and M. Portais (Paris: ORSTOM, 1989), 185.

harvest labour.[37] By the late eighteenth century, over 75 per cent of the Andean population in both Cuenca and Loja consisted of emigrants – registered either as *forasteros* or as *coronas*.[38] The bark trade also changed the face of the south sierra in terms of the value and tenure of land. By the late 1700s and early 1800s, cinchona trees still prospered only in their natural habitat: at elevations from roughly 1,600 to 2,400 metres on the eastern slopes of the Andes,[39] in rough, stony and uneven terrain[40] that was 'entirely impassable on horseback by its nature (*por naturaleza*), and hardly less so barefoot (*a pie desnudo*)'.[41] These hitherto unsolicited, mountainous and inhospitable territories, of little or no agricultural productivity,[42] gradually came to be valorized, and coveted, with the onset of the cinchona export boom.[43] The territories' sudden valorization not only brought on a wave of litigations about land ownership, demarcation and licensing in the area.[44] The very expansion of large, landed estates observable during the late 1700s and early 1800s, at the expense of previously ownerless land, or smaller agrarian properties (*minifundios*) held by less wealthy, or

[37] Alongside cinchona harvest, stock-raising and mule-rearing allowed the region to attract a certain degree of in-migration, albeit a modest one. Ibid., 183–84. Other historians have argued, too, that the Loja bark trade brought about in-migration. Petitjean and Saint-Geours, 'La ecomomía de la cascarilla,' 18.

[38] Andrien, *The Kingdom of Quito*, 41; 48. On migration and the category of the *forastero* in Andean colonial history, see Sarah Albiez-Wieck, 'Indigenous Migrants Negotiating Belonging: Peticiones de cambio de fuero in Cajamarca, Peru, 17th–18th Centuries', *Colonial Latin American Review* 26, no. 4 (2017), 483–508.

[39] For a discussion of the divergent contemporary opinions on the trees' minimum and maximum height, from Francisco José de Caldas to Alexander von Humboldt, see Adolph Pleischl, *Über die Nothwendigkeit, Fürsorge zu treffen, dass der leidenden Menschheit der nöthige Bedarf an Chinarinden und an den daraus bereiteten chemischen Präparaten auch in der Folge sichergestellt werde* (Wien: Druck von Carl Gerold's Sohn, 1857), 27.

[40] For contemporary voices on the terrain, see Pedro Cevallos, 'En el Pueblo de Malacatos en dies y siete dias del mes de Septiembre de mill setecientos setenta y tres años para la Informacion dicha hise compareser à Pedro Cevallos Natural del Pueblo de Bilcabanbba,' *Archivo General de Indias*, Quito 239, Malacatos, 1773-09-17.

[41] 'El Presidente Guzman acompaña una representación,' fs. 120–21.

[42] Moya, *Auge y Crisis de la Cascarilla*, 242.

[43] Cevallos, 'En el Pueblo de Malacatos,' 125. See also Moya, *Auge y Crisis de la Cascarilla*, 115.

[44] On licence disputes, see 'Licencias de Cascarilla,' *Archivo Nacional de la Historia*, Quito, Fondo General, Serie Cascarilla, Caja 4, Expediente 3, Quito, 1793-05-22 – 1793-08-27. For a litigation, see, for example, 'Juan Camacho, en el articulo que sige, contra Dn. Joseph Sanchez Barragan, mi suegro, sobre el derecho de cascarilla, que por accion de dn. Baleriano Guapulena, cacique del pueblo de San Miguel de Chimbo, tengo adquirida posecion en los Montes de Iluicachiuru pertenecientes a la Parzialidad de sus Yndios, paresco ante VM, por la persona que tiene mi Poder, que presento y juro,' *Archivo Nacional de la Historia*, Quito, Fondo General, Serie Cascarilla, Caja 2, Expediente 2, n.p., 1778-06-23.

politically less influential, citizens and Indian communities, was motivated, as historians have argued, by the landowners' attempts to lay claim to bark-growing forests.[45] And indeed, whereas the individuals implicated in the harvest were 'generally the poor', as Sánchez de Orellana y Riofrío put it rather candidly in his 1776 report[46] – Indians, in the majority, but also 'mestizos', whites and, occasionally, 'mulattos'[47] – the five or six principal Loja cinchona merchants were, at the same time, among the foremost landowners in the area.[48] In Cuenca, too, many of the principal landowners played an important part in the cinchona export trade.[49] As Miguel García de Cáceres denounced in 1770, the wealthy and 'powerful of the Provinces' (los poderosos de la Provincias) bought 'all the cinchona forests, that have been and are being discovered [...] at very low prices, [...] with the design of harvesting on their own account the bark they produce'.[50] The Loja and Cuenca merchants and, even more so, the merchants based in the port cities of Piura as well as Lima export traders invariably reaped the largest share of the profits.[51] By 1775, at a time when merchants paid the harvesters 22 reales for an arroba, the bark was re-sold at Piura or Paita at 36 reales per arroba, and at 88 reales per arroba in Lima.[52] What is more, the same merchants also came to seize some of the principal bark-growing territories and with it the bulk of the bark trade, reducing the Indians to 'labourers that work for low wages' (peones que trabajan por un vil jornal).[53]

---

[45] Moya, Auge y Crisis de la Cascarilla, 124–29.

[46] Sánchez de Orellana y Riofrío (Marqués de Villa de Orellana), 'Dictamen sobre la conveniencia, ó perjuicio, que estancandosse la Quina pueda resultar a la causa Publica, y comercio de estas Provincias,' 952.

[47] See, for instance, the sales lists composed between March 2 and April 20, 1780, by Sebastián José López Ruiz (1741–1832) – the official charged with the abbroachment of cinchona for the Royal Pharmacy in the Santa Fé Audiencia between 1778 and 1787, which record the sellers' caste. Sebastián José López Ruiz, 'Razon de los sugetos, que desde el dia 2 de Marzo, asta el 20 de Abril del año corriente de 1780, me han vendido Quina, y se la hé comprado a cuenta de la Rl. Hazienda en esta ciudad; con expresion de sus arrobas, y libras, el justo valor, que, por ellas les he satisfecho, de los nombres, y apellidos de los que me la han trahido; los Pueblos, y territorios, de donde me han asegurado ser vecinos; y de los Montes, de que igualmente me han dicho que la han sacado,' Archivo General de Indias, Indiferente 1554, Santa Fé, 1780-03-02 / 1780-04-20.

[48] Moya, Auge y Crisis de la Cascarilla, 80–83; 156. See also Anda Aguirre, La Quina o Cascarilla en Loja, 58.

[49] Andrien, The Kingdom of Quito, 96.

[50] Report by Miguel García de Cáceres, cited in Jaramillo Baanante, 'El comercio de la cascarilla,' note 67.

[51] Ibid., 87. 'Few' were the 'merchants engaged in this commerce, as one contemporary observer noted. Santisteban, 'Copia de Carta,' 787.

[52] Jaramillo Baanante, 'El comercio de la cascarilla,' 68.

[53] Report by Miguel García de Cáceres, cited in ibid., 79. Before 1768, the bark was harvested without restriction in the highland forests. The trees grew mostly on hills

The Spectre of Extinction

The most 'injurious abuse' the harvest workers were culpable of, wrote
Francisco José de Caldas (1768–1816) in his 1808 memorandum on the
state of the cinchona trees, 'in general, and in particular those of Loja',
was that of removing the trees' bark by means of 'barbarous', destructive
techniques that caused them to wither and die, entirely 'careless about
the future' of the tree populations, such that by that time one found but
with great difficulty a cinchona sapling, or plant, in the area around Loja.
Many of the harvesters, Caldas grieved,

> excorticate the tree, break the branches in the most rustic and gross manner [to] take
> the bark and [thereby] render that individual unusable forever, for, thus mistreated,
> it inevitably dries up. Others, the first thing they do is to fell the tree at the base, a
> mindless practice, though less detrimental than the previous one. The stump
> regenerates into two, three or sometimes, five sprouts (*el tronco principal arroja dos,
> tres, y algunas veces, cinco renuevos*). It is to this beneficial natural regeneration that we
> owe the trees that provide his Majesty, and our pharmacies [with the bark]. Without
> it, they might perhaps [already] have extinguished the species.[54]

Francisco José de Caldas was an exceptionally well-connected and well-
travelled man – he had worked as a peddler in the Andes before training
as a botanist – and his observation that the harvesting practices adopted
by bark cutters were destructive, that their carelessness, greed and ignorance were to be blamed for the trees' destruction,[55] echoed what was
conventional wisdom at the time, among colonial officials and bark
merchants as well as 'experienced harvesters'. Bark cutters did indeed
either, to be able to excorticate more conveniently, fell the trees directly
or decorticate the entire trunk of cinchona trees, and that practice contributed, as another contemporary phrased it, to 'crippl[ing] the [tree]
forever' and, almost inevitably caused it to dry up, wither and die.[56]
As one observer noted in 1779, but a very small minority of trees, 'those

that 'had no known owner' and as such belonged to the King (*realengos*). Jijon y León,
'Recomendaciones.' On the ownership of cinchona-growing hills, see Moya, *Auge y
Crisis de la Cascarilla*, 113–15.

[54] Caldas, *Memoria*, 14.

[55] While Caldas blamed the 'the brute bark-cutter' (*bárbaro cascarillero*), others criticized
the workers' 'greed' (*codicia*). Ibid; Sobre la conservacion de Montes de Cascarilla,' fs.
2–3.

[56] 'Sobre la conservacion de Montes de Cascarilla.' See also Pedro Xavier de Valdivieso y
Torre, in a report to José García de Leon y Pizarro, dated as of December 22, 1779, in
'Expediente sobre el corte de cascarilla en los Montes de Loxa,' f. 22. On destructive
harvesting practices, see also Ruiz López, *Quinología*, 14. On the same destructive
harvesting practices – the cinchonas' 'wasteful exploitation', as contemporary observers
phrased it – see also Moya, *Auge y Crisis de la Cascarilla*, 134; Petitjean and Saint-Geours,
'La ecomomía de la cascarilla,' 32–33.

that had escaped standing, and with some part of their bark', survived a first harvest to regrow a second layer of bark.[57]

The exact number of trees that withered and died as a consequence of destructive harvesting practices eludes us, of course, but we can approximate the scale and magnitude of the destruction. If we are to believe contemporaries' estimates, a 30- to 40-year-old cinchona tree at the time yielded around 3 to 4 *arrobas* – that equals some 40 to 50 kilograms – of bark. A younger one – say, 5 or 6 years old – bore only some 12 to 14 *libras*, that is, some 6 kilograms.[58] Calculating with these upper and lower bounds, and with a moderate estimate of 400 tons of free, legal exports per annum, these would have occasioned the destruction of between 8,000 and 70,000 trees. These figures apparently escalated at times, owing to changing consumer tastes and fashions. Miguel García Cáceres (fl. 1779–1785), the governor of the Jaén de Bracamoros *gobierno*, lamented in 1785 that the current preference for fine, thin bark in commerce was to be held accountable for excessive tree mortality. With a robust, regular-sized tree yielding on average some 5 *libras* of thin bark, García Cáceres calculated, it was 'necessary to fell 100,000 large trees to collect 20,000 *arrobas* of thin (*fina*) cinchona', some 230 tons. If thick bark, or crust, were also used it would only have required 34,000 trees, and it would not have been necessary, García Cáceres grieved, 'to miserably destroy 66,000 just to satisfy the whim of the merchants of Europe'.[59] Not only were tens of thousands of cinchona trees destroyed every year, close observers soon apprehended that the recovery of tree populations – the natural regeneration Caldas pinned his hopes on – might be slower than anticipated. As Pedro Xavier de Valdivieso y Torre alerted in 1779, one was not to expect a tree to mature 'in the short period or two or three years', as did some of his contemporaries, or even to 'yield marketable bark' (*se sazona su corteza*) in five or six years. Rather, according to his observations, it took saplings more than two decades to become full-grown trees.[60] Also, while felled trees did often regenerate

---

[57] Pedro Xavier de Valdivieso y Torre, in a report to José García de Leon y Pizarro, dated as of December 22, 1779, in 'Expediente sobre el corte de cascarilla en los Montes de Loxa.'

[58] These estimates roughly agree with those given by Pedro Javier Valdivieso y Torres, in his 1779 report to José García de Leon y Pizarro, who wrote that a 'perfectly seasoned' (*su perfecta sazón*) tree yielded 'one or two *arrobas*'. Ibid., 22. See also Caldas, *Memoria*, 16.

[59] Cited in Moya, *Auge y Crisis de la Cascarilla*, 195–96. See also ibid., 54. On changing preferences for thin and thick, bark, see also Crawford, *Empire's Experts*, 179–80.

[60] Pedro Xavier de Valdivieso y Torre, in a report to José García de Leon y Pizarro, dated as of December 22, 1779, in 'Expediente sobre el corte de cascarilla en los Montes de Loxa,' f. 22.

into sprouts, harvesters were frequently censured for not allowing the sprouts time to grow and thicken, but instead decorticating young trees, 'not sparing even the most tender sapling' (*no perdonan el mas tierno arbolillo con el trato*).[61] That practice was considered not only unprofitable – ten young trees taken together yielded but one or two *libras* – but detrimental to the recovery of tree populations.[62] Deforestation was, like many other ecological changes, unanticipated and unintended by humans. It was a result of the partnership humans formed with the trees' material potentiality, their fragility, that pushed them in directions they neither envisioned nor intended.[63]

Though cinchona trees had, by all accounts, abounded for centuries on the cloud-swept, precipitous eastern slopes of the Andes from the late seventeenth century onwards, contemporaries occasionally observed and expressed unease about alterations in the distribution and abundance of cinchona tree populations on the hills around the city of Loja – the first, and, for over a century, practically the only harvest site.[64] Observations of the trees' finiteness and a sense of their impending scarcity were unobtrusive and rare, however, until the mid-eighteenth century. As Antonio de Ulloa (1716–1795), a member of the Charles-Marie de La Condamine expedition (1735–1745), who stayed in Loja in 1736, noted, 'there were dense and extensive forests' of cinchona trees, even though it was sometimes acknowledged that there was 'much diminution', since too little care was bestowed on sowing young trees, and 'the number of those that regenerate[d] d[id] not correspond to the number that [was] felled'.[65] Loja's cinchona forests continued to be seen by most as 'dense and extensive' into the 1750s, a time of veritable gold rush, when hundreds of bark cutters, responding to high market demand and exorbitant bark prices, went to the bark-growing regions and often brought back hundreds of kilograms each in one season. Back then, the bark was not even 'measured in *arrobas* but in *petacas* of six *arrobas* and a half', as one contemporary phrased it, and local legend had it that some of the boldest harvesters gathered together up to 'fifty *petacas*, all from

---

[61] Pedro Xavier de Valdivieso y Torre, in a report to José García de Leon y Pizarro, dated as of February 21, 1782, in ibid., f. 2–3.

[62] For this practice in the Loja area, see 'Sobre la conservacion de Montes de Cascarilla,' fs. 2–3. See also Moya, *Auge y Crisis de la Cascarilla*. For the same practice in Huánuco, see Petitjean and Saint-Geours, 'La ecomomía de la cascarilla,' 32–33.

[63] For the idea that materiality often pushed humans in directions they neither envisioned nor intended, see Timothy LeCain, 'Against the Anthropocene. A Neo-Materialist Perspective,' *International Journal for History, Culture and Modernity* 3, no. 1 (2015).

[64] Boumediene, *La colonisation du savoir*.

[65] Antonio de Ulloa, *Viaje a la América meridional*, ed. Andrés Saumell (Madrid: Historia 16, 1990 (1748)), 416.

Uritusinga' – 3.7 tons of bark – while 'other rural folk received money for four to six *petacas*', 300 to 450 kilograms.[66] With the onset of the export boom and massive exploitation from the 1750s, however, the depletion of cinchona tree populations progressed quickly. By the 1760s, colonial officials, harvesters as well as merchants noted the depletion of some particular mountain slopes. Fernando Calderón, a servant at a local hacienda who had been ordered to explore the mountainside of Las Palmas near Loja in 1769, related to officials he had not seen 'one single mature cinchona tree (*que fuese de buena sason*), only some piles of cut and decorticated branches', and several young cinchona trees, not yet fit for harvesting. There were no more hills that had 'good cinchona adjacent to this city [of Loja]', Calderón told Loja officials, and 'all the bark that people bought, and sold in it [nowadays], was from [the hills of] Cajunuma, and Uritusinga'.[67] Deforestation advanced with the harvesters. By 1782, it was 'commonly known' (*consta, por Vos comun*) that cinchona trees were 'very scarce' on the hills of Uritusinga.[68] Soon, Spanish colonial officials began to express unease about a decline in exports, owing not to a decrease in demand but to an actual shortage of supply. By 1779, the hills of 'Boquerón, Ama, Yunza, Solomaco, Guaycopamba, and San Miguel de la Tuna' in the Loja province – 'beyond comparison larger than Uritusinga' – taken together yielded less than a hundred *quintales* – some 4.6 tons – per annum.[69] By the late 1770s and early 1780s, there were still some merchants, landowners and officials who claimed that cinchona supplies were abundant and inexhaustible, but an awareness of the vulnerability, scarcity and fragility of cinchona tree populations had already come to supersede and supplant the earlier vision of a land of plenty in official as well as popular discourse.[70]

[66] Pedro Xavier de Valdivieso y Torre, in a report to José García de Leon y Pizarro, dated as of December 22, 1779, in 'Expediente sobre el corte de cascarilla en los Montes de Loxa,' f. 22.

[67] 'Expediente tramitado desde Febrero de 1768 hasta Febrero de 1769 por el Virrey de Nueva Granada Don Pedro Mesia de la Zerda, el Secretario de Camara y Gobierno del Virreynato Don Francisco Silvestre, el Presidente de Quito Don Jose Diguja, el Corregidor y Justicia Mayor de Loja Manuel Daza y Fominaya, el Cabildo de Loja, y Don Pedro Javier de Valdivieso y Torres sobre la clasificacion de la mejor cascarilla para enviarla a la Real Botica de Su Majestad,' *Archivo Nacional de la Historia*, Quito, Fondo Especial / Fondo Presidencia de Quito, Caja 24, Volumen 66, Documento 2858, Loja, 1768-02 / 1769-02, 111.

[68] Ibid., f. 111.

[69] 'Sobre la conservacion de Montes de Cascarilla de la Prov.a de Loxa y proveymiento de este Genero para la Real Botica.' By the 1780s, the amount of bark in the annual shipments to the Royal Pharmacy was on the decline owing to the trees' scarcity. 'Expediente sobre el corte de cascarilla en los Montes de Loxa.'

[70] Pedro Xavier de Valdivieso y Torre, in a report to José García de Leon y Pizarro, dated as of December 22, 1779, in ibid., 22.

Too notorious was the 'decline of the hills where the fine cinchona grows', as a group of prominent Loja residents phrased it in 1782, 'owing to the quantities that had been extracted from them'.[71] The most disquieting and disturbing reports stressing resource limits came from the harvesters themselves. Echoing the testimonies of Mariano Villa and his fellow prisoners about how they had spent many days in the forests without discovering a single cinchona tree, about how the hills had been 'lumbered and destroyed entirely',[72] many bark cutters complained about the ever greater difficulty and laboriousness implicated in searching for the bark 'in the roughness of the hills'.[73] Of the several hundred men who entered the forests every year, many returned without having found a single plant.[74] By 1788, Tadeo Celis, the priest of Malacatos and Vilcabamba, the parish that comprised some of the principal cinchona harvest areas – Cajunuma, Uritusinga and Boquerón – stated that 'all investigations of this branch [of trade] had served to demonstrate' the same: the 'near annihilation' of all of the cinchona forests of Loja.[75] By 1809, preoccupation with the trees' scarcity had developed a sense of urgency and finality intense enough for Caldas to evoke even the spectre of the trees' extinction. Within some 16 to 20 square *leguas*, some 67.2 to 84 square kilometres, of Loja – Malacatos, Vilcabamba and Gonzanamá – Caldas intimated, there was now hardly a 'single [cinchona] tree'.[76]

From the late 1700s onwards, Creole and Spanish contemporaries became increasingly eloquent on the necessity of measures responding to the threat of the cinchonas' extinction. Whereas some proposed forest restoration – the replenishing of tree populations by the systematic

---

[71] Report by the Loja Cabildo, dated as of March 27, 1782, in 'Sobre la conservacion de Montes de Cascarilla,' 15.

[72] Testimony of Melchor Bera, Loja, April 27, 1782, in ibid.

[73] Spanish archival documents retain the voices of bark cutters since Crown officials occasionally conducted questionings among them in their efforts to contain commerce, increase revenue or locate tree populations. Alexandro Toledo, 'En el Pueblo de Malacatos en dicho dia mes y año para la dicha Informacion, comparecio Alexandro Toledo,' *Archivo General de Indias*, Quito 239, Malacatos, 1773-09-17, 124.

[74] 'Testimonio de los Autos en que se comprehenden varios Informes y Diligencias practicadas en virtud de Real Cedula, sobre si sera, o no combeniente el Estanco de la Cascara Quina,' *Archivo Nacional de la Historia*, Quito, Fondo General, Serie Cascarilla, Caja 1, Expediente 11, Santa Fé, 1777-12-10, f. 26.

[75] Report by Tadeo Celis, dated as of May 16, 1788, cited in 'Expediente obrado por el Corregidor de Loxa en satisfaccion de la Real Orden de 24 de Noviembre de 1787 acreditando la imposivilidad de cumplimentarla por la suma escasez de Cascarilla de aquellos Montes que no permiten llevar a devido efecto la Ynstruccion formada a 17 de Marzo de 1773,' *Archivo Nacional de la Historia*, Quito, Fondo General, Serie Cascarilla, Caja 3, Expediente 9, 10-V-1788, Malacatos, 1788-05-16 / Loja, 1788-07-12, 4.

[76] Caldas, *Memoria*, 4; 13.

planting of saplings – other colonial officials forwarded conservationist schemes, the restriction of clearance in endangered forest areas, enforced with the help of special guards, to 'avoid the cutting of saplings' (*evitar el corte de la tierna*).[77] Plans for the setting up of a cinchona plantation system to halt the plant's 'ruination' and lessen the 'immense difficulties' posed by its harvest were frequently voiced, but like forest restoration or conservationist measures, plantations were never, it would seem, successfully implemented on a significant scale.[78] The chief, though by all accounts temporary, solution pursued by the Crown and merchants was to identify new harvest areas,[79] and in several of these – in the Cuenca *gobernación*, the environs of Santa Fé and part of the Huánuco province – the cycle of boom, overexploitation and scarcity repeated itself.[80] It did so even in the territories of the Viceroyalty of Brazil where, in a short-lived mini-boom, trees taken for cinchona were found to 'grow wild' (*cresce [...] espontaneamente*) and where 'people ran off to harvest wherever they could' (*correrão os Povos a tirala aonde cada hum podia*)', such that the forests were soon at risk of being 'ruined'.[81] By the late 1700s and early 1800s there was considerable controversy both over the boundary and confines of cinchona via-à-vis other plants and over the varieties cinchona was to encompass – the plants that belonged to 'the said species

---

[77] 'Sobre la conservacion de Montes de Cascarilla,' fs. 2–3; 13–14.

[78] Caldas was persuaded that only the systematic 'planting' or, rather, establishing of 'plantations' (*plantíos*) of cinchona trees would 'halt the ruination, or at least immense difficulties' encountered in its gathering, but lamented that all efforts undertaken in this vein had so far failed, owing to the 'ignorance' that prevailed among 'the inhabitants of Loja'. Caldas, *Memoria*, 15. For a similar proposal, see Ruiz López, *Quinología*, 15. Matthew Crawford has argued that another reason for the failure to implement a plantation system was royal officials' persuasion that the trees were not amenable to cultivation and a broader understanding of the Andes as a 'space of extraction rather than cultivation'. Crawford, *The Andean Wonder Drug*, 74.

[79] Crawford, *Empire's Experts*, 187.

[80] Several of the principal Cuenca harvest areas were declared exhausted after 1783 to 1784. Moya, *Auge y Crisis de la Cascarilla*, 132–33; 78–79; Petitjean and Saint-Geours, 'La ecomomía de la cascarilla,' 38. By 1787, upon their second visit to Huánuco, Hipólito Ruiz López and José Antonio Pavón reported on the imminent destruction of the cinchona forests. Crawford, *The Andean Wonder Drug*, 134. On the exhaustion of the cinchona forests in the environs of Santa Fé, see Salvador Vázquez, 'Las quinas del norte de Nueva Granada,' 425.

[81] This was the case in the captaincy of Piauí, where trees taken for cinchona were discovered in 1778. By 1798, more than a thousand *arrobas* – some 11 to 12 tons – had been gathered, and according to the governor, the forests were at risk of being 'ruined'. D. João de Amorim Pereira. 'CARTA do (governador do Piauí), D. João de Amorim Pereira, á rainha (D. Maria I).' *Arquivo Histórico Ultramarino*, 016 – PIAUÍ – CATÁLOGO DE DOCUMENTOS MANUSCRITOS AVULSOS / Cx. 21, D. 1074, Piauí, 1798.

cinchona' (la d[ic]ha especie de cascarilla)[82] or, in the eyes of other contemporaries, the kinds and number of species that were to be contained in the genus Cinchona[83] – and, accordingly, also great disagreement and dispute about the extension of the tree's habitat.[84] In all of this, variations in how contemporaries evaluated the urgency of scarcity and the possibility of extinction closely correlated with uncertainty about the confines, solidity and definiteness of the species, as well as its potential areas of growth. The concept and possibility of extinction are contingent both on accurate botanical knowledge and either certainty about a species' endemism or the ability to contextualize globally. As a matter of fact, the greatness of poorly explored territory, where supposedly extinct species might still be found undetected, was – other than ideas about the mutability of species – a key argument against the reality of extinction around 1800[85] (FIGURE 5.2).

We tend to associate unsustainable harvest practices, the overexploitation of vegetable raw materials and anxieties about species' extinction with the late modern era, but, as historians have argued for some time now, at least some Prussian and French, as well as, it would seem, Spanish and Portuguese American naturalists of the late eighteenth and early nineteenth centuries, were acutely aware not only of the possibility of alterations in the distribution and abundance of vegetation but also of that of plant species in particular, as a result of economic forces acting upon environments.[86] As a matter of fact, by the time Caldas invoked the spectre of the extinction of cinchona as a species, the very possibility of species loss and extinction – the conception that the world of plants was fragmented into a series of discrete, fixed and stable ontic unities that could appear or vanish forever – had only just arisen.[87]

---

[82] 'Sobre el acopio de la Quina de los Montes de Loxa Callysalla,' ff. 34–36; 'Expediente sobre el corte de cascarilla en los Montes de Loxa,' f. 1; 'Don Juan Gavilanes de este vecindario, pobre de solemnidad declarado por la soberania de S.A. segun dro. ante V. paresco y digo,' Archivo Nacional de la Historia, Quito, Fondo General, Serie Cascarilla / Caja 5, Expediente 11, Quito, 1809-04-19; Informe de la Contaduría, 835v–36r.

[83] See, for instance, Mutis, Instrucción formada por un facultativo existente por muchos años en el Perú, relativa de las especies y virtudes de la quina, 416–17; López Ruiz, 'Representación,' fls. 117–18.

[84] See, for instance, Caldas on the subject of the limiting factors – altitude and degrees of latitude, in his view – regarding the trees' growth. Caldas, Memoria, 7–9.

[85] Grove, Green Imperialism, 350; 55.   [86] Ibid.

[87] Ibid., 245; Barrow, Nature's Ghosts, 23–26. Georges Cuvier (1769–1832) between 1796 and 1806 completed widely publicized studies of living and extinct elephants that provided almost irrefutable evidence for the reality of extinction. Ibid., 40–41; Martin J. S. Rudwick, The Meaning of Fossils. Episodes in the History of Palaeontology (Chicago: University of Chicago Press, 1976). The question of extinction was linked to that of the fixity of species, a doctrine upheld by Georges Cuvier, against several of his

Figure 5.2 Watercolour of a cinchona tree (*cascarilla*, in colloquial Spanish), by a native artist. Baltasar Jaime Martínez Compañón (1782–1789), *Trujillo del Perú or Colección original inédita de mapas relativos al obispado de Perú; retratos en colores y dorados de Arzobispos, Vireyes y otros personajes del Perú; planos de ciudades; cuadros sobre lenguas indígenas*, Estampa IX, Vol. III
© *Patrimonio Nacional (España)*.

Caldas's lamentation certainly had little in common with our present-day understanding of species loss, deforestation and extinction as catastrophic, wanton events that signify a potentially permanent loss to a standing, coveted, even normative biological diversity, or injury to the tropical forest as wilderness, a place of intrinsic, romantic or ecological value.[88]

contemporaries. David Sepkoski, 'Extinction, Diversity, and Endangerment,' in *Endangerment, Biodiversity and Culture*, ed. Fernando Vidal and Nélia Dias (New York: Routledge, 2015), 63–64.

[88] Sepkoski, 'Extinction, Diversity, and Endangerment,' 63–64. On ideas about the tropical forest as 'wilderness', a place of intrinsic, 'romantic' or ecological, value, see William Cronon, 'The Trouble with Wilderness; or, Getting Back to the Wrong Nature,' in *Uncommon Ground: Rethinking the Human Place in Nature*, ed. William Cronon (New York: W. W. Norton, 1995); José Augusto Pádua, 'Tropical Forests in Brazilian Political Culture. From Economic Hindrance to Endangered Treasure,' in *Endangerment, Biodiversity and Culture*, ed. Fernando Vidal and Nélia Dias (London: Routledge, 2016); José Augusto Pádua, 'European Colonialism and Tropical Forest Destruction

Rather, he resorted to the emergent concepts of plant rarity and closeness to extinction in a utilitarian reading, one tied to specific medical and economic purposes assigned to the bark – as singular[89] remedy, source of income and tax revenue.[90] Still, Caldas shared with Spanish officials, Creole merchants and Indian bark cutters an acute awareness of, and preoccupation with, alterations in the distribution and abundance of the cinchona tree, an anxiety with it becoming scarce, or even becoming extinct. The bark's shortage in Loja was quite as notorious as Mariano Villa had declared it to be in 1782.[91]

The Bark Cutters

By the late 1700s, in contrast to the earlier colonial period, in both the Quito and Santa Fé *Audiencias*, the official, general policy was that Indian labour was free – that is, that the Indians were allowed to sell their labour at will and be paid for it.[92] And indeed, in the late 1700s and early 1800s, many bark cutters, in Cuenca, Loja and Santa Fé alike, would go into the forests of their own accord and, on their return, sell the bark either in their home villages or on the streets and markets of the capital cities to the highest bidder – to official cinchona commissioners or one of the 'countless' 'small traders' (*traficantes compradores*) who made money out of buying and reselling the bark.[93] As merchants and Crown officials often lamented, these harvesters were under no obligation to anyone, but 'free (*suelta*), to work when they chose to, and for whom it pleased them (*a quien es de su agrado*)'.[94] From the sales lists and account books compiled by cinchona commissioners and merchants that record the harvesters' names, the amount of cinchona they sold and, occasionally, their home villages and caste, it will seem that many of these men, and

in Brazil. Environment Beyond Economic History,' in *Environmental History. As If Nature Existed*, ed. John R. McNeill, José Augusto Pádua and Mahesh Rangarajan (Oxford: Oxford University Press, 2010).
[89] Lambert, *A description of the genus Cinchona*, 1.   [90] Caldas, *Memoria*, 11–13.
[91] Testimony of Mariano Villa, dated as of April 27, 1782, in 'Sobre la conservacion de Montes de Cascarilla,' fs. 18–19.
[92] Juan A. Villamarin and Judith E. Villamarin, *Indian Labor in Mainland Colonial Spanish America* (Newark: University of Delaware Press, 1975), 19–20.
[93] This quote is taken from a report by Sebastián José López Ruiz (1741–1832) relating to the Santa Fé *Audiencia*. Sebastián José López Ruiz, 'Carta,' *Archivo General de Indias*, Indiferente 1557, Santa Fé, 1804-08-13, f. 656v. On free harvest labour in Malacatos and Vilcabamba, see Petitjean and Saint-Geours, 'La ecomomía de la cascarilla,' 22. On the prevalence of free harvest labour, see also Moya, *Auge y Crisis de la Cascarilla*, 76; 94.
[94] Pedro Xavier de Valdivieso y Torre, in a report to José García de Leon y Pizarro, dated as of December 22, 1779, in 'Expediente sobre el corte de cascarilla en los Montes de Loxa.'

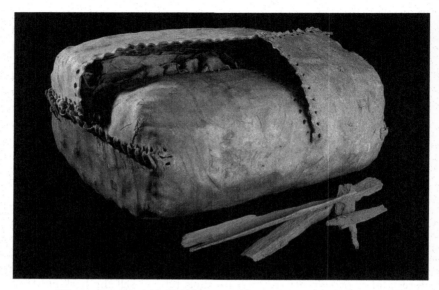

Figure 5.3 Rawhide bag for storing cinchona bark, collected by the Botanical Expedition to the Viceroyalty of Peru (1778–1816), under the command of Hipólito Ruiz López and José Antonio Pavón.
© The Science Museum / Science and Society Picture Library.

some few women, went into the forests sporadically to come back with as much as they found in a day and were able to carry (FIGURE 5.3). As a matter of fact, most of the harvesters lived in areas within or adjoining the mountain ranges where cinchona trees grew. In the sales lists compiled between March 2 and April 20, 1780, by Sebastián José López Ruiz (1741–1832) – the official charged with the abbroachment of cinchona for the Royal Pharmacy in the Santa Fé *Audiencia* between 1778 and 1787 – for instance, bark cutters from Facatativá, to the north-west of Bogotá, sold López Ruiz bark from Matima mountain,[95] while those from Bojacá harvested bark on Mount Zipacón, a two to four hours' walk from their homes.[96] These bark cutters would usually sell López Ruiz less than one *arroba*, sometimes – as in the case of Jacinta Pays, an Indian woman from Pasca, a village in what is today Cundinamarca – less

[95] Felipe Hernandez and Francisco Guzmán, both *vecinos* from Facatativá, sold López Ruiz cinchona from Matima. López Ruiz, 'Razon de los sugetos.'
[96] Bartolomé Orejuela, a *vecino* from Bojacá, brought back three *arrobas* and three *libras* from Zipacón. Ibid.

than 2 kilograms.[97] For Jacinta Pays, like many other harvesters, bark cutting, since it was seasonal work, would have been one of a variety of sources of income, most of which would not have resulted from wage labour. They presumably cut bark for a couple of days or weeks to add to their income derived from farming, sharecropping or trade of their own home-produced commodities.[98]

As in other sectors of the colonial economy, however, free labour and various forms of corvée coexisted in the extraction of cinchona, with varying emphases over the last decades of the eighteenth and the early decades of the nineteenth centuries.[99] For a variety of reasons – the need to secure more and stable supplies, the reputed shortage of labour and presumably also rising competition among traders – merchants and Crown officials increasingly instituted the system of advance payment in which Mariano Villa and his companions had become involved. Within that system, harvesters willing to collect bark for the Crown had to present themselves before the governor (corregidor) 'to ask for and receive money',[100] that is, to be paid in advance for a specified amount of cinchona they were expected to deliver upon their return from the mountains. Francisco Caldas described the procedure in the area around Loja in 1809 as follows:

Fifty labourers (peones) called cascarilleros, all of them residing in Malacatos, Vilcabamba and Gonzanamá. In the month of June, the Corregidor assigns to each of them the quantity in arrobas they are to deliver in December, depending on their robustness, agility and experience. They are paid in advance at a rate of twenty reales per arroba (Se le adelanta el valor á razón de

---

[97] Jacinta Pays, one out of two women who had dealings with López Ruiz, sold him four libras, or 1.8 kilograms. Ibid.

[98] See also Moya, Auge y Crisis de la Cascarilla, 113–15. Andean bark cutters deviated in many aspects from the classic definition of the 'labourer' formulated in the North Atlantic tradition – individuals who live exclusively by selling their labour power to one entrepreneur for a wage – in that they often had several incomes, most of which did not result from wage labour. On changing definitions of the 'labourer', see, for instance, Marcel van der Linden, 'The "Globalization" of Labour and Working-Class History and Its Consequences,' in Global Labour History. A State of the Art, ed. Jan Lucassen (Bern: Peter Lang, 2006), 25.

[99] In the mines of Potosí, for instance, at the end of the colonial period, nearly half the labourers were mitayos and the rest 'free'. Raquel Gil Montero, 'Free and Unfree Labour in the Colonial Andes in the Sixteenth and Seventeenth Centuries,' International Review for Social History 56 (2011), 313. On Spanish imperial labour policy, see also Sanchez-Albornoz, 'El trabajo indigena en los Andes: teorías del siglo XVI,' 179–80.

[100] 'Expediente del Dr. Dn. Antonio Marin y Parra Abogado de esta Rl. Audiencia sobre que en los Montes de su Hacienda no se introduzgan a sacar Cascarilla,' Archivo Nacional de la Historia, Quito, Fondo General, Serie Cascarilla, Caja 5, Expediente 1, n.p., 1793-06-28.

veinte reales por arroba). Each *cascarillero* provides himself with meat and other victuals in June; in August he goes into the forest, from whence he returns with the assigned portion; in December he takes it to Loja, where it is encased and remitted to Piura in January and passes into the hands of those Royal Officials who embark it for Callao.[101]

Cinchona merchants from Loja, Cuenca or Piura proceeded similarly, though, in a variant of the *reparto* or *repartimiento de mercancías* system – the forced purchase of merchandise by Indian communities[102] – they often advanced goods rather, or other, than money. As was common practice also in cotton, cochineal[103] and indigo cultivation across Spanish America, merchant-financiers supplied funds or commodities and were to be repaid in produce by the harvesters. Don Miguel de Armestar, one of the principal Piura merchants who settled in Cuenca and sold his merchandise in Lima, for instance, 'distributed [...] clothes and other goods' among the local residents, 'expecting to be repaid in cinchona from the forests of this jurisdiction'. To the Indians from Ayabaca alone de Armestar declared to have advanced 30,000 *pesos*.[104] By 1787, upon Hipólito Ruiz López's and José Antonio Pavón's second visit to Huánuco, they observed the same system of advance payment, in which 'those who engage[d] in this business' advanced money and merchandise – 'scandalously priced' – to the bark cutters, who, upon their return from the mountains, paid their debts to various merchants by

---

[101] Caldas, *Memoria*, 16.

[102] The *reparto* system – or, as it was also called, *repartimiento de mercancías* or *repartimiento de comercio* – which allowed *corregidores* to force Indian communities to purchase merchandise advanced to them, often at inflated prices, was legalized in 1751 and entered into effect in 1756 (sanctioned until 1783) in the Viceroyalty of Peru. See Scarlett O'Phelan Godoy, *Rebellions and Revolts in Eighteenth Century Peru and Upper Peru* (Köln: Böhlau Verlag, 1985), 99. See also Kenneth J. Andrien, *Andean Worlds. Indigenous History, Culture, and Consciousness under Spanish Rule, 1532–1825* (Albuquerque: University of New Mexico Press, 2001), 203–04. The system was never officially, or nominally, imposed in the Kingdom of Quito, but variants and echoes of it – other than in cinchona extraction, in the production of woollens, for instance – were prevalent nonetheless. Andrien, *The Kingdom of Quito*, 119; Brian R. Hamnett, 'Between Bourbon Reforms and Liberal Reforma: The Political Economy of a Mexican Province – Oaxaca, 1750–1850,' in *The Political Economy of Spanish America in the Age of Revolution, 1750–1850*, ed. Kenneth J. Andrien and Lyman L. Johnson (Albuquerque: University of New Mexico Press, 1994), 245.

[103] On the organization of labour in cochineal production, see Carlos Marichal, 'Mexican Cochineal, Local Technologies and the Rise of Global Trade from the Sixteenth to the Nineteenth Centuries,' in *Global History and New Polycentric Approaches*, ed. Manuel Perez Garcia and Lucio De Sousa (Singapore: Palgrave Macmillan, 2018).

[104] 'Sobre el acopio de la Quina de los Montes de Loxa Callysalla,' fs. 11–12. On Miguel de Armestar and other cinchona merchants who advanced goods, see Moya, *Auge y Crisis de la Cascarilla*, 165–69.

presenting the assigned bushels of bark.[105] Some religious orders proceeded similarly. As the historian Carlos Pérez has shown for the region
of Caupolicán, in the second half of the eighteenth century, Franciscan
missionaries controlled the trade between their Indian charges and itinerant merchants in various forest products, including cinchona. Priests
advanced goods to the Indians in exchange for the bark.[106] A debt
system was common both along the cinchona commodity chain – peninsular merchants advanced goods to Lima merchants, who passed them
on to Piura or Guayaquil merchants, and their agents and factors in the
harvest areas[107] – and across different sectors of the eighteenth-century
Spanish American labour market. It certainly had its advantages for
harvesters, allowing them to provide themselves with provisions and
securing them payment. It also created a system that not only prevented
the harvesters from selling bark to other agents, however, but also got
them into serious difficulties if – as was all too often the case with
communities subjected to *reparto*[108] – they were ever unable to deliver
the agreed amounts.

Mariano Villa and his fellow inmates were not the only ones who,
despite having 'searched all over the forest where the bark was cut',[109]
had found only a fraction, or, worse still, no part of the amounts they had
committed themselves to deliver. With the rapid advance of deforestation
in the principal harvest areas – first in the Loja *corregimiento* and, later, in
the Cuenca *gobernación*, the environs of Santa Fé and the Huánuco
province – it became common for bark cutters in these parts 'to wander
many days without finding so much as a single tree'.[110] As Ruiz López
and Pavón observed, several of the bark cutters were 'continually in
debt for one to two hundred bushels of bark' to cinchona merchants.[111]

[105] *The Journals of Hipólito Ruiz, Spanish Botanist in Peru and Chile, 1777–1815* (Portland:
Timber Press, 1998), 306. See also Crawford, *The Andean Wonder Drug*, 134.

[106] Carlos Pérez, 'Quinine and Caudillos: Manuel Isidoro Belzu and the Cinchona Bark
Trade in Bolivia, 1848–1855' (unpublished PhD dissertation, University of California,
1998), 35.

[107] 'Sobre la conservacion de Montes de Cascarilla,' fs. 18–19.

[108] As historians have shown, indigenous communities often found it difficult to meet their
reparto quotas, partly because prices for goods were frequently inflated. Andrien,
Andean Worlds, 203.

[109] Testimony of Mariano Villa, dated as of April 27, 1782, in 'Sobre la conservacion de
Montes de Cascarilla,' fs. 18–19.

[110] See, for instance, the statement made by Alexandro Toledo, who in 1773 had 'cut bark
for the Royal Pharmacy since the year [17]69'. Toledo, 'En el Pueblo de Malacatos
en dicho dia mes y año para la dicha Informacion, comparecio Alexandro Toledo,'
f. 124 v/r. On the increasing difficulty for harvesters to satisfy their obligations with
deforestation, see also Moya, *Auge y Crisis de la Cascarilla*, 97.

[111] Crawford, *The Andean Wonder Drug*, 134.

Afraid of falling into debt peonage, or of 'being jailed' (*verse reducidos a prisión*) as debtors of bark, harvesters who had received advance payment took greater risks to find and decorticate trees.[112] Cinchona trees frequently 'prospered in the rockiest parts of the hills', in rough and uneven terrain, and some bark cutters risked their lives to get at them.[113] Seeking to reach trees growing on rocky outcrops and suspensions, they lowered themselves down 'on a cordage, or rope, which they fix at the edge of the rock'.[114] Consequently, many came home not just empty-handed but injured, or 'seriously ill' (*mui enfermo*), while others died in their attempts to explore new harvesting areas, in accidents or of exposure.[115] Again other bark cutters who found less than the amounts they had committed themselves to deliver, 'escaped into the mountains' (*se han ausentado a Montañas*), or left the province to circumvent detention as debtors.[116]

Even where the harvesters found enough bark, the costs they bore came to exceed their reward. Not only were they exposing themselves to greater dangers to find trees, but the monetary reward for it became less advantageous. Even though the prices paid by merchants varied,[117] between 1768 and 1815 the Crown paid a set price of 24 *reales*, that is, 3 *pesos*, for an *arroba* of genuine cinchona from the Loja area,[118] and proportionately lower prices – roughly between 2 *pesos* and 20 *reales* – for Loja barks of lesser quality, that is, of divergent texture, taste, consistency or colour.[119] While the bark prices were set, the costs for harvesters increased with

[112]   Manuel Vallano y Cuesta, 'Auto para manifestar el deplorable estado en que se hallan los Montes que producen la Quina selecta que se ha remitido a la Real Botica de S.M. En la ciudad de la Concepción de Loxa à diez días del mes de Mayo de mil setecientos ochenta y ocho años el señor Don Manuel Vallano y Cuesta Corregidor, Juez Conservador, y especial comisionado del ramo de cascarilla de esta ciudad por S.M,' *Archivo General de las Indias*, Malacatos, 1788-05-16, 1485–86.

[113]   Cevallos, 'En el Pueblo de Malacatos,' 125.

[114]   Testimony by Tadeo Celis, dated as of May 10, 1788, in Vallano y Cuesta, 'Auto,' f. 1485–86.

[115]   Testimony by Doña Micaela del Castillo, *vecina* and widow of Don Fernando de la Vega, dated as of April 29, 1782, in 'Sobre la conservacion de Montes de Cascarilla,' fs. 13–15.

[116]   Testimonies of Mariano Villa, Luis Yaguana, and Thomas Lapo, Loja, dated as of April 27, 1782. Ibid.

[117]   By 1785, merchants paid Loja bark cutters some 7 *pesos* for an *arroba* of 'genuine' cinchona and between 3 and 5 *pesos* for an *arroba* of barks of lesser 'quality'. Petitjean and Saint-Geours, 'La ecomomía de la cascarilla,' 36. By 1795, according to José Garcia de Leon y Pizarro, merchants paid the bark cutters 5 *pesos* for an *arroba* of cinchona, for which the Crown paid 2.5 *pesos*. Garcia de Leon y Pizarro, 'El Conde de Casavalencia informa difusamente,' 178.

[118]   Petitjean and Saint-Geours, 'La ecomomía de la cascarilla,' 36–37.

[119]   See, for instance, the report by Matias de Salazar, a landowner from Vilcabamba commissioned with overseeing the extraction of cinchona from the hills around

deforestation.[120] When the bark 'was first discovered there were corpulent trees, often several of them together that yielded up to three or four *arrobas*', observers related, but it had since become 'rare to find two or three plants together'.[121] As a consequence, in the principal harvest areas – first in Loja and later also in Cuenca – harvesters had to cover ever longer distances and spend more time roaming 'through the thicket, and among precipices' in search of cinchona trees.[122] Tadeo Celis related that, according to his parishioners who told him of their grievances, 'formerly, when this specific abounded', 'one or two days sufficed for one person to extract an *arroba* or more'.[123] By the 1780s, harvesters covered distances of 'three, or four *leguas* every day' – between 17 and 24 kilometres – venturing 'deeper and deeper into the mountains in search of batches of cinchona'.[124] At times they walked for fifteen, twenty or more days, spending all their advances on their provisions, and 'returned home afterwards without having found a single tree'.[125] As Melchor Bera, a Loja bark cutter, put it in 1782, it now took him two or three months to gather together one *arroba*, and that only after 'several entrances and gleanings'.[126] Other harvesters, too, related how they hurried back to the transfer sites after 'having collected one or two *arrobas* in two months' to deliver their allotted portions.[127] By 1792, as Don Pablo Miguel de Larreategui y Beamudo put it, it was 'publicly known that the hills where trees with fine cinchona grow were utterly destroyed, such that the men employed in its harvest complained much about the increased labour it takes them to get at an *arroba*'.[128] Labour in the cinchona harvest had been a highly profitable and worthwhile endeavour when Andeans made 3 *pesos* in one or two days – the annual salary for adult male Indians who worked for wages in the Quito *Audiencia* was between 14 and 18 *pesos*[129] – but it

Malacatos, who paid bark cutters between 2 and 2.5 *pesos* for every *arroba*. Matias de Salazar, 'Zertificacion de Don Matias de Salazar, Vessino de la Ciudad de Loxa, residente como hasendado en el Pueblo de Villcabamba, y Comissionado para selar la extraccion de Cascarilla de los serros desde Pais de Malacatos,' *Archivo General de Indias*, Quito 239, Loja, 1773.
[120] Moya, *Auge y Crisis de la Cascarilla*, 241.
[121] 'Testimonio de los Autos en que se comprehenden varios Informes y Diligencias,' f. 26.
[122] Ibid. On Cuenca, see Moya, *Auge y Crisis de la Cascarilla*, 133.
[123] Testimony by Tadeo Celis, Malacatos, dated as of May 16, 1788, in 'Expediente obrado por el Corregidor de Loxa.'
[124] Ruiz López, *Quinología*, 14.   [125] Ibid.
[126] Testimony of Melchor Bera, Loja, April 27, 1782, in 'Sobre la conservacion de Montes de Cascarilla.'
[127] Cevallos, 'En el Pueblo de Malacatos.'
[128] Testimony of Don Pablo Miguel de Larreategui y Beamudo, dated as of April 29, 1782, in 'Sobre la conservacion de Montes de Cascarilla.'
[129] Mark van Aken, 'The Lingering Death of Indian Tribute in Ecuador,' *Hispanic American Historical Review* 6, no. 3 (1981), 433.

quickly became far less so when it came to take them several months to make the same amount of money.

Predictably, in those areas where the harvesting of cinchona became less profitable, and more hazardous for the harvesters, ever fewer Andeans set off to cut bark or offered to hire themselves out as labourers. As José Garcia de León y Pizarro (1770–1835), Count of Casavalencia, then President of the Quito *Audiencia*, put it in 1795, earlier 'the Indians [had] volunteered to be employed' in the extraction of cinchona, 'and not only the Indians but also the mestizos and other people of colour of this province, because they found they received a just reward for their work', but now there was 'no one to be found willing to extract it'.[130] As Coronel don Pedro Romero (fl. 1780s), who sought to recruit bark cutters and to advance payments around 1782, observed, 'the Indians resisted and refused to take the money', and those who eventually did 'were unwilling to accept [money] for two or three *arrobas*, as had been common previously, and wished only to take over half an *arroba*'.[131] The Loja city council noted that bark collectors, because of the 'very great labour, danger and cost necessary to extract one, or two *arrobas*', though they had 'formerly (*antes*) [...] presented themselves to ask for, and receive money, voluntarily', today went to harvest only when 'forced to (*lo hacen forzados*), [...] by royal authority', and that it was 'necessary for the *Corregidor* of the Malacatos Valley to go in person to call and persuade its inhabitants so they enter into these mountains to harvest the specific for our Catholic Monarchs'.[132] By 1794, Tomás Ruiz Gómez de Quevedo, royal cinchona commissioner between 1790 and 1816, reported that almost all of those employed in the harvesting of cinchona were Indians and that they had been coerced into it 'with full force' (*con toda fuerza se les obliga à ello*).[133] It was not deforestation alone that was responsible for labour shortage and the various forms of corvée it entailed, to be sure. Other businesses – like sugar mills – had long vied with cinchona traders for the insufficient workforce.[134] From the late

[130] Report dated as of November 14, 1795. Garcia de León y Pizarro, 'El Conde de Casavalencia informa difusamente,' 178.
[131] Testimony of Don Pablo Miguel de Larreategui y Beamudo, dated as of April 29, 1782, in 'Sobre la conservacion de Montes de Cascarilla.' For more examples of Indians who refused to accept money or were unwilling to accept money only for very small accounts, see Moya, *Auge y Crisis de la Cascarilla*, 102.
[132] 'Expediente obrado por el Corregidor de Loxa,' f. 4.
[133] 'El Presidente Guzman acompaña una representación.'
[134] On labour shortage in the south Quito *Audiencia*, see Minchom, 'The Making of a White Province,' 32. On competition with other businesses, see Petitjean and Saint-Geours, 'La ecomomía de la cascarilla,' 34. The months of bark cutting also coincided with the harvest season, when men were required to stay home 'to make the most of

1700s the rapid expansion of plantation agriculture, particularly cacao production in the coastal province of Guayaquil, drew ever more discontented Andeans from the south sierra, who left to seek employment along the coast.[135] It was environmental degradation, together with the opening up of other opportunities for employment, that entailed both increasing scarcity of harvesters and, in the eyes of officials, the necessity of force.

Force had many faces to it. Some bark traders sent their own, dependent servants to work in the mountains. There were 'so few Indian or servile men (*gente India ó servil*) in that province, and especially in the villages that supply cinchona', Ruiz Gómez de Quevedo lamented, 'that even though the *vecinos* have only few contracted servants, they will force those to go harvest the specific [for them], because they cannot find free men for that labour'. That practice caused 'the landowners significant injuries, and losses' since 'their estates and crops' suffered neglect in the absence of the workers.[136] Clergymen relied on their own methods of coercion to ensure compliance. As Sebastián José López Ruiz reported in an 1804 letter from the Santa Fé harvest areas, 'the priests, at the altar [...] during mass and from the pulpit, exhort and threaten their parishioners, if they sell their cinchona to anyone but their priests' (*exhortan e amenazan á sus feligreses, si las quinas que estos sacan se las venden á otros, sino á sus párrocos*).[137] Throughout the Newgranadine and Peruvian harvest areas, priest–bark traders were also accustomed to using their own parishioners as muleteers and harvesters. Two clerics from the province of Chachapoyas, 'brothers called Villavicencios', the presbytery don Josef Quadra, or the curate of San Felipe, in the province of Jaén de Bracamoros, were but some of the many clergymen who were known to make 'their Indians' harvest bark for them and caravan it on the back of their mules out of the highlands.[138]

As in other imperial settings, where environmental disadvantages and degradation often entailed increased labour coercion and lesser labour

---

their crops'. Pedro Xavier de Valdivieso y Torre, in a report to José García de Leon y Pizarro, dated as of December 22, 1779, in 'Expediente sobre el corte de cascarilla en los Montes de Loxa.'

[135] Andrien, *The Kingdom of Quito*, 43; 50–51; 54.
[136] 'El Presidente Guzman acompaña una representación,' f. 136–37.
[137] López Ruiz, 'Carta,' f. 656v.
[138] On the 'brothers called Villavicencios', see Miguel de García de Cáceres, 'Proyecto sobre quina,' *Archivo General de Indias*, Indiferente 1554, Guayaquil, 1779-03-16, f. 1072r. On Josef Quadra and the curate of San Felipe, see Intendencia de Cuenca, 'Testimonio,' *Archivo General de Indias*, Indiferente 1555, Gualeco, 1787-11-28, 1166.

compensation to increase or make up for low yields,[139] in the Andes, deforestation and the shortage of hands entailed the return to a labour draft system. It will seem that in the Quito *Audiencia* at least, merchants and Crown officials opted to fall back on the *mita* – a rotational labour draft system, which obligated adult male tributaries to work periodically for a set period in the service of the colonial state or, in some areas, Spanish *hacendados*, mine or factory owners.[140] Although the *mita* was in decline by the late 1700s in the south Quito *Audiencia*, adducing the south's alleged chronic labour shortage,[141] *hacendados*, mine owners and Crown officials occasionally resorted to small-scale use of a reduced number of *mitayos* – sometimes only one or two.[142] In 1791, Tomás Ruiz Gómez de Quevedo and Vicente Olmedo y Rodríguez (1763–1854), the botanist appointed co-director of the royal cinchona reserve in Loja in 1790,[143] suggested employing 50 *mitayos* from the Cuenca *corregimiento* at the expense of the Royal Treasury for two or three years – always replacing the 'dead, refugees and the useless (*inútiles*)' when need be – to employ them in bark cutting and in the making of inroads into the 'most remote forests'.[144] Though other officials like José Garcia de Leon y Pizarro criticized the proposal on the grounds that it exceeded the accorded timeframe for the *mita* – which required an annual shift – and thus was in breach of the laws that protected the Indians from both 'violence and compulsion',[145] it would seem that a cinchona *mita* was eventually implemented nonetheless. As the historian Luz del Alba Moya

---

[139] According to Philip W. Porter and others, for instance, in tropical Africa planters compensated for comparatively low cotton yield, attributable to differences in light duration, through the stricter control of labour. Philip W. Porter, 'Note on Cotton and Climate: A Colonial Conundrum,' in *Cotton, Colonialism, and Social History in Sub-Saharan Africa*, ed. Allen F. Isaacman and Richard L. Roberts (London / Portsmouth N.H.: James Currey / Heinemann, 1995), 49. See also Corey Ross, *Ecology and Power in the Age of Empire: Europe and the Transformation of the Tropical World* (New York: Oxford University Press, 2017), 55. Along those lines, historians have also written about the connection between ecology – especially adverse climates – and the imperial division of labour. See, for instance, Peder Anker, *Imperial Ecology: Environmental Order in the British Empire, 1895–1945* (Cambridge, Mass.: Harvard University Press, 2001), 167.

[140] On the *mita*, see, for instance, Villamarin and Villamarin, *Indian Labor in Mainland Colonial Spanish America*, 18. The *mita* found varying expressions and entailed different obligations across colonial Spanish America. On the eighteenth-century history of the *mita*, see, for instance, Enrique Tandeter, *Coercion and Market: Silver Mining in Colonial Potosí, 1692–1826* (Albuquerque: University of New Mexico Press, 1993).

[141] According to some contemporary observers, lamentations of labour shortage were a mere myth to justify coercion. Garcia de Leon y Pizarro, 'El Conde de Casavalencia informa difusamente,' 178.

[142] Minchom, 'The Making of a White Province,' 32.

[143] On Vicente Olmedo, see chapter 6 in Crawford, *The Andean Wonder Drug*.

[144] Garcia de Leon y Pizarro, 'El Conde de Casavalencia informa difusamente,' 175–76.

[145] Ibid.

has found, on the grounds of ever more Andeans' refusal to contract themselves out as bark cutters, a bark *mita*, implicating 50 *peones*, was effected towards 1790.[146]

Another means of forcing Andeans into cinchona harvest was through the tribute system, as historians have shown.[147] The obligation to pay tribute had been placed upon the native population – any adult male registered as Indian, with the exception of the sick and disabled, as well as village officeholders, caciques and their eldest sons – soon after the conquest and continued to be exacted throughout the colonial period and, in fact, beyond.[148] Tribute not only provided a large portion of treasury revenues, that is, more than one-third of all tax collections in the Quito *Audiencia* on the eve of independence. As in other parts of Spanish America, from the point of view of the elites and the state, it served to expand the legal grounds on which coercion was exercised and compel the Indian population to enter the monetized economy as labourers.[149] Indeed, as is evident from official reports and court files about licences, many of those registered as Indians engaged in harvest labour solely to procure cash 'to pay the royal tribute'.[150] As the historian Luz del Alba Moya has found, some Indian communities were granted special licences to work in privately owned harvest areas with the proviso that they would pay their tribute with it. The Indian communities of Chancaymina, Huambacola and Colambo were given permission to exploit the cinchona forests on the estate of Manuel Mora in the Malacatos Valley, so that they could pay their tribute, while the Indian community of Chillanes, in the jurisdiction of Guaranda, was granted the licence to exploit royal territories so they would be in a position to pay the royal tribute and maintain themselves and their families.[151] On the same principle, extenuation was granted if some tributary had infringed upon the successive royal prohibitions that restricted cinchona harvest on the hills around Loja after 1751. The Indian Martín Banda, for instance, accused

---

[146] Moya, *Auge y Crisis de la Cascarilla*, 76; 107.   [147] Ibid., 94.

[148] On the long history of tribute in the Quito *Audiencia*, and exemptions from paying it, see Aken, 'The Lingering Death of Indian Tribute in Ecuador,' 433.

[149] Villamarin and Villamarin, *Indian Labor in Mainland Colonial Spanish America*, 19; Aken, 'The Lingering Death of Indian Tribute in Ecuador,' 431. Nathan Wachtel and, later, Carlos Sempat Assadourian were the first to propose that it was the monetarization of tribute in the wake of the 1570s that forced the Andean population to enter the colonial economy by selling either their goods or their labour. Nathan Wachtel, *Sociedad y ideología: ensayos de historia y antropología andinas* (Lima: Instituto de Estudios Peruanos, 1973), 79–158; Carlos Sempat Assadourian, 'La despoblación indígena en Perú y Nueva España durante el siglo XVI y la formación de la economía colonial,' *Historia Mexicana* 38, no. 3 (1989), 427.

[150] See, for instance, Cuenca, 'Testimonio,' 1158–59.

[151] Moya, *Auge y Crisis de la Cascarilla*, 115–16.

of having harvested cinchona in violation of the royal prohibition, would not refute the charge. He said in his defence, however, that 'the weight of the royal tribute' had impelled him to extract cinchona in the service of don Joaquin Crespo and don Francisco Ortega, both of whom had assured him, in addition, that they were in possession of special permits.[152] Though conceived originally as a levy of Indian specie and labour, the obligation to pay tribute extended at times, and in certain instances and areas of colonial Spanish America, also to other, non-native castes, and 'mestizos'.[153] Indeed, some of those who asked for a special licence to collect cinchona in order to enable them, 'like other wretched ones (*miserables*)', to 'satisfy the royal tribute' they found themselves 'obliged to pay, and to maintain a poor, numerous family', described themselves in their solicitations as 'mestizos'.[154] Tribute rates varied according to the taxpayer's status – as *quinto* or *forastero* Indian – and from province to province, but generally ranged, by the late 1700s and early 1800s, from 3 to 6 *pesos* in the Cuenca government and the Loja *corregimiento*[155] – an amount bark cutters could, for much of the period under consideration, realistically have earned in one harvest season. Not all cinchona-based tribute payments were transacted in cash, however. Some *corregidores* directly collected tribute in specie rather than cash, presumably in order to resell the bark more advantageously. The *corregidor* in charge of collecting tribute in the village of Chillanes, for instance, asked the Indians in his jurisdiction to pay their tribute in cinchona. A 1792 legal file retains the dispute that ensued when he had

---

[152] Cuenca, 'Testimonio,' 1158–59.
[153] Miscegenation between the native population and European and African arrivals had long produced a bewildering plethora of men and women whose existence not only belied the legal fiction of the 'two republics' – Indian and Spanish – but necessitated a re-evaluation of the existing tribute scheme. On non-native tribute, see Cynthia E. Milton and Ben Vinson III, 'Counting Heads: Race and Non-Native Tribute Policy in Colonial Spanish America,' *Journal of Colonialism and Colonial History* 3, no. 3 (2002); Ronald Escobedo Mansilla, 'El tributo de los zambaigos, negros y mulatos libres en el Virreinato peruano,' *Revista de Indias*, no. 1 (1981); Chantal Caillavet and Martin Minchom, 'Le métis imaginaire: idéaux classificatoires et stratégies socio-raciales en Amérique latine (xvie–xxe siècle),' *L'Homme* 32, no. 122–124 (1992).
[154] See the application made by Josef Vosques, who described himself as 'mestizo (*Mestiso*) resident in the village of San José de Chimbo (*Vecino del Pueblo de San Jose de Chimbo*)'. 'Licencias de Cascarilla,' f. 59–60.
[155] According to Jacques Poloni-Simard, in Cuenca *Indios quintos* paid 6 pesos, *Indios forasteros* three. Poloni-Simard, *El Mosaico Indígena*, 264. In the decades around 1800, in the Loja province, *Quinto* Indians paid 5 *pesos* and 2 *reales*, while *forastero* Indians only paid 3 *pesos*. Minchom, 'The Making of a White Province,' 29. In the first decades of the eighteenth century, adult male Indians had paid between 7 and 10 *pesos* in the Quito *Audiencia*, but tribute rates declined towards the close of the century. See Aken, 'The Lingering Death of Indian Tribute in Ecuador,' 432.

several of the tributaries arrested for non-payment, 'even though they had money to pay it, because they did not give it in cinchona (*por que no lo davan en Cascarilla*)'.[156] Non-payment of tribute was not treated lightly, and gives us a vivid idea of the force in play. Corporal punishment – whippings or torture – was a common reprimand for non-payment of tribute, as was imprisonment.[157] Somewhat paradoxically, then, even though the bark trade altered nearly every facet of the central and northern Andes – its landscape of possession, commerce and demographics, the distribution and abundance of vegetation, the livelihood, health and the fate of so many men and women – it did not affect these societies' basic socio-economic structure. On the contrary: the elites were not only those who reaped the largest share of the profits,[158] they used traditional colonial elements of subjugation – the tribute, servitude or the *mita* – to do so.[159]

★★★

Temporarily, at least, cinchona exports came to a halt with the independence movements. The struggle for independence severely affected the central and northern Andes[160] and, by 1814, the responsible commissioners reported to the President of the Quito *Audiencia*, Toribio Montes (r. 1812–1817), that for three years there had been no remittance of cinchona for the Royal Pharmacy because of 'the circumstances of the times'.[161] Contraband soared with the disruptions of the period. Since the beginnings of the insurrection Crown officials lamented that they had been unable to keep contraband in check. People were beginning to say that 'there was no King, and that Spain did not exist', and officials found themselves unable to get the harvesters to hand over their portions, for they 'poked fun at those who gave orders'.[162] The disruption of trade

---

[156] 'Autos seguidos por Da. Rosa Orosco contra la Protecturia sobre los perjuicios que recibe en los Montes de Quina de su Hazienda del Naranjal – Secretaria del Capitan Dn. Luis Cifuentes, Relator el Dr. dn. Ramon Ybarguren,' *Archivo Nacional de la Historia*, Quito, Fondo General, Serie Cascarilla, Caja 4, Expediente 1, Riobamba, 1792-08-18, 5.

[157] Villamarin and Villamarin, *Indian Labor in Mainland Colonial Spanish America*, 19; Aken, 'The Lingering Death of Indian Tribute in Ecuador,' 433.

[158] Moya, *Auge y Crisis de la Cascarilla*, 241.   [159] Ibid., 243.

[160] Vega Ugalde, 'Cuenca en los movimientos independentistas,' 28.

[161] 'Relatibo a la remision que han hecho a Espana los Comisionados por el acopio de Quina electa de la Ciudad de Loxa de diez mil novecientas libras de este especifico,' *Archivo Nacional de la Historia*, Quito, Fondo General, Serie Cascarilla, Caja 5, Expediente 13, Loja, 1814-09-26. See also Moya, *Auge y Crisis de la Cascarilla*, 121–22.

[162] Cited in Crawford, *The Andean Wonder Drug*, 176.

patterns during the dislocations of the independence era contributed to an overall prolonged commercial decline of the area,[163] and cinchona export trade in South America was never to recover its exalted, monopolistic position entirely. Soon after the collapse of the Spanish Empire, during the 1830s, Dutch and British agents smuggled cinchona seedlings out of the Andes, and some years later, botanists successfully created cinchona plantations – in Dutch Java in 1835 and in British India in 1861 – effectively undermining the South American monopoly on the bark.[164] By the second half of the nineteenth century, North America and northern Europe were the main producers and manufacturers of the bark's derivatives. Andean South America had become a periphery of the trade in cinchona.

[163] Moya, *Auge y Crisis de la Cascarilla*, 200.
[164] Andrien, *The Kingdom of Quito*, 50; 161.

Conclusion
A Plant of the World

Millions of men and women in societies rimming or tied to the Atlantic basin – be they Ottoman courtiers, Hamburg paupers or Andean villagers – had, by the late 1700s and early 1800s, come to assign medicinal properties as well as particular therapeutic practices and cultural imaginaries to dried shreds of bark removed from the cinchona tree. From Philadelphia to Rome, from Nagasaki to Rio de Janeiro, people had encountered and espoused that singular remedy, as well as techniques and understandings attendant to its consumption, through exposure to the written word, medical practice or word of mouth. Though coming from a single, limited source in the central and northern Andes, the bark would have been available to them through a variety of mercantile conduits and from a wide range of purveyors the world over: on offer from a Frankfurt apothecary, attainable from a Jesuit dispensary in Macao and buyable over the counter of a Smyrna market stand. Many of the bark's users, traders and supporters would have been conversant with and, in some measure, tempted into its consumption or advocacy by stories about its discovery by 'wild Indians' – 'natural men' of whom, in their unlettered simplicity, closeness to nature and 'divine folly', mankind could reasonably expect the revelation of nature's most coveted secrets, its most excellent remedies. Many sufferers and medical practitioners would also have shared recipes and medical practices calculated to render cinchona more effectual and agreeable, with its therapeutic properties, powdery texture and nauseous taste entailing a necessity for solvents, sweeteners and the presence of kindred febrifuges and aligning preparations of the bark the world over. Gentlewomen in the Scottish Lowlands, physicians at the Ottoman Porte and Mexican hospital orderlies alike would have known how to concoct, or where to purchase, the classic bittersweet febrifugal lemonades and aromatic compound wines of the bark. Many consumers would have shared not only fanciful narratives about the bark's discovery and formulae to concoct preparations of it that were palatable, and efficacious, but also diagnostic experience, expertise in indications and an understanding of the environs

179

and ailments in which the use of the Peruvian bark would be most beneficial. From the fenlands of Cambridgeshire to the river valleys of Lima, sufferers and practitioners from all walks of life, be they sailors, householders or seasonal workers, would most commonly have trusted the bark's protection and curative properties in fevers, occasioned by insalubrious situations and climates. Knowledge, *A Singular Remedy* holds, is not bound ineluctably to just one place or locality. As the case of the Peruvian bark shows, by the late 1700s and early 1800s, men and women had come to share expertise and experiences across imperial and social divides at a level as fundamental as that of their body, of sickness and cure.

A Singular Remedy pushes back against a long-standing emphasis on locality in the history of science in that it holds knowledge to be movable. It does not, however, deny or disregard the relevance of place. Scholarship on early and late modern globalization, with its liquid language of elusive flows and unconstrained circulation, has come under criticism for evoking an idea of movement as erosive and anti-thetical to place,[1] and for 'assuming coherence and direction instead of probing [the] causes and processes' of movement.[2] Indeed, to the Peruvian bark, and knowledge of it, the world was neither an even nor an indifferent medium of transmission. Both moved primarily along the routes of Spanish and Portuguese American, Iberian, English, Dutch and French commerce, imperialism or proselytizing, with their course defined by the situation of these societies' trade entrepôts, military outposts and diaspora communities. They covered vast distances, but with their movement ever limited and measured by the fragmentation, standstills and relays intrinsic to maritime and overland conveyance – the speed and range of sailing vessels, horse-drawn vehicles or human legs – and to the, just as yet, polycentric configuration of the early modern world. For the bark's – Iberian, Spanish American, French, Dutch or British – suppliers submitted and catered to the ways of Levantine or Cantonese trading networks, and to the desires of the Tokugawa shogunate or the Moroccan 'Alawi court.[3] The bark, and

---

[1] For critiques of the 'liquid' language of scholarship on modern and early modern 'globalization', see, for instance, Rockefeller, 'Flow'; Jürgen Osterhammel, 'Globalizations,' in *The Oxford Handbook of World History*, ed. Jerry H. Bentley (Oxford: Oxford University Press, 2014). See also Gänger, 'Circulation.'

[2] Frederick Cooper, 'What Is the Concept of Globalization Good For? An African Historian's Perspective,' *African Affairs* 100, no. 189–213 (2001).

[3] On the limits of early modern globalization, see Jan de Vries, 'The Limits of Globalization in the Early Modern World,' *Economic History Review* 63, no. 3 (2010). Instead of a single political centre, in the early modern era, there were several centres with their own

knowledge of it, rarely strayed far from the veins of its passage in ways that point both to 'connections and their limits'[4] in the Age of Revolution. Cinchona was available to a wide range of men, women and children, so long as they lived or moved in places tied to the wider Atlantic World: in convents or adjoining bazaars, near seaports or in towns, areas with a peculiar material culture that often diverged from that of the surroundings. Indeed, the principal vehicles and veins for the – printed, penned or verbal – communication of medical understandings and practices attendant to the bark's administration were social relations between settings tied together by Atlantic trade, proselytizing and imperialism: between enlightened physician-authors and their readership, soldiers and enemy civilians, planters and slaves. Differences in status, wealth and welfare within consumer societies, furthermore, tended to guide the bark and knowledge of it to upper- and middle-class households and royal courts. Charity or – political, economic or military – expediency would, in turn, lead it to poor-houses, slave plantations or navy vessels, in ways that further revise our assumptions about the social depth of processes of convergence and integration in the realm of medicine and the participation of non-elite sufferers and practitioners in them. Whereas shared understandings of bodily health, particularly across the period's Christian and Islamic empires, certainly facilitated the bark's passage across continents and bodies of water by the late 1700s and early 1800s, the medical topography which those shared understandings entailed, with their distinct contours of insalubrious, febrile environments, also confined the bark's use to particular situations: the world's low-lying marshlands, the sickly air of close, crowded spaces – the narrow confines of ships, army camps and rapidly growing urban spaces – and the hot and humid climates of the tropics that a growing number of American, Portuguese, Ottoman, or Dutch soldiers, colonists and merchants encountered in a period of global imperialism, warfare and settlement. Indeed, the very possibility, scale and volume of the bark's worldwide distribution hinged on its source in the vulnerable human and physical ecology of the central and northern Andes – a volatile labour force and a fragile, thinning forest cover – that regulated, retarded or delimited supplies.

peripheries, a relative balance of power between the Spanish Empire, the Dutch and English trading companies, China and the gunpowder empires of the Islamic world. On the debate about the 'Great Divergence', see Pomeranz, *The Great Divergence*; Rosenthal and Wong, *Before and Beyond Divergence: The Politics of Economic Change in China and Europe*.
4 Cooper, 'What Is the Concept of Globalization Good For?,' 213.

Place and locality did not just guide and ground the movement of bark knowledge. They also introduced contingency and variability into it. It was men and women from across the Atlantic World – Spanish physicians and French housewives, Angolan surgeons and Portuguese merchants – who, engaging with and relying on one another's beliefs and practices, had apprehended that substance's admirable effects, directed its exploitation and jointly contrived registers, and crafted routines of medication attendant to its administration, in ways that undermine the very notion of origins, and any 'dichotomy of local [...] creation', and transferral outwards.[5] The religious beliefs, peculiar culinary lore or economic possibilities, all of them brought to bear in consumption practices, in turn, introduced variation and contingency into cinchona formulae. Men and women often tinkered with the particulars of recipes for preparations of the bark until they satisfied their palate or suited their medical needs or drinking habits. They insisted that the wine in recipes be replaced with vinegar, favoured capillaire syrup over cane sugar and substituted Spanish oranges for Portuguese lemons. They subtly adapted modes of administration to their own constitution, the obstinacy of their condition and the season or climate they sought to shield themselves from. They rendered its name in their own language and refashioned its history – imaginatively elaborating on, or not attending much to, tropes of the Indians' secrecy and closeness to nature – in ways evocative of the linguistic and cultural idiosyncrasies that set the Atlantic World apart. Indeed, it was often their ability to appropriate, and situate that substance, to ensconce it in their culinary habits, cultural imaginary or medical locality that made bark knowledge both varied and ubiquitously appealing. By the late 1700s and early 1800s, knowledge and use of the Peruvian bark was common across a wide range of geographically disperse and socially diverse societies within, or tied to, the Atlantic World, in part because of its ability to 'delicately adjust to context'.[6] Bark knowledge did not travel to distant societies because it was transcendent, or of universal validity, as the positivist founders of the discipline of the history of science would have it. Indeed, it was not. Rather, knowledge of the Peruvian bark was made to acquire validity elsewhere, in laboursome

---

[5] Secord, 'Knowledge in Transit,' 661. This takes inspiration from historians of science who advocated a 'synchronic account of the intimate ways in which all sites of knowledge making and technique were implicated in networks of skilled practice and of mobile collaboration'. Lissa Roberts and Simon Schaffer, preface to *The Mindful Hand. Inquiry and Invention from the Late Renaissance to Early Industrialisation*, ed. Lissa Roberts, Simon Schaffer and Peter Dear (Amsterdam: Koninklijke Nederlandse Akademie van Wetenschappen, 2007), xx.

[6] Daston, 'Introduction. Speechless,' 18.

acts of 'adaptive appropriation'[7] and calibrated replication. Not only could movement be the very making of knowledge, as the case of cinchona shows. Knowledge moved in ways that were as contingent, as much tied to circumstance and situation as those long associated with its making.

[7] Schaffer et al., introduction, xx.

Bibliography

List of Archives

Spain
General Archive of the Indies (*Archivo General de Indias*), Seville
General Archive of Simancas (*Archivo General de Simancas*)
Archive of the Royal Botanical Garden (*Archivo del Real Jardín Botánico*), Madrid
Archive of the Royal Palace (*Archivo del Palacio Real*), Madrid

Portugal
Portuguese Overseas Archive (*Arquivo Histórico Ultramarino*), Lisbon
National Library of Portugal (*Biblioteca Nacional de Portugal*), Manuscripts Section, Lisbon
National Archive of Torre do Tombo (*Arquivo Nacional da Torre do Tombo*), Lisbon

Peru
Archive of the Riva Agüero Institute (*Archivo Histórico, Instituto Riva Agüero*), Lima
National Library of Peru (*Biblioteca Nacional del Perú*), Manuscripts Section, Lima

Ecuador
National Historical Archive (*Archivo Nacional de la Historia*), Quito

France
Academy of Medicine Library (*Bibliothèque de l'Académie de Médicine*), Paris
Central Library of the National Museum of Natural History (*La Bibliothèque centrale du Muséum national d'histoire naturelle*), Paris

Switzerland
Archive of the Institute for the History of Medicine, University of Zurich (*Archiv für Medizingeschichte, Medizinhistorisches Institut, Universität Zürich*)

United Kingdom
Wellcome Library, Archives and Manuscripts Collection, London

Manuscripts and Archival Sources

Abascal y Sousa, José Fernando de (Marqués de la Concordia), Virrey del Perú. 'Expediente que la Real Hacienda envía al Marqués de La Concordia, virrey del Perú incluyendo relación de las medicinas proporcionadas por el farmacéutico Luis de Montes, para el auxilio de los escuadrones de guerra, quien solicita el pago por este servicio.' *Biblioteca Nacional del Perú*, Sección Manuscritos – XPB/D27, Lima, 1816.

'ALVARÁ (CÓPIA), do príncipe regente (D. João), ordenando que a Junta do Protomedicato ficasse responsável pela preparação das 'águas de Inglaterra' (quina) para consumo das Armadas Reais e nos domínios ultramarinos.' *Arquivo Histórico Ultramarino*, CU 076 – REINO RESGATE 20121023 / Cx. 30-A, Pasta 16, Queluz, 1799-06-24.

Amorim Pereira, D. João de. 'CARTA do (governador do Piauí), D. João de Amorim Pereira, á rainha (D. Maria I).' *Arquivo Histórico Ultramarino*, 016 – PIAUÍ – CATÁLOGO DE DOCUMENTOS MANUSCRITOS AVULSOS / Cx. 21, D. 1074, Piauí, 1798.

'Aprovado con la advertencia de que si se pudiese despachar toda la quina en rama seria mas sencilla.' *Archivo General de Indias*, 1557, Madrid, 1807-01-28 / 1807-02-03.

Arriaga, Julián de. 'Carta.' *Archivo General de Indias*, Indiferente 1552, Madrid, 1753-01-23.

'Autos seguidos por Da. Rosa Orosco contra la Protecturia sobre los perjuicios que recibe en los Montes de Quina de su Hazienda del Naranjal – Secretaria del Capitan Dn. Luis Cifuentes, Relator el Dr. dn. Ramon Ybarguren.' *Archivo Nacional de la Historia*, Quito, Fondo General, Serie Cascarilla, Caja 4, Expediente 1, Riobamba, 1792-08-18.

Azara, Nicolás de. 'Carta a Pedro de Lorena.' *Archivo General de Simancas*, Legajo 961/2, Rome, 1789-10-09.

Blad, Juan Pablo de. 'Carta a Salvador Rizo.' *Archivo del Real Jardín Botánico*, Real Expedición Botánica del Nuevo Reino de Granada (1783–1816). José Celestino Mutis. Correspondencia. Correspondencia a Salvador Rizo, RJB03/0001/0003/0149, Neiva (Colombia), 1789-10-27.

'A booke of divers receipts.' *Wellcome Library*, Archives and manuscripts, MS.1322/54, n.p., c. 1660–1750.

Bueno, Cosme. 'Carta al S. Visitador y Superior Intendente General.' *Biblioteca Nacional del Perú*, Expediente sobre el Estanco de la Cascarilla, Sección Manuscritos – C388, Lima, 1785-09-19.

Caballero y Góngora, Antonio. 'Carta.' *Archivo General de Indias*, Indiferente 1555, Santa Fé, 1786-10-19.

'Carta al Marquès de Sonora.' *Archivo General de Indias*, Indiferente 1554, Santa Fé, 1786-10-19.

'Copia de Carta Reservada.' *Archivo del Palacio Real*, Papeles del Almacén de la Quina, Caja 22283 / Expediente 2, Turbaco, 1788-05-28.

'El arzobispo Virrey da cuenta.' *Archivo General de Indias*, Indiferente 1554, Santa Fé, 1786-09-24.

'El Arzobispo Virrey de Santa Fé Informa à V.E. el testimonio indirecto que ha tenido de las clandestinas extracciones de Quina que hacen los extranjeros en las costas septentrionales del Reyno.' *Archivo General de Indias*, Indiferente 1554, Carta 315, Santa Fé, 1782–1789.

Calbo y Lopez, Ramón. 'Experimentos [sobre las] propiedades de la Quina o cascarilla nuevamente descubierta.' *Archivo General de Indias*, Indiferente 1557, n.p., 1816–7.

Casas, Simon de las. 'Carta a Miguel de Muzquiz.' *Archivo General de Simancas*, Legajo 907, Vienna, 1772-10-03.

Castro, D. José Luis de. 'OFÍCIO do (vice-rei do Estado do Brasil), conde de Resende, (D. José Luis de Castro), ao (secretário de estado da Marinha e Ultramar), D. Rodrigo de Sousa Coutinho.' *Arquivo Histórico Ultramarino*, 017 – RIO DE JANEIRO – CATÁLOGO DE DOCUMENTOS MANU-SCRITOS AVULSOS / Cx. 165, D. 12275, Rio de Janeiro, 1798-05-04.

Cevallos, Pedro. 'En el Pueblo de Malacatos en dies y siete dias del mes de Septiembre de mill setecientos setenta y tres años para la Informacion dicha hise compareser à Pedro Cevallos Natural del Pueblo de Bilcababa.' *Archivo General de Indias*, Quito 239, Malacatos, 1773-09-17.

'Collection of medical receipts and prescriptions: in Italian, by various hands.' *Wellcome Library*, Archives and manuscripts, Closed stores WMS 4, MS.4105, n.p., n.d.

'Collection of medical receipts by several hands: in Italian.' *Wellcome Library*, Archives and manuscripts, Closed stores WMS 3, MS.4110, n.p., c. 1800.

'Collection of medical receipts, in Italian. Perhaps originally part of a Pharmacist's Prescription book. With indented alphabet.' *Wellcome Library*, Archives and manuscripts, Closed stores WMS 4, MS.4106, n.p., n.d.

'Collection of medical receipts, preceded by some domestic receipts, receipts for liqueurs.' *Wellcome Library*, Archives and manuscripts, Closed stores WMS 4, MS.4082, n.p., c. 175–1775.

'Collection of medical receipts, with a few cookery receipts: by several hands.' *Wellcome Library*, Archives and manuscripts, Closed stores WMS 3, MS.4057, n.p., n.d.

'Collection of medical receipts, with a few household and veterinary receipts: in French.' *Wellcome Library*, Archives and manuscripts, Closed stores WMS 4, MS.4087, n.p., n.d.

'Colonie de la Louisiane.' *Archivo del Palacio Real*, Suministros de Medicinas y Papeles de la Real Botica 1671–1781, Caja 22284, Madrid, n.d.

Cornejo, Juan. 'Informe de Don Juan Cornejo Ministro de V.M. en Genova.' *Archivo General de Simancas*, Legajo 961/2, Geno, 1785.

Cuenca, Intendencia de. 'Testimonio.' *Archivo General de Indias*, Indiferente 1555, Gualeco, 1787-11-28.

'Decretos do príncipe regente.' *Arquivo Histórico Ultramarino*, 076 – REINO RESGATE 20121023 / Cx. 30-A, Pasta 18, Queluz, 1804-09-22.

Diguja, Joseph. 'El Excelentíssimo Señor Bailio Frey Don Julian de Arriaga, con fecha de siete de Mayo de este presente año, y de R. Orden, me previno.' *Archivo General de Indias*, Indiferente 1554, Quito, 1773-12-20.

'Don Juan Gavilanes de este vecindario, pobre de solemnidad declarado por la soberania de S.A. segun dro. ante V. paresco y digo.' *Archivo Nacional de la Historia*, Quito, Fondo General, Serie Cascarilla / Caja 5, Expediente 11, Quito, 1809-04-19.

'Doutor Juiz de Fora Prezidente, Vereadores e Procurador da Camera abaixo asignados da Cidade de São Filippe de Benguella, e Sua Capitania por Sua Alteza Real o PRÍNCIPE REGENTE Nosso Senhor, que Deus Guarde.' *Arquivo Nacional da Torre do Tombo*, Ministério do Reino / Negócios diversos do Físico-Mor / Maço 469, Caixa 585, 22, São Filipe de Benguela, 1800-09-02.

Dundas, Lady Eleanor. 'Collections of medical and cookery receipts in English, by several hands.' *Wellcome Library*, Archives and manuscripts, MS.2242, Falkirk / London, c. 1785.

Dunham, Maria. 'Maria Dunham Her Book.' *Wellcome Library*, Archives and manuscripts, Closed stores WMS 4, MS.8301, n.p., 1781.

'El Cura Párroco, el Alcalde, Regidores, y Procurador Síndico del Lugar de Vargas del Arzobispado de Toledo Hacen presente à V.M. que la mayor parte de su vecindario son pobres jornaleros.' *Archivo General de Indias*, Legajo 961, 1786-08-09.

'El Presidente Guzman acompaña una representación del Corregidor de Loxa, y del Botanico, encargado del acopio de la Quina (Tomás Ruiz Gómez de Quevedo y Vicente Olmedo y Rodríguez) en que tratan prolijamente del Estado de su Comision.' *Archivo General de Indias*, Indiferente 1556, Quito, 1794-11-21.

'El Sub.do de la Real Hacienda de Lima contesta a la Real Orden de remisión de Quina.' *Archivo General de Indias*, Indiferente 1554, Lim, 1787.

'Expediente del Dr. Dn. Antonio Marin y Parra Abogado de esta Rl. Audiencia sobre que en los Montes de su Hacienda no se introduzgan a sacar Cascarilla.' *Archivo Nacional de la Historia*, Quito, Fondo General, Serie Cascarilla, Caja 5, Expediente 1, n.p., 1793-06-28.

'Expediente obrado por el Corregidor de Loxa en satisfaccion de la Real Orden de 24 de Noviembre de 1787 acreditando la imposivilidad de cumplimentarla por la suma escasez de Cascarilla de aquellos Montes que no permiten llevar a devido efecto la Ynstruccion formada a 17 de Marzo de 1773.' *Archivo Nacional de la Historia*, Quito, Fondo General, Serie Cascarilla, Caja 3, Expediente 9, 10-V-1788, Malacatos, 1788-05-16 / Loja, 1788-07-12

'Expediente sobre el corte de cascarilla en los Montes de Loxa.' *Archivo Nacional de la Historia*, Quito, Fondo General, Serie Cascarilla, Caja 2, Expediente 5, Loja, 1779-08-19.

'Expediente tramitado desde Febrero de 1768 hasta Febrero de 1769 por el Virrey de Nueva Granada Don Pedro Mesia de la Zerda, el Secretario de Camara y Gobierno del Virreynato Don Francisco Silvestre, el Presidente de Quito Don Jose Diguja, el Corregidor y Justicia Mayor de Loja Manuel Daza y Fominaya, el Cabildo de Loja, y Don Pedro Javier de Valdivieso y Torres sobre la clasificacion de la mejor cascarilla para enviarla a la Real Botica de Su Majestad.' *Archivo Nacional de la Historia*, Quito, Fondo Especial / Fondo Presidencia de Quito, Caja 24, Volumen 66, Documento 2858, Loja, 1768-02 / 1769-02.

'Expediente y cartas de José García de Leon y Pizarro.' *Archivo General de Indias*, Indiferente 1554, Quito, 1782-08-18.

Fernandez Alexo, Fray Josef. 'Fray Josef Fernandez Alexo Religioso observante de San Francisco y Comisario de los Santos Lugares de Jerusalén expone.' *Archivo General de Simancas*, Legajo 960, Jerusalem, 1781.

Finger, Mrs, and Anna Maria Reeves. 'Collection of medical, cookery, and household receipts: with additions by several hands. With numerous inserted receipts and cuttings from newspapers, etc. pasted in.' *Wellcome Library*, Archives and manuscripts, MS.2363, n.p., c. 1750–1775.

García de Cáceres, Miguel de. 'Proyecto sobre quina.' *Archivo General de Indias*, Indiferente 1554, Guayaquil, 1779-03-16.

Garcia de Leon y Pizarro, José. 'El Conde de Casavalencia informa difusamente sobre la representación del corregidor de Loxa y del Botanico encargados del acopio de Quina.' *Archivo General de Indias*, Indiferente 1556, Madrid, 1795-11-14.

'Informe.' *Archivo General de Indias*, Indiferente 1556, Madrid, 1706-01-07.

Gauché, Joseph. 'Description d'un Quinquina indigène á St. Domingue, par Joseph Gauché, habitante, concessionnaire et administrateur des eaux thermales de Boynes, membre du Cercle des Philadelphes du Cap Français. Mémoire lu à l'Académie des Sciences, le 24 juillet 1787.' *La Bibliothèque centrale du Muséum national d'histoire naturelle*, Ms 1275, n.p., c. 1787.

Gomes, Bernardino António. 'OFÍCIO de Bernardino António Gomes ao (secretário de estado da Marinha e Ultramar), visconde de Anadia (João Rodrigues de Sá e Melo Meneses e Souto Maior).' *Arquivo Histórico Ultramarino*, 017 – RIO DE JANEIRO – CATÁLOGO DE DOCUMENTOS MANUSCRITOS AVULSOS / Cx. 243, D. 16604, Lisboa, 1807-03-20.

Gomes da Silva, Vicente. 'OFÍCIOS (14) de oficiais da marinha e comandantes de embarcações, ao (secretário de Estado dos Negócios da Marinha e Ultramar), visconde de Anadia, (D. João Rodrigues de Sá e Melo Meneses e Souto Maior).' *Arquivo Histórico Ultramarino*, 076 – REINO RESGATE 20121023 / Cx. 302-A, Pasta 5, Rio de Janeiro, 1806-02-04.

Grimaldi, Marques de. 'Carta a Miguel de Muzquiz.' *Archivo General de Simancas*, Legajo 907, Rome, 1779-10-21.

Hernandez de Gregorio, Manuel. 'Dn. Manuel Hernandez de Gregorio, Boticario de Camara presenta una memoria compuesta de 37 artículos, queriendo persuadir las grandes conveniencias de la estancación general, y parcial de la Quina en beneficio de la salud publica, y del interés del Real Erario, detallando las reglas gubernativas para su administración.' *Archivo General de Indias*, Indiferente 1556, Madrid, 1804.

Hueber, Johannes. 'Handbüchlein angefangen den 6ten Mey 1727 in Coullioure ein Meerport in der Provinz Roussillon, in den Pireneischen Gebirgen. Von Johannes Hueber Chirurgien in dem löblichen Regiment de D'Hemel Schweitzer, zu diensten dero Königl. Majeßtet in Franckreich.' *Archiv für Medizingeschichte*, Universität Zürich, Rezeptbücher, MS J 4, Collioure, 1727-05-06.

'Informe de la Contaduría de 9 de Julio de 1774 y respuesta del Sor Fiscal de 30 de Agosto del mismo año.' *Archivo General de Indias*, Indiferente 1554, Madrid, 1774-07-09 / 1774-08-30.

'Informe del Ministro de V.M. en Roma.' *Archivo General de Simancas*, Legajo 961/2, Roma, 1785.

'Instancia solicitando quina para los vasallos de la encomienda del Infante Duque de Parma.' *Archivo General de Simancas*, Legajo 907, n.p., 1775-07-13.

Jesus Maria, João de. 'Colleção medica de receitas p[ar]a quasi todos os achaques a q[ue] está sugeita a natureza humana, e muitas uezes exprimentadas com bom sucesso pellos melhores medicos, cirurgiõis, e quimicos deste reino, e estrangeiros: e muitas dellas tidas por segredos quasi infaliveis de cazas illustrissimas, e medicos grandes deste reino em muitos annos.' *Biblioteca Nacional de Portugal*, Manuscritos Reservados, COD. 5077, Lisboa, 1760.

'Juan Camacho, en el articulo que sige, contra Dn. Joseph Sanchez Barragan, mi suegro, sobre el derecho de cascarilla, que por accion de dn. Baleriano Guapulena, cacique del pueblo de San Miguel de Chimbo, tengo adquirida posecion en los Montes de Iluicachiuru pertenecientes a la Parzialidad de sus Yndios, paresco ante VM, por la persona que tiene mi Poder, que presento y juro.' *Archivo Nacional de la Historia*, Quito, Fondo General, Serie Cascarilla, Caja 2, Expediente 2, n.p., 1778-06-23.

'Kräuterrezepte und Hausmittel.' *Archiv für Medizingeschichte*, Universität Zürich, Rezeptbücher, MS J 29, Solothurn, c. 1767.

Laganes, Manuel de. 'El Guardián y Comunidad de este Convento de San Francisco de Descalzos de la V. de Ciempozuelos.' *Archivo del Palacio Real*, Copias de Ordenes comunicadas por el Ministro de Hacienda al Señor Sumiller de Corps. Real Botica, Reinados Carlos III / Legajo 197, 3, Ciempozuelos, 1775-07-06.

'Licencias de Cascarilla.' *Archivo Nacional de la Historia*, Quito, Fondo General, Serie Cascarilla, Caja 4, Expediente 3, Quito, 1793-05-22 – 1793-08-27.

'Livre des remedes de Madame la presidante de Maliverny seulement de sus dont ie fait le pruve tan de sus que l'on ma donne que sus que ie pris aux livres.' *Wellcome Library*, Archives and manuscripts, Closed stores WMS 4, MS.3409, Aix[-en-Provence], 1719–1739.

López Ruiz, Sebastián José. 'Carta.' *Archivo General de Indias*, Indiferente 1557, Santa Fé, 1804-08-13.

'Para los efectos que convengan remite copia de una representación que ha dirigido al Virrey en que refiere los abusos y desordenes que se cometen en la colectación beneficio y comercio de las quinas de aquel reyno.' *Archivo General de Indias*, Indiferente 1557, Santa Fé, 1804-08-13.

'Razon de los sugetos, que desde el dia 2 de Marzo, asta el 20 de Abril del año corriente de 1780, me han vendido Quina, y se la hé comprado a cuenta de la Rl. Hazienda en esta ciudad; con expresion de sus arrobas, y libras, el justo valor, que, por ellas les he satisfecho, de los nombres, y apellidos de los que me la han trahido; los Pueblos, y territorios, de donde me han asegurado ser vecinos; y de los Montes, de que igualmente me han dicho que la han sacado.' *Archivo General de Indias*, Indiferente 1554, Santa Fé, 1780-03-02 / 1780-04-20.

'Representación.' *Archivo General de Indias*, Indiferente 1554, n.p., 1779-11-22.

'Los felices efectos que ha hecho la Quina en los exercitos franceses.' *Archivo del Palacio Real*, Papeles del Almacén de la Quina, Caja 22283 / Expediente 7, Madrid, 1797–1798.

Losada, Duque de. 'Nota de varios encargos que hace el Rey de Marruecos.' *Archivo del Palacio Real*, Real Botica, Reinados Carlos III / Legajo 197, 3, Aranjuez, 1771-04-20.

Martínez de Beltrán, Luis. 'Oficio de D. Luis Martínez de Beltrán a Manuel Muzquiz, comunicándole que cuando lleguen los dos cajones de quina regalada a la Emperatriz Reina de Hungría, los hará seguir a su destino.' *Archivo General de Simancas*, Legajo 907, Genova, 1771-04-27.

'Medicin Buch von der verstorbenen Frauen Anna Katharina Haydtin.' *Wellcome Library*, Archives and manuscripts, MS.2797, n.p., 1756.

'Medicina Primitiva ou Colecção de Remedios Escolhidos e aprovados por experiências constantes, Para o uso das Pessoas do campo, dos Ricos e Pobres, traduzida do inglez, de Wesley sobre a décima terceira edição, revista, e aumentada consideravelmente, e Agora traduzida do Franzes na Lingua Portugueza.' *Arquivo Nacional da Torre do Tombo*, Manuscritos da Livraria, PT/TT/MSLIV/0134, London, 1760-11-10.

Molinari, Danielle. 'Ricete.' *Archiv für Medizingeschichte*, Universität Zürich, Rezeptbücher, MS J 8, Graubünden/ Chur, c. 1740–1762.

Moñino y Redondo, José (Conde de Floridablanca). 'El Procurador Sindico general del lugar de la Moraleja [...] ha representado [...] que teniendo cerca de quatrocientos tercianarios, para cuya curación carecen de los auxilios necesarios, está expuesto todo el Pueblo a un contagio perlífero, si no se le socorre con una competente porción de Quina para su pronto remedio.' *Archivo del Palacio Real*, Copias de Ordenes comunicadas por el Ministro de Hacienda al Señor Sumiller de Corps. Real Botica, Reinados Carlos III / Legajo 197, 3, San Ildefonso, 1788-09-13.

'El Rey ha mandado que á la Villa de Santa María la Real de Nieva se de una porción de Quina para mas de 300 enfermos pobres que tiene.' *Archivo del Palacio Real*, Copias de Ordenes comunicadas por el Ministro de Hacienda al Señor Sumiller de Corps. Real Botica, Reinados Carlos III / Legajo 197, 3, San Ildefonso, 1788-09-17.

'En carta confidencial que me ha escrito el capital general de Castilla la Vieja me dice que continúan las crueles tercianas de los Pueblos de aquella provincia.' *Archivo del Palacio Real*, Copias de Ordenes comunicadas por el Ministro de Hacienda al Señor Sumiller de Corps. Real Botica, Reinados Carlos III / Legajo 197, 3, San Lorenzo, 1788-11-01.

'Regalo de Alhajas y Joyas hecho por S.M. y A.A. al embiado de la Puerta Otomana Ahmet Vasif Effendi.' *Archivo General de Simancas*, Secretaría de Hacienda / Legajo 906, El Pardo, 1788-02-29.

Montes, Luis de. 'Relación de las Medicinas, y utensilios que he entregado al físico del Equadron de Usares, Dn. Geronimo Maria del Aguila, de las Recirculas del Profesor de Farmacia Dn. José Solano, y Razón a continuación de las existentes, en mi poder en estado de exclusión.' *Biblioteca Nacional del Perú*, Expediente que la Real Hacienda envía al Marqués de La Concordia, virrey del Perú incluyendo relación de las medicinas proporcionadas por el farmacéutico Luis de Montes, para el auxilio de los escuadrones de guerra, quien solicita el pago por este servicio, Sección Manuscritos - 1816 - XPB/D27, Lima, 1816.

Mutis, José Celestino. 'Borrador del oficio de José Celestino Mutis al virrey Pedro Mendinueta y Muzquis.' *Archivo del Real Jardín Botánico*, Real Expedición Botánica del Nuevo Reino de Granada (1783–1816). José Celestino Mutis. Correspondencia, RJB03/0002/0002/0172, Santa Fé de Bogotá (Colombia), 1801-10-24.

'Borrador y copia de un oficio de José Celestino Mutis al arzobispo virrey Antonio Caballero y Góngora comunicándole que ha redactado un "Plan de curación para las enfermedades que padecen las tropas del rey establecidas en el Darién" que consisten en diversas clases de calenturas y fiebres. Le notifica que ha enviado un cargamento de quina al hospital de estas tropas. Adjunto, borrador manuscrito de Mutis y copia del citado plan de curación, ambos incompletos.' *Archivo del Real Jardín Botánico*, Real Expedición Botánica del Nuevo Reino de Granada (1783–1816). José Celestino Mutis. Documentación oficial. Oficios de José Celestino Mutis, RJB03/0002/0002/0069, Mariquita (Colombia), 1786-05-18.

Muzquiz, Miguel de. 'Carta a Julián de Arriaga.' *Archivo General de Indias*, Indiferente 1554.

'El Rey ha resuelto que se embie al Señor Infante Duque de Parma un cajón de Quina.' *Archivo del Palacio Real*, Copias de Ordenes comunicadas por el Ministro de Hacienda al Señor Sumiller de Corps. Real Botica, Reinados Carlos III / Legajo 197, 3, San Lorenzo, 1772-12-22.

'El Rey quiere que se remitan a la Señora Infanta Duquesa de Parma dos arrobas de quina selecta.' *Archivo del Palacio Real*, Copias de Ordenes comunicadas por el Ministro de Hacienda al Señor Sumiller de Corps. Real Botica, Reinados Carlos III / Legajo 197, 3, El Pardo, 1783-03-30.

'En 31 de Diciembre del año próximo anterior me pasaron el Gefe de la Real Botica, y el primer ayuda de dicho Real Oficio la cuenta adjunta del consumo que en todo el año de 1777 hubo de quina.' *Archivo del Palacio Real*, Copias de Ordenes comunicadas por el Ministro de Hacienda al Señor Sumiller de Corps. Real Botica, Reinados Carlos III / Legajo 197, 3, n.p., 1778-09-12.

'En los años de 1752, 54 y 56 se sirvió el Rey mandar entregar varias porciones de Quina selecta para la curación de los terzados y enfermos del Real Hospital de las Minas de Almaden.' *Archivo del Palacio Real*, Copias de Ordenes comunicadas por el Ministro de Hacienda al Señor Sumiller de Corps. Real Botica, Reinados Carlos III / Legajo 197, 3, San Ildefonso, 1776-09-12.

'Entre otras cosas que el Rey envia al Papa quiere S.M. remitirle una arroba de quina.' *Archivo del Palacio Real*, Copias de Ordenes comunicadas por el Ministro de Hacienda al Señor Sumiller de Corps. Real Botica, Reinados Carlos III / Legajo 197, 3, San Ildefonso, 1771-07-27.

'Queriendo el Rey regalar a Su Hermana la Señora Electriz Viuda de Baviera veinte y cinco libras de quina de la mas selecta, lo participo a V.E. de orden de S.M.' *Archivo del Palacio Real*, Copias de Ordenes comunicadas por el Ministro de Hacienda al Señor Sumiller de Corps. Real Botica, Reinados Carlos III / Legajo 197, 3, El Pardo, 1780-02-18.

'Neste Secretaria da Junta do Proto-Medicato da Repartiçao de Medicina se achaõ enformes que deram os Medicos d'esta Corte, e Provincias, relativamente a os Queitos a que se lhes mandou responder por Ordem da mesma Junta dos quaes o seu theor he o Seguinte.' *Arquivo Nacional da Torre do Tombo*, Ministério do Reino / Negócios diversos do Físico-Mor, Maço 469/ Caixa 585, 4, Lisboa, 1799-04-17 / Tavira, 1799-05-06.

'Nota del cacao de Soconurco, Polbillo de Oaxaca, Vaynillas Tavaco y Quina que se ha distribuido a los sujetos y parages que irán mencionados.' *Archivo General de Indias*, Indiferente 1552, Cádiz, 1753-10-09.

'Notas de las Corachas de Cacao soconusco, Botes de Tabaco de Tabaco Havano, de media arroba cada uno; de Sevilla y negrillo de a seis libras; y Quina que su majestad envía de regalo a la Corte de Napoles en este año de 1773.' *Archivo General de Simancas*, Legajo 907, n.p., 1773-07-20.

'OFÍCIO (3) ao (secretário de Estado dos Negócios da Marinha e Ultramar) Martinho de Melo e Castro.' *Arquivo Histórico Ultramarino*, 076 – REINO RESGATE 20121023 / Cx. 30-A, Pasta 5, Faro (Lisboa), Caldas da Rainha, 1784-08-24 / 1787-05-08.

'Pareceres de los médicos sobre los efectos de la Quina de Santa Fé.' *Archivo del Palacio Real*, Papeles del Almacén de la Quina, Caja 22283 / Expediente 1, Madrid, 1784-12-19 / 1784-12-18.

Pinto de Miranda Montenegro, Caetano. 'OFÍCIO do governador e capitão-general da capitania de Mato Grosso Caetano Pinto de Miranda Montenegro ao (secretário de estado da Marinha e Ultramar) Rodrigo de Sousa Coutinho.' *Arquivo Histórico Ultramarino*, 010 – MATO GROSSO – CATÁLOGO DE DOCUMENTOS MANUSCRITOS AVULSOS / Cx. 34, D. 1791, Vila Bela, 1798-06-14.

Pombo, José Ignacio de. 'Carta de José Ignacio de Pombo a José Celestino Mutis explicando su informe sobre las quinas.' *Archivo del Real Jardín Botánico*, Real Expedición Botánica del Nuevo Reino de Granada (1783–1816). José Celestino Mutis. Correspondencia, RJB03/0001/0001/0288, Cartagena, 1805-10-30.

'Quina y Tabaco para los Regalos acostumbrados de las Cortes de Napoles y Toscana, y para Venecia.' *Archivo General de Indias*, Legajo 907, n.p., 1777-05-24.

'Quina y Tabacos para los Regalos que S.M. regala en este año a Napoles, Toscana y Venecia.' *Archivo General de Simancas*, Legajo 907, n.p., 1783-07-19.

'Real Cedula.' *Archivo General de Indias*, Indiferente 1552, Madrid, 1751-08-27.

'Receipts copied from Miss Myddleton's Book, August 15th, 1785. With many added receipts for remedies by various later hands, extracts, and pasted-in cuttings from newspapers, etc.'. *Wellcome Library*, Archives and manuscripts, Closed stores WMS 4, MS.3656, n.p., c. 1785–1818.

'Receitas medicinais para diversas enfermidades.' *Biblioteca Nacional de Portugal*, Manuscritos Reservados, COD. 6177//3, n.p., 1759.

'Regalo al Papa.' *Archivo General de Simancas*, Legajo 907, n.p., 1774-08-30.

'Regalos hechos a la Corte de Constantinopla con motivo de la Paz, concluida en este año.' *Archivo General de Simancas*, Secretaría de Hacienda / Legajo 906, n.p., 1785.

'Relacion puntual de los Regalos de cacao soconusco, tabaco habano de Sevilla y Quina, que S.M. envía por la via de Alicante, y Genova, a sus Magestades

Sicilianas, y demás particulares de la Corte de Napoles en el presente año de 1786, como se hizo en el 1785.' *Archivo General de Simancas*, Legajo 907, n.p., 1786.

'Relatibo a la remision que han hecho a Espana los Comisionados por el acopio de Quina electa de la Ciudad de Loxa de diez mil novecientas libras de este especifico.' *Archivo Nacional de la Historia*, Quito, Fondo General, Serie Cascarilla, Caja 5, Expediente 13, Loja, 1814-09-26.

'Remedios singulares usados por los Misioneros de tierra firme, q[u]e se allan en el Apendice I° de la H[istori]a del Orinoco del Abate Felipe Salvador Gilij. tomo 3° en Roma año 1782.' *Wellcome Library*, Archives and Manuscripts, Closed stores WMS/Amer.8, n.p., 1782.

Rieux, Luis de. 'Carta a Miguel Cayetano de Soler.' *Archivo General de Indias*, Indiferente 1557, Aranjuez, 1800-05-14.

Rosuela, Joseph. 'Pertenece este al uso de Frai Joseph Rosuela.' *Wellcome Library*, Archives and manuscripts, WMS/Amer.22-24, San Diego, 1771-03-22.

Saavedra, Francisco de, and Felipe Carrillo. 'Informe sobre una porción de Quina pedida al Rey por el Sindico Procurador gral. de la Isla de Tenerife.' *Archivo del Palacio Real*, Caja 22283 / Expediente 7, San Lorenzo, 1797-12-06 / Canarias, 1797-10-13.

Salazar, Matias de. 'Zertificacion de Don Matias de Salazar, Vessino de la Ciudad de Loxa, residente como hasendado en el Pueblo de Villcabamba, y Comissionado para selar la extraccion de Cascarilla de los serros desde Pais de Malacatos.' *Archivo General de Indias*, Quito 239, Loja, 1773.

San Martin Cueto, Miguel de. 'Razon de las libras de cascarilla que se han extrahido para fuera del reyno.' *Archivo General de Indias*, Indiferente 1554, Cádiz, 1776-10-25.

Sánchez de Orellana y Riofrío (Marqués de Villa de Orellana), Clemente. 'Dictamen sobre la conveniencia, ó perjuicio, que estancandosse la Quina pueda resultar a la causa Publica, y comercio de estas Provincias.' *Archivo General de Indias*, Indiferente 1554, n.p., 1776.

Santisteban, Miguel de. 'Copia de Carta.' *Archivo General de Indias*, Indiferente 1554, Santa Fé, 1753-06-04.

'Séance du Mardi 30 Juin. La Société m'a chargé de porter sur ses plumitifs le résumé suivt. concernant les différentes especes de quinquina qui ont été soumises á son examen.' *Bibliothèque de l'Académie de médecine*, Procès-verbaux des séances de la Société Royale de la Médicine, Ms 11/11, Paris, 1789-06-30.

'Secretos medicos, y chirurgicos.' *Wellcome Library*, Archives and manuscripts, Closed stores WMS / Amer. 22, Mexico, n.d.

'Sobre el acopio de la Quina de los Montes de Loxa Callysalla y otros que la produzcan de buena calidad, y su envio a Espana de cuenta de la Rl. Hazienda.' *Archivo Nacional de la Historia*, Quito, Fondo General, Serie Cascarilla, Caja 3, Expediente 13, Cuenca, 1790-08-26.

'Sobre la conservacion de Montes de Cascarilla de la Prov.a de Loxa y proveymiento de este Genero para la Real Botica.' *Archivo Nacional de la Historia*, Quito, Fondo General, Serie Cascarilla, Caja 2, Expediente 11, Loja, 1782-02-21 / 1783-04-27.

Sor María Margarita. 'La Abadesa y Religiosas del Convento de capuchinas de la ciudad de Huesca en el Reyno de Aragon con el debido respetto a los Rs. Ps. de V.M. exponen.' *Archivo del Palacio Real*, Reinados Carlos III / Legajo 197, 3, Huesca, 1759–1788.

'Tabaco, y Quina para la Señora Infanta Duquesa de Parma.' *Archivo General de Simancas*, Legajo 907, n.p., 1786-01-22.

'Tacabo y Quina para la Señora Infanta Duquesa de Parma.' *Archivo General de Simancas*, Legajo 907, n.p., 1783.

'Tasación hecha por Manuel Seminario de la botica que Juan Lucas Camacho legó al Hospital de la Caridad. Se incluye inventario de los bienes existentes en dicha botica, indicando precios.' *Archivo Histórico / Instituto Riva Agüero* PUCP, Colección Maldonado, A-I-42, Lima, 1759-09-13.

Tavares, Francisco, José Correa Pieanço, and José Francisco Oliveira. 'Sendo S. Magestade Servida Determinar por Avizo expedido pela Secretaria d'Estado dos Negocios do Reyno em data de seis d'Abril d'este corrente anno que a Junta do Proto-Medicato naõ comprehenda a André Lopes de Castro no processo e execuçao annunciados no Edital de 15 de Março; e que antes suspendendo todo o procedimento, e conservando tudo procedimento, e conservando tudo o que respeita a esta dependencia de que trata a representaçaó no seu primeiro estado, faça presente a S. Mag[estad]e em Consulta o que á Junta occorrer para S. Mag(estad)e revolver a que fosse do seu Serviçio.' *Arquivo Nacional da Torre do Tombo*, Ministério do Reino / Negócios diversos do Físico-Mor, Maço 469/ Caixa 585, Lisbon, 1799-08-12.

'Testimonio de los Autos en que se comprehenden varios Informes y Diligencias practicadas en virtud de Real Cedula, sobre si sera, o no combeniente el Estanco de la Cascara Quina.' *Archivo Nacional de la Historia*, Quito, Fondo General, Serie Cascarilla, Caja 1, Expediente 11, Santa Fé, 1777-12-10.

'Testimonio de los Autos, sobre que se establezcan el estanco de la Quina, o Cascarilla, en virtud de la real orden de S. M.' *Archivo General de Indias*, Indiferente 1554, Cuenca, 1776-07-29 / Villa Orellana, 1776-08-18.

Toledo, Alexandro. 'En el Pueblo de Malacatos en dicho dia mes y año para la dicha Informacion, comparecio Alexandro Toledo.' *Archivo General de Indias*, Quito 239, Malacatos, 1773-09-17.

Valdez, Antonio. 'El Presbítero Don Antonio María Chacón, Administrador del Real Hospital de San Fernando del Sitio de San Ildefonso, ha hecho presente que con motivo de haberse aumentado considerablemente el vecindario de aquel pueblo, ha crecido también el numero de los pobres enfermos.' *Archivo del Palacio Real*, Copias de Ordenes comunicadas por el Ministro de Hacienda al Señor Sumiller de Corps. Real Botica, Reinados Carlos III / Legajo 197, 3, San Lorenzo, 1787-11-21.

Vallano y Cuesta, Manuel. 'Auto para manifestar el deplorable estado en que se hallan los Montes que producen la Quina selecta que se ha remitido a la Real Botica de S.M. En la ciudad de la Concepción de Loxa à diez días del mes de Mayo de mil setecientos ochenta y ocho años el señor Don Manuel Vallano y Cuesta Corregidor, Juez Conservador, y especial comisionado del ramo de cascarilla de esta ciudad por S.M.' *Archivo General de las Indias*, Malacatos, 1788-05-16.

'Varios Papeles pertenecientes á la Quina del Péru.' *Archivo del Palacio Real,* Papeles del Almacén de la Quina, Caja 22282 / Expediente Número 6, Madrid, 1773-02-12.

Villaluenga, Juan José. 'Copia del oficio de Juan José Villaluenga del 18 de julio de 1786 al arzobispo virrey Antonio Caballero y Góngora comentando una donación anual de quina a fray Juan Antonio Gago, prior del Hospital de San Juan de Dios, de Cassa Plaza, para su administración a enfermos pobres.' *Archivo del Real Jardín Botánico,* Real Expedición Botánica del Nuevo Reino de Granada (1783–1816). José Celestino Mutis. Documentación oficial. Oficios varios, RJB03/0002/0003/0097, Quito/Turbaco, 1786-10-08.

Volmar, Paul. 'Gehört Paulus Volmar. Med. Pract.' *Archiv für Medizingeschichte,* Universität Zürich, Rezeptbücher, MS J 4, Winterthur, 1772.

White, Mary, and Mary Downing Cartwright. 'Reciepts [sic] Physick & Chirugery.' *Wellcome Library,* Archives and manuscripts, Closed stores WMS 4, MS.8300, n.p., 1715–1719.

Printed Sources

Alsinet, Joseph, *Nuevas utilidades de la quina, demostradas, confirmadas y añadidas por el Doctor Don Josef Alsinet, Medico de la Familia de su Magestad, y Jubilado del Real Sitio de Aranjuez. Se manifiesta el modo cómo cada uno en su casa podrá quitar el amargor á la Quina, sin perjuicio de su virtud febrífuga* (Madrid: Don Miguél Escribano, 1774).

Anon., 'El Medico verdadero. Prontuario singular de varios selectisimos remedios, para los diversos males à que està expuesto el Cuerpo humano desde el instante que nace. Compuesto por un curioso, para el alivio de todos los que se quieran curar con èl,' in *La medicina popular peruana,* ed. Hermilio Valdizán and Angel Maldonado (Lima: Imprenta Torres Aguirre, 1922 (1777)), 417–87.

'Recitario eficaz para las familias. Medicamentos caseros,' in *La medicina popular peruana,* ed. Hermilio Valdizán and Angel Maldonado (Lima: Imprenta Torres Aguirre, 1922 [n.d.]), 107–316.

Aydüz, Salim, and Esma Yildirim, 'Bursalı Ali Münşî ve Tuhfe-i Aliyye. Kına Kına Risâlesi Adlı Eserinin Çevirisi.' *Yeni Tıp Tarihi Araştırmaları* 8 (2002), 85–105.

Badus, Sebastianus, *Anastasis corticis Peruviae; seu chinae chinae defensio, Sebastiani Badi Genvensis patrij vtriusque Nosochomij olim medici, et Publicae Sanitatis in Ciuitate Consultoris. Contrà ventilationes Joannis Jacobi Chifletii, gemitusque Vopisci Fortunati Plempii, Illustrium Medicorum* (Genoa: Typis Petri Joannis Calenzani, 1663).

Bailey Thomas, Robert, *The Farmer's Almanack, Calculated on a New and Improved Plan, for the Year of our Lord 1801. Fitted to the Town of Boston, but will serve for any of the adjoining states* (Boston: Manning & Loring, 1800).

Baker, George, 'Observations on the Late Intermittent Fevers to which is added a Short History of the Peruvian Bark.' *Medical Transactions* III (1785), 141–216.

Baldinger, E. G., 'Geschichte der Chinarinde und ihrer Wirkungen.' *Magazin vor Aerzte* 7 (1778), 993–1030; 49–67.

Baumes, Jean Baptist Timothee, *Traité des fièvres rémittentes et des indications qu'elles fournissent pour l'usage de quinquina*, 2 vols., vol. 2 (Montpellier: Sevalle, 1821).

Bergen, Heinrich von, *Versuch einer Monographie der China* (Hamburg: Hartwig & Müller, 1826).

Bigelow, Jacob, *A treatise on the materia medica: intended as a sequel to the Pharmacopoeia of the United States: being an account of the origin, qualities and medical uses of the articles and compounds, which constitute that work, with their modes of prescription and administration* (Boston: Ewer, 1822).

Blair, William, *The Soldier's Friend: or, the Means of Preserving the Health of Military Men; Addressed to the Officers of the British Army* (London: Mr. Longman et al., 1798).

Blane, Gilbert, *A Short Account of the Most Effectual Means of Preserving the Health of Seamen, particularly in the Royal Navy, to the Flag-Officers and Captains of his Majesty's Ships of War on the West-India Station* (Sandwich, off Antigua: – 1780).

Brown, George, 'The concentrated tincture of yellow Peruvian bark: prepared by George Brown, chemist, at no. 79, St. Paul's Church Yard: a certain cure for the intermittents, and very useful in all complaints which require large doses of the bark.' *Wellcome Library*, Med. ephemera, Drug Advertising: Pre-1850: Box 1, EPH381G. 19, London, 1805.

Bruce, James, *Travels to discover the source of the Nile, in the years 1768, 1769, 1770, 1771, 1772, and 1773*, vol. 5 (Edinburgh / London: J. Ruthven / J. Robinson, 1790).

Buchan, William, *Domestic Medicine: Or, a treatise on the prevention and cure of diseases* (London: W. Strahan, 1774).

Domestic Medicine, or, the Family Physician: Being an Attempt To Render the MEDICAL ART more generally useful, by shewing people what is in their own power both with respect to the PREVENTION and CURE of Diseases. CHIEFLY Calculated to recommend a proper attention to REGIMEN and SIMPLE MEDICINES (Edinburgh: Balfour, Auld, and Smellie, 1769).

Medicina doméstica (Madrid: Imprenta Real, 1785).

Medicina domestica, ou Tratado de prevenir, e curar as enfermidades com o regimento, e medicamentos simplices, escripto em inglez pelo D.r Guilherme Buchan, Socio do Collegio dos Medicos de Edimburgo, Traduzido em Portuguez, com varias notas, e observações concernentes ao Clima de Portugal, e do Brazil, com o receituario correspondente, e hum Appendice sobre os Hospitaes Navaes, Cura, e Dieta dos Enfermos dos mesmos Hospitaes, por Manoel Joaquim Henriques de Paiva, 4 vols. (Lisboa: Typografia Morazziana, 1787).

Bueno, Cosme, *El conocimiento de los tiempos, Efemeride del Año de 1796. Bisiesto, en que van puestos los principales Aspectos de la Luna con el Sol, calculados para el Meridiano de esta muy Noble, y muy Leal Ciudad de LIMA, Capital y Emporio de esta América Meridional* (Lima: Imprenta Real, 1795).

El conocimiento de los tiempos; efemeride del año de 1788. Bisiesto, en que van puestos los principales Aspectos de la Luna con el Sol, calculados para el Meridiano de esta muy Noble, y muy Leal Ciudad de LIMA, Capital y Emporio de esta América Meridional (Lima: Imprenta Real, 1787).

Caldas, Francisco José de, *Memoria sobre el estado de las quinas en general y en particular sobra la de Loja*, ed. Federico Gonzalez Suarez, Anales de la Universidad Central de Quito (Quito: Tipografia y Encuadernación Saletiana, 1907 (1809)).

Campenhausen, Balthasar von, *Bemerkungen über Rußland, besonders einige Provinzen dieses Reiches und ihre Naturgeschichte betreffend* (Leipzig: Friedrich Christian Dürr, 1807).

Canals, Tomas, *Tratado doméstico de algunas enfermedades bastante comunes en esta capital* (Lima: Imprenta Real del Telegrafo Peruano, 1800).

Castiglioni, Luigi, *Storia delle piante forastiere le piu importanti nell'uso medico, od economico* (Milano: Stamperia di Giuseppe Marelli, 1791).

Citte, Francois, 'De l'usage du quinquina et des règles de son application dans les fièvres intermittentes,' in *Collection des thèses soutenues a l'école de médicine de Montpellier pendant l'an XII*, ed. L'École de Médicine de Montpellier (Montpellier: G. Izar et A. Richard, 1804), 3–23.

Clark, James, *A Treatise on the Yellow Fever, as it appeared in the Island of Dominica, in the Years 1793-4-5-6; to which are added, Observations on the Bilious Remittent Fever, on Intermittents, Dysentery, and Some Other West Indian Diseases* (London: J. Murray & S. Highley, 1797).

Clark, John, *Observations on Diseases which prevail in Long Voyages to Hot Countries, Particularly on those in the East Indies, and on the Same diseases as they appear in Great Britain*, 2 ed. (London: J. Murray, 1792).

Clarke, Edward Daniel, *Travels in Various Countries of Europe, Asia and Africa. Part the First: Russia, Tahtary and Turkey*, 4 ed., 2 vols., vol. 2 (London: T. Cadell and W. Davies, 1817).

Cobo, Bernabé, *Inca Religion and Customs*, ed. John Howland Rowe, trans. Roland Hamilton (Austin: University of Texas Press, 1990 (1653)).

Cockburn, William, *The Present Uncertainty in the Knowledge of Medicines in a Letter to the Physicians in the Commission for Sick and Wounded Seamen* (London: Benj[amin] Barker, 1703).

Colingwood, T., 'Observations on the Peruvian Bark.' *Medical Commentaries* X (1785).

Collingwood, Thomas. 'Observations on the Peruvian Bark.' *Medical Commentaries* X (1785), 265–75.

Crespo Nolasco, Pedro. 'Carta apologética de la quina o cascarilla.' *Mercurio Peruano (Lima)* 8 (1795 [1861]).

Crest de Saint-Aubin, Stéphanie Félicité du, Countess de Genlis, *Zuma, ou la découverte du quinquina* (Paris: Maradan, Libraire, 1818).

Curtis, Samuel, *A Valuable Collection of Recipes, Medical and Miscellaneous. Useful in Families, and Valuable to Every Description of Persons* (Amherst, N.H.: Elijah Mansur, 1819).

Curvo Semedo, João, *Polyanthea Medicinal. Noticias galenicas, echymicas. Repartidas em tres Tratados, dedicadas a's saudisas memorias, e veneradas cinzas do eminentissimo senhor Cardenal de Sousa, Arcebispo de Lisboa, Capellam Mor do Serenissimo Senhor Rey Dom Pedro II & seu Conselheyro de Estado* (Lisboa: Antonio Pedrozo Galram, 1727).

Dancer, Richard, *Dissertatio Medica Inauguralis de Cinchona; quam annuente summo numine, ex auctoritate reverendi admodum viri, D. Georg II Baird, SS.T.P. Academiae Edinburgenae Praefecti; Necnon Amplissimi senatus academici consensu; et nobilissimae facultatis medicae decreto; pro gradu doctoris, summisque in medicina honoribus ac privilegiis rite et legitime consequendis; eruditorum examini subjicit Ricardus Dancer* (Edinburgh: C. Stewart, 1809).

Dancer, Thomas, *The Medical Assistant; or Jamaica Practice of Physic: Designed chiefly for the Use of Families and Plantations* (Kingston, Jamaica: Alexander Aikman, 1801).

Dávalos, José Manuel de, 'Specimen Academicum de morbis nonnullis Limae grassantibus, ipsorumque Therapeia.' *Journal de Médicine* (1787), 5–136.

Davis, John B., *A scientific and popular view of the fever of Walcheren, and its consequences, as they appeared in the British troops returned from the late expedition; with an account of the morbid anatomy of the body, and the efficacy of drastic purges and mercury in the treatment of this disease* (London: Samuel Tipper, 1810).

Documentos régios que authorizão a verdadeira Agoa de Inglaterra, da composição do doutor Jacob de Castro Sarmento, manipulada presentemente por José Joquim de Castro, na sua Real Fabrica, por decreto de sua Alteza Real o Príncipe Regente nosso Senhor. Com huma relação dos professores de medicina, e cirurgia deste reino de Portugal e seus Dominios, que tem attestado a excellencia da dita Agoa de Inglaterra (Lisboa: Impressão Régia, 1809).

Dufau, Julien, *Essai sur l'application du quinquina dans le traitement des fièvres intermittentes, Collection des theses soutenues a l'ecole de médicine de Paris* (Paris: Didot Jeune, 1805).

Esteyneffer, Juan de, *Florilegio medicinal de todas las enfermedades, sacado de varios, y clasicos autores, para bien de los pobres, y de los que tienen falta de Medicos, en particular para las Provincias remotas, en donde administran los RR. PP. Missioneros de la Compañia de Jesus* (Mexico: Herederos de Juan Joseph Guillena Carrascoto, 1712).

Ewell, James, *The planter's and mariner's medical companion: treating, according to the most successful practice, I. The diseases common to warm climates and on ship board. II. Common cases in surgery, as fractures, dislocations, &c. &c. III. The complaints peculiar to women and children. To which are subjoined, a dispensatory, shewing how to prepare and administer family medicines, and a glossary, giving an explanation of technical terms* (Philadelphia: John Bioren, 1807).

Fernández de Oviedo y Valdés, Gonzalo, *Primera parte de la historia natural y general de las indias, yslas e tierra firme del mar oceano* (Sevilla: Iuam Cromberger, 1535).

Ferreira Furtado, Júnia, ed. *Erário mineral* (Rio de Janeiro: Editora FIOCRUZ, 2002 (1735)).

Fodéré, François-Emmanuel, *Recherches expérimentales, faites à l'hôpital civil et militaire de Martigues, sur la nature des fièvres á périodes et sur la valeur des différens remèdes substitués au Quinquina,* (Marseille: Jean Mossy, 1810).

Gil y Alveniz, Manuel, *Colección de Memorias Médicas: Contiene una memoria premiada por la Real Academia de Medicina práctica de Barcelona, sobre las epidemias generales de España en los años de 1803 y 4. La práctica de la vacunación en cinco memorias presentadas y aprobadas por la Real Junta Superior Gubernativa de Medicina. Y un tratado sobre la pronta, fácil, segura y económica curación de las calenturas intermitentes por medio de la quina* (Madrid: Iearra, 1820).

Gonzalez, Pedro Maria, *Tratado de las enfermedades de la gente de mar, en que se exponen sus causas, y los medios de precaverlas* (Madrid: Imprenta Real, 1805).

Gray, John, William Arrot and Phil Miller, 'An Account of the Peruvian or Jesuits Bark.' *Philosophical Transactions* 40 (1737/38), 81–86.

Gunn, John C., *Gunn's Domestic Medicine: Or Poor Man's Friend, in the Hours of Affliction, Pain and Sickness: This Book Points Out, in Plain Language, Free from Doctor's Terms, the Diseases of Men, Women, and Children, and the Latest and Most Approved Means Used in Their Cure and is Expressly Written for the Use of Families in the Western and Southern States* (Springfield: John M. Gallagher, 1835).

Helvétius, Claude-Adrien, *Manière de donner le quinquina aux pauvres* (Versailles: Imprimerie François Muguet, 1686).

Henriques de Paiva, Manuel Joaquim, *Farmacopéa Lisbonense, ou Collecçao dos simplices, preparaçoes e composiçoes mais efficazes e de major uso* (Lisboa: Officina de Filippe da Silva e Azevedo, 1785).

Memoria sobre a excellencia, virtudes, e uso medicinal da verdadeira agua de Inglaterra da invenção do Dr. Jacob de Castro Sarmento, Membro do Real Collegio de Medicos de Londres, e Socio da Sociedade Real. & c. Actualmente preparada por José Joaquim de Castro, na sua Real Fabrica, por Decretos de Sua Alteza Real o Principe Regente N.S. ordenada por M. J. H. de P. Impressa na Bahia na Typog. de Manoel Antonio da Silva Serva, no anno de 1815, com as licenças necessarias (Lisboa: Impressão Regia, 1816).

Hernández de Gregorio, Manuel, ed. *El arcano de la quina. Discurso que contiene la parte médica de las cuatro especies de quinas oficinales, sus virtudes eminentes y su legítima preparación. Obra póstuma del doctor D. José Celestino Mutis, Director u Gefe de la espedición botánica de Santa Fé de Bogotá en el nuevo reyno de Granada* (Madrid: Ibarra, Impresor de Cámara de S. M., 1828).

Hill, John, *A History of the Materia Medica: Containing Descriptions of all the Substances used in Medicine; their Origin, their Characters when in Perfection, the Signs of their Decay, their Chymical Analysis, and an Account of their Virtues, and of the several Preparations from them now used in the Shops* (London: T. Longman, C. Hitch and L. Hawes, 1751).

Humboldt, Alexander von, *Ensayo político sobre el reino de la Nueva España*, trans. Vicente Gonzalez Arnao, vol. 4 (Paris: Rosa, 1822).

Ideen zu einer Geographie der Pflanzen: Nebst einem Naturgemälde der Tropen-länder: auf Beobachtungen und Messungen gegründet, welche vom 10ten Grade nördlicher bis zum 10ten Grade südlicher Breite, in den Jahren 1799, 1800, 1801, 1802 und 1803 angestellt worden sind (Tübingen / Paris: F. G. Cotta / F. Schoell, 1807).

Hunter, John, *Observations on the Diseases of the Army in Jamaica: And on the Best Means of Preserving the Health of Europeans, in that Climate* (London: Printed for G. Nicol, Pall-Mall, Bookseller to His Majesty, 1788).

Huth, Georg Leonhart, *Sammlung verschiedener die Fieberrinde betreffender Abhan-dlungen und Nachrichten* (Nürnberg: Seeligmann, 1760).

Huxham, John, D. *Johann Huxham, Mitglieds der Königlichen Gesellschaft in London, Abhandlung von Fiebern, welche von der Beschaffenheit des Geblütes herrühren* (München: Merz u. Mayer, 1756).

Essai sur les différentes espèces de fièvres, avec des dissertations sur les fievres lentes, nerveuses, putrides, pestilentielles & pourprées (Paris: D'Houry, Impr. Lib. de Mgr. le Duc D'Orléans, 1776).

An Essay of Fevers, and Their Various Kinds: As Depending on Different Constitutions of the Blood: with Dissertations on Slow Nervous Fevers; on Putrid, Pestilential, Spotted Fevers; on the Small-pox; and on Pleurisies and Peripneumonies (London: S. Austen, 1750).

Irving, Ralph, *Experiments on the Red and Quill Peruvian Bark: with Observations on its History, Mode of Operation, and Uses* (Edinburgh: C. Elliot, 1785).

Jackson, Robert, *An Outline of the History & Cure of Fever, Endemic and Contagious; More Expressly the Contagious Fevers of Jails, Ships, & Hospitals, the Concentrated Endemic, Vulgarly Called the Yellow Fever of the West Indies* (Edinburgh: John Meir, 1808).

Jijon y León, Miguel de, 'Recomendaciones para la explotación y comercialización real de la cascarilla.' *Revista del Archivo Nacional de Historia, Sección del Azuay* 6 (1986 (1776)), 127–45.

Juan XXI (Pedro Julião / Pedro Juliano), *Libro de medicina llamado Tesoro de pobres: en que se hallaràn remedios muy aprouados para la sanidad de diuersas enfermedades* (Sevilla: Nicolas Rodriguez, 1655).

Kentish, Richard, *Experiments and observations on a new species of Bark, shewing its great efficacy in very small doses* (London: J. Johnson, 1784).

La Condamine, Charles-Marie de, 'Sur l'arbre du quinquina.' *Mémoires de l'Academie Royale* MDCCXL (1738 (1737)), 226–44.

Lambert, Aylmer B., *An illustration of the genus Cinchona: Comprising Descriptions of all the Officinal Peruvian Barks, incl. Several New Species. Baron de Humboldt's Account of the Cinchona Forests of South America and Laubert's Memoir of the Different Species of Quinquina* (London: Searle, 1821).

Lambert, Aylmer Bourke, *A description of the genus Cinchona, comprehending the various species of vegetables from which the Peruvian and other barks of a similar quality are taken. Illustrated by figures of all the species hitherto discovered. To which is prefixed Professor Vahl's dissertation on this genus, read before the Society of natural history at Copenhagen* (London: B. and J. White, 1797).

Larruga, Eugenio D., *Memorias políticas y económicas sobre los frutos, comercio, fábricas y minas de España*, vol. XXXV (Madrid: Por don Antonio Espinosa, 1795).

Lewis, William, and John Rotheram, *The Edinburgh new dispensatory: with the additions of the most approved formulae, from the best foreign pharmacopoeias; the whole interspersed with practical cautions and observations; and enriched with the latest discoveries in natural history, chemistry, and medicine; with new tables of elective attractions of antimonial and mercurial preparations, &c.; and several copperplates of the most convenient furnaces, and principal pharmaceutical instruments; being an improvement of the New dispensatory by Dr. Lewis* (Walpole, Newhampshire: Thomas & Andrews, 1796).

Lind, James, *An Essay on Diseases incidental to Europeans in hot Climates with the Method of Preventing their fatal Consequences. To which is added, an Appendix Concerning Intermittent Fevers, and, a simple and easy Way to render Sea Water fresh, and to prevent a Scarcity of Provisions in long Voyages at sea* (London: Printed for J. Murray, 1788).

Long, Edward, *The History of Jamaica. Or, General Survey of the Antient and Modern State of that Island: With Reflections on Its Situation, Settlements, Inhabitants, Climate, Products, Commerce, Laws, and Government*, 3 vols., vol. 2 (London: T. Lowndes, 1774).

Lyall, Robert, *The Character of the Russians, and a Detailed History of Moscow* (London: T. Cadell, in the Strand, and W. Blackwood, 1823).

Mallet, M., *Sur le Quinquina de la Martinique, connu sous le nom de Quinquina-Piton* (Paris: 1779).

Masdevall, Joseph, *Bericht über die Epidemien von faulen und bösartigen Fiebern welche in den letzten Jahren im Fürstenthum Catalonien geherrscht haben nebst der glücklichen, geschwinden und sichern Heilmethode dieser Krankheiten*, trans. E. H. Spohr (Braunschweig: Schulbuchhandlung, 1792).

Medicamens, et précis de la méthode de Mr. Masdevall docteur médecin du Roi d'Espagne Charles IV, pour guérir toutes les maladies epidémiques, putrides & malignes, fiévres de différents genres &c. & pour en préserver. Divisés en paragraphes & en numéros correspondans, à l'usage des familles dépourvues de médecins (New Orleans: Chez Louis Duclot, 1796).

Relacion de las epidemias de calenturas pútridas y malignas que en estos últimos años se han padecido en el Principado de Cataluña y principalmente de la que se descubrió el año pasado de 1783 en la ciudad de Lérida, Llano de Urgel y otros muchos corregimientos y partidos, con el método feliz, pronto y seguro de curar semejantes enfermedades (Madrid: Imprenta Real, 1797).

Relación de las epidemias de calenturas pútridas y malignas, que en estos últimos años se han padecido en el Principado de Cataluña; y principalmente de la que se descubrió el año pasado de 1783 en la ciudad de Lérida, Llano de Urgel y otros muchos Corregimientos y Partidos, con el método feliz, pronto y seguro de curar semejantes enfermedades (Barcelona: Imprenta Real, 1786).

Relazione dell'epidemie sofferte nel principato di Catalogna principalmente nell'anno 1783. Scritta in lingua spagnuola dal nobile sig. dottor Giuseppe Masdevall, attuale medico di camera del re cattolico: In cui si espone il suo nuovo metodo specifico per guarire le febbri putride, maligne, ed altre simili malattie: Tradotta, ed illustrata di nuove osservazioni, e memorie sino all'anno 1788 dall'abate Pietro Montaner (Ferrara: Per gli eredi di Giuseppe Rinaldi, 1789).

Mendes, José Antonio, *Governo de Mineiros mui necessário para os que vivem distantes de professores, seis, oito, dez e mais léguas, padecendo por esta causa os seus domésticos e escravos, queixas que pela dilação dos remédios se fazem incuráveis, e as mais das vezes mortais* (Lisboa: Officina de Antônio Rodrigues Galhardo, 1770).

Monardes, Nicolás, *Primera y segunda y tercera partes de la historia medicinal, de las cosas que se traen de nuestras Indias Occidentales, que siruen en Medicina* (Sevilla: Alonso Escriuano, 1574).

Moreau de Maupertuis, Pierre Louis, *Lettre sur le progrès des sciences* (Berlin: 1752).

Motherby, George, *A New Medical Dictionary; Or, General Repository of Physic Containing an Explanation of the Terms, and a Description of the Various Particulars Relating to Anatomy, Physiology, Physic, Surgery, Materia Medica, Pharmacy &c. &c. &c.* (London: J. Johnson, 1775).

Murray, Johan Andreas, *Johan Andreas Murray's Vorrath von einfachen, zubereiteten und gemischten Heilmitteln, zum Gebrauche praktischer Aerzte bearbeitet*, ed. Ludwig Christoph Althof, 2 vols., vol. 1 (Göttingen: Johann Christian Dieterich, 1793).

Mutis, Josè Celestino, 'Carta testimonial de Mutis al dr. Francisco Martínez Sobral,' in *José Celestino Mutis*, ed. Federico Gredilla (Bogotá: Academia Colombiana de Historia, 1982 (1789)), 89–92.

Instrucción formada por un facultativo existente por muchos años en el Perú, relativa de las especies y virtudes de la quina (Cádiz: Don Manuel Ximenez Careño, 1792).

Nunes Ribeiro Sanches, António, *Tratado da conservaçam da saude dos povos. Obra util, e igualmente necessaria aos Magistrados, Capitaens Generaes, Capitaens de Mar, e Guerra, Prelados, Abbadessas, Medicos, e Pays de familias. Com hum appendix: Consideraçoens sobre os terremotos, com a noticia dos mais consideraveis, de que faz mençaõ a Historia, e deste ultimo, que se sentio na Europa no 1 de Novembro de 1755* (Lisboa: Na Officina de Joseph Filippe, 1757).

O'Ryan, Michael, *A Letter on the Yellow Peruvian Bark, Containing an Historical Account of the first Introduction of that Medicine into France, and a Circumstantial Detail of Its Efficacy in Diseases, Addressed to Dr. Relph, Physician to Guy's Hospital* (London: J. Nunn, 1794).

'Op Woensdag den 19 Juny 1793, 's morgens ten 11 uren precies, zal men, te Leverpool, publyk veilen en verkoopen: de GANTSCHE LADING, van het Schip Le Federatif, van St. Domingo komende en naar Bourdeaux gedestineert gewest zynde.' *Rotterdamsche Courant N° 70*, 1793-06-11, 3.

Padréll et Vidal, Joseph, 'Dissertation sur l'usage et l'abus du quinquina dans le traitement des fièvres intermittentes; présentée et soutenue á l'École de Médicine de Montpellier le 23 Prairial an 10 (de la République),' in *Collection des thèses soutenues a l'École de Médicine de Montpellier*, ed. L'École de Médicine de Montpellier (Montpellier: Imprimerie de G. Izar e A. Ricard, 1802).

Percival, Thomas. 'Experiments on the Peruvian Bark.' *Philosophical Transactions* LVII (1767).

Pereyra, Luis Joseph, *Tratado completo de Calenturas: fundado sobre las leyes de la Inflamacion, y Putrefaccion, que constantemente observaron los mayores, y mas ilustrados medicos del mundo* (Madrid: Imprenta de Antonio Marin, 1768).

Pharmacopoea Rossica (St Petersburg: Academiae Scientiarum, 1778).

Pinckard, George, *Notes on the West Indies written during the expedition under the command of the late General Sir Ralph Abercromby: including observations on the island of Barbadoes, and the settlements captured by the British troops, upon the coast of Guiana; likewise remarks relating to the Creoles and slaves of the western colonies, and the Indians of South America: with occasional hints, regarding the seasoning, or yellow fever of hot climates*, 3 vols., vol. 3 (London: Longman, Hurst, Rees, and Orme, 1806).

Pinkerton, John, *Modern Geography. A Description of the Empires, Kingdoms, States, and Colonies; with the Oceans, Seas, and Isles; in all the Parts of the World: Including the most Recent Discoveries and Political Alterations*, 2 vols., vol. 2 (London: T. Cadell and W. Davies, Strand, 1817).

Pinto de Azeredo, Jose, *Ensaios sobre algumas enfermidades d'Angola* (Lisboa: Na Regia Officina Typografica, 1799).

Piso, Guilherme, *História natural do Brasil Ilustrada*, trans. Alexandre Correia (São Paulo: Companhia Editora Nacional, 1948 (1648)).

Pleischl, Adolph, *Über die Nothwendigkeit, Fürsorge zu treffen, dass der leidenden Menschheit der nöthige Bedarf an Chinarinden und an den daraus bereiteten chemischen Präparaten auch in der Folge sichergestellt werde* (Wien: Druck von Carl Gerold's Sohn, 1857).

Pomme, Pierre, 'Mémoire et observations cliniques sur l'abus du Quinquina,' in *Supplément au traité 'Les affectations vaporeuses des deux sexes', ou Maladies nerveuses* (Paris: Chez Cussac, 1804), 5–173.

Pringle, John, *Observations on the diseases of the Army* (London: A. Millar; D. Wilson; T. Durham, T. Payne, 1764).

Observations on the Nature and Cure of Hospital and Jayl-Fevers in a Letter to Doctor Mead, Physician to his Majesty (London: A. Millar and D. Wilson, 1750).

Quack, Jan de, *Zuma, of De ontdekking van den kina-bast: tooneelspel in vier bedrijven* (Amsterdam: J.C. van Kesteren, 1819).

Rambach, Johann Jacob, *Versuch einer physisch-medizinischen Beschreibung von Hamburg* (Hamburg: Carl Ernst Bohn, 1801).

Ramel, M.F.B., *De l'Influence des marais et des étangs sur la santé de l'homme, ou Mémoire couronné par la ci-dev. Société Rle de médecine de Paris* (Marseille: J. Mossy, 1801).

Relph, John, *An Inquiry into the Medical Efficacy of a new Species of Peruvian Bark, Lately Imported into this Country under the Name of Yellow Bark: Including Practical Observations Respecting the Choice of Bark in General* (London: James Phillips, 1794).

'Remèdes éprouvés contre les Fiévres tierces, double tierces, & autres Fiévres intermittentes.' *Journal Oeconomique, ou Memoires, Notes et Avis sur l'Agriculture, les Arts, le Commerce, & tout ce qui peut avoir rapport à la santé, ainsi qu'à la conservation & à l'augmentation des biens des Familles, & c.* (1764), 221–22.

Renard, Johann Claudius, *Die inländischen Surrogate der Chinarinde in besonderer Hinsicht auf das Kontinent von Europa* (Mainz: F. Kupferberg, 1809).

'Review of "Medicaments, et Precis de la Methode de Mr. *Masdevall*, &c &c. – That is, Prescriptions, and a Sketch of the Method of Mr. Masdevall, Physician of Charles the Fourth, King of Spain, for curing all epidemic, putrid, and malignant Distempers, Fevers of different Kinds, &c. &c. with the Means of Prevention. Divided into Paragraphs, for the Use of Families who are unable to procure Physicians. New-Orleans. Duclot. 1796. pp. 47 8vo,"' in *The Medical Repository*, ed. Samuel L. Mitchill, Edward Miller and Elihu H. Smith (New York: T. & J. Swords, 1800), 211–15.

Rigby, Edward, *An Essay on the Use of the Red Peruvian Bark in the Cure of Intermittents* (London: J. Johnson, 1783).

Roberts, John, *Observations on Fevers, wherein the Different Species, Nature and Method of Treating those Diseases, are Represented in New and Interesting Points of View. The Whole written in a simple and concise Manner, divested of the Terms of Art, and adapted to common Capacities, on the plain Principles of*

Common Sense. Being designed for the Use of those, who may be afflicted with those Diseases, as well as the Medical Reader (London: Rozea et al., 1781).

Robertson, Robert, *Directions for Administering Peruvian Bark, in a Fermenting State, in Fever and Other Diseases in Which Peruvian Bark is Proper; and More Especially in Such Cases as the Usual Formulae of the Bark are Rejected by the Stomach, or Nauseated by the Sick; with some Experiments to Ferment the Peruvian Bark with Different Sweets* (London: Marche and Teape, 1799).

Directions for Preparing and Administering Peruvian Bark Gingerbread as a Preventive and Cure of Tertian and Remitting Fever, Extracted from Dr. Robertson's Synopsis Morborum (London: D.N. Shury, 1812).

Observations on Fevers and Other Diseases: Which Occur on Voyages to Africa and the West Indies (Cambridge: Cambridge University Press, 2011 (1792)).

Rosen de Rosenstein, Nils, *Anweisung zur Kenntniss und Cur der Kinderkrankheiten* (Göttingen: J.C. Dieterich, 1768).

Rousseau, Jean-Jacques, *A Discourse on Inequality*, trans. Maurice Cranston (London: Penguin, 1984 (1755)).

A Discourse upon the Origin and Foundation of the Inequality among Mankind (London: R. an J. Dodsley, 1761 [1755]).

Ruiz López, Hipólito, *Quinología O Tratado del Arbol de la Quina o Cascarilla, con su descripción y la de otras especies de quinos nuevamente descubiertas en el Perú, del modo de beneficiarla, de su elección, comercio, virtudes, y extracto elaborado von cortezas recientes* (Madrid: La viuda é hijo de Marin, 1792).

The Journals of Hipólito Ruiz, Spanish Botanist in Peru and Chile, 1777–1815 (Portland: Timber Press, 1998).

Ruiz López, Hipólito, and José Antonio Pavón Jiménez, *Suplemento a la quinologia, en el qual se aumentan las Especies de Quina nuevamente descubiertas en el Perú por Don Juan Tafalla, y la Quina naranjada de Santa Fé con su estampa* (Madrid: Imprenta de la viuda e hijo de Marín, 1801).

Russell, John L., 'The Botanical and Horticultural Literature of the Olden Times, with Remarks on the Species and Sorts.' *The Magazine of Horticulture, Botany and all Useful Discoveries and Improvements in Rural Affairs* 24 (1858), 168–76.

Salazar, Thomas de, *Tratado del uso de la quina* (Madrid: Viuda de Ibarra, 1791).

Sánchez y Sánchez, Juan, *Disertacion quimico-medica sobre la opiata antifebril, inventada por el ilustre señor doctor D. Josef de Masdevall, Medico de Camara con Exercicio de S.M. Católica, & c. Explicase quimicamente la naturaleza de esta convinacion, sus propiedades, virtudes, y maravilloso modo de obrar en toda clase de calenturas, especialmente putridas y malignas con un tratado muy curioso sobre el Ayre fixo y la diferencia entre èste y el Gàs mefitico productivo de las epidemias* (Malaga: D. Felix de Casas y Martinez, 1794).

Schall, Johann Eberhard Friedrich, *Handbuch für Leute die keine Aerzte sind zur Beförderung nützlicher und angenehmer Kenntnisse. Dritter und letzter Theil* (Riga: Johann Friedrich Hartknoch, 1781).

Skeete, Thomas, *Experiments and Observations on Quilled and Red Peruvian Bark: Among Which are Included Some remarkable Effects arising from the Action of Common Bark and Magnesia upon each other, with Remarks on the Nature and*

Mode of Treatment of Fevers, Putrid Sore-Throat, Rheumatism, Scrophula, and other Diseases; in order to ascertain the Cases in which Bark may be administered – either alone, or combined with other Remedies – to the best Advantage (London: J. Murray, 1786).

Skinner, Joseph, *The Present State of Peru: Comprising its Geography, Topography, Natural History, Mineralogy, Commerce, the Customs and Manners of its Inhabitants, the State of Literature, Philosophy, and the Arts, the Modern Travels of the Missionaries in the Heretofore Unexplored Mountain Territories, &c. &c. the Whole Drawn from Original and Authentic Documents, Chiefly Written and Compiled in the Peruvian Capital; and Embellished by Twenty Engravings of Costumes, &c.* (London: Richard Phillips, 1805).

Sloane, Hans, *A voyage to the islands Madera, Barbados, Nieves, S. Christophers and Jamaica, with the Natural History of the Herbs and Trees, Four-footed Beasts, Fishes, Birds, Insects, Reptiles, &c. of the last of those islands; to which is prefix'd, an Introduction, Wherein is an Account of the Inhabitants, Air, Waters, Diseases, Trade, &c. of that Place, with some Relations concerning the Neighbouring Continent, and Islands of America*, 2 vols., vol. 1 (London: B.M., 1707).

Spilsbury, Francis, *The friendly physician: A new treatise: containing rules, schemes, and particular instructions, how to select and furnish small chests with the most approved necessary medicines; to which are added many excellent receipts for particular disorders* (London: Sold by Mr. Wilkie, in St. Paul's Church yard; Mr. Stevenson, Newry, in Ireland; Mr. Swinney, Birmingham; Mr. Gilbert, Dublin; Mr. Crowse, Norwich; Mr. Saint, Newcastle on Tyne; and by the author, at No. No. [sic] 5, in Mount Row, near Westminster Bridge, 1773).

Suarez de Ribera, Francisco, *Ilustracion, y publicacion de los diez y siete secretos del Doctor Juan Curvo Semmedo, confirmadas sus virtudes con maravillosas observaciones* (Madrid: Imprenta de Domingo Fernandez de Arrojo, 1723).

Swieten, Gerard, Freiherr van, *The diseases incident to armies: with the method of cure* (Philadelphia: R. Bell, 1776).

Tavares, Francisco, *Observações, e reflexões sobre o uso proveitoso, e saudavel da quina na gota* (Lisbon: Regia Oficina Typografica, 1802).

Pharmacopeia Geral para o reino, e dominios de Portugal, publicada por ordem da Rainha Fidelissima D. Maria I, 2 vols. (Lisboa: Na Regia Officina Typografica, 1794).

Tegut, Enrico, *Le mirabili virtú della Kinakina, con la maniera di servirsene in qualunque sorte di Febbre, e complessione: con un'aggiunta di un nuovo metodo dalle osservazioni del signor barone di Wansvieten, e del cavaliere Eislingeu, Membro dell' Accademia di Londra, per servirsi di questo rimedio senza prenderlo per bocca* (Venice: Presso Antonio Zatta, e Figli, 1785).

Terne, C., *Verhandelingen over de Vraage, in hoe verre zou men, by gebrek van de Apotheek, uit kelder en keuken de vereischte Geneesmiddelen, ook tegen de zwaarfte ziekten en kwaalen, zo uit- als inwendig, kunnen bekomen, mits uitzondere de volgende middelen, Kina, Kwik, Opium, Staal, Delfzuuren, Rhabarber en Ipecacoanna* (Amsterdam: Petrus Conradi, 1788).

Theden, Johann Christian Anton, *Unterricht für die Unterwundärzte bey Armeen, besonders bey dem königlich-preußischen Artilleriecorps* (Berlin: Friedrich Nicolai, 1782).

Theobald, John, *Every Man His Own Physician: Being, a Complete Collection of Efficacious and Approved Remedies, for Every Disease Incident to the Human Body. With Plain Instructions for Their Common Use* (London / Boston: Griffin / Cox and Berry, 1767).

Thomson, James, *A treatise on the diseases of negroes, as they occur in the island of Jamaica: with observations on the country remedies* (Jamaica: Alex. Aikman, jun., 1820).

Thornton, Robert John, *New Family Herbal: Or Popular Account of the Natures and Properties of the Various Plants Used in Medicine, Diet and the Arts* (London: Richard Phillips, 1810).

Tissot, Samuel Auguste André, *Anleitung für das Landvolk in Absicht auf seine Gesundheit* (Zürich: Heidegger und Compagnie, 1763).

Avis au peuple sur sa santé (Lausanne: J. Zimmerli, 1761).

Tissot, Simon André, *Aviso al pueblo acerca de su salud ó Tratado de las enfermedades mas frequentes de las gentes del campo*, trans. Juan Galisteo y Xiorro (Madrid: Imprenta de Pedro Marin, 1790).

Townsend, Joseph, *A Journey through Spain in the Years 1786 and 1787, with particular attention to the Agriculture, Manufactures, Commerce, Population, Taxes, and Revenue of that Country* (London: C. Dilly, 1791).

Ulloa, Antonio de, *Viaje a la América meridional*, ed. Andrés Saumell (Madrid: Historia 16, 1990 (1748)).

Unanue, José Hipólito, 'Botánica. Introducción a la descripción científica de las plantas del Perú, Vol. II, núms. 43 y 44 (29 de mayo y 2 de junio de 1791),' in *El Mercurio Peruano: 1790–1795. Antología*, ed. Jean-Pierre Clément (Frankfurt am Main: Vervuert, 1998 (1791)), 93–114.

Velloso, Jose Mariano, *Quinografia Portugueza ou Colleccao de varias memorias sobre vinte e duas especies de quinas, tendentes ao seu descobrimiento nos vastos dominios do Brasil, copiada de varios authores modernos, enriquecida com cinco estampas de Quinas verdadeiras, quatro de falsas, e cinco de Balsameiras. E colligada de ordem de sua alteza real o principe do Brasil nosso senhor* (Lisboa: Impressor da Santa Igreja Patriarcal, 1799).

Villalba, Joaquin de, *Epidemiologia española, ó, Historia cronológica de las pestes, contagios, epidemias y epizootias que han acaecido en España desde la venida de los cartagineses hasta el año 1801. Con noticia de algunas otras enfermedades de esta especie que han sufrido los españoles en otros reynos, y de los autores nacionales que han escrito sobre esta materia, así en la peninsula como fuera de ella* (Madrid: Fermin Villalpando, 1803).

Villalobos, Baltasar de, *Método de curar tabardillos, y descripción de la fiebre epidemica, que por los años de 1796 y 97 afligio varias poblaciones del partido de Chancay: Escrito de órden de este Superior Gobierno y Real Acuerdo de Justicia* (Lima: Imprenta Real del Telégrafo Peruano, 1800).

Woyts, Johann Jacob, *Abhandlung aller innerlichen und äußerlichen Kranckheiten, in zwei Theilen, in welchen jedwede Kranckheit deutlich beschrieben, und zur Kur die bewährtesten Arzney-Mittel aus denen Schrifften derer berühmtesten Aerzte and die Hand gegeben werden* (Leipzig: Friedrich Lanckisches Erben, 1753).

Wylie, James, *Pharmacopoeia castrensis Ruthena* (St Petersburg: Typographia Medica, 1808).

Secondary Sources

Achan, Jane, Ambrose O. Talisuna, Annette Erhart, Adoke Yeka, James K. Tibenderana, Frederick N. Baliraine, Philip J. Rosenthal, and Umberto D'Alessandro, 'Quinine, an Old Anti-malarial Drug in a Modern World: Role in the Treatment of Malaria.' *Malaria Journal* 10, no. 144 (2011), 1–12.

Achim, Miruna, 'From Rustics to Savants: Indigenous *Materia medica* in Eighteenth-Century Mexico.' *Studies in History and Philosophy of Biological and Biomedical Sciences* 42 (2011), 275–84.

Lagartijas medicinales. Remedios americanos y debates científicos en la Ilustración (Mexico: Conaculta/UAM-C, 2008).

Agrawal, Arun, 'Dismantling the Divide between Indigenous and Scientific Knowledge.' *Development and Change* 26 (1995), 413–39.

Aken, Mark van, 'The Lingering Death of Indian Tribute in Ecuador.' *Hispanic American Historical Review* 6, no. 3 (1981), 429–59.

Albala, Ken, 'Medicinal Uses of Sugar,' in *The Oxford Companion to Sugar and Sweets*, ed. Darra Goldstein (Oxford: Oxford University Press, 2005), 439–40.

Alberola Romá, Armando, and David Bernabé Gil, 'Tercianas y calenturas en tierras meridionales valencianas: una aproximación a la realidad médica y social del siglo XVIII.' *Revista de Historia Moderna* 17 (1998–1999), 95–112.

Albiez-Wieck, Sarah, 'Indigenous Migrants Negotiating Belonging: Peticiones de cambio de fuero in Cajamarca, Peru, 17th–18th Centuries', *Colonial Latin American Review* 26, no. 4 (2017), 483–508.

Alegre Pérez, María Esther, 'La asistencia social en la Real Botica durante el último cuarto del siglo XVIII.' *Boletín de la Sociedad Española de Historia de la Farmacia* 35, no. 139 (1984), 199–210.

Alexander, John T., *Bubonic Plague in Early Modern Russia: Public Health and Urban Disaster* (New York: Oxford University Press, 2003).

Amaro, Ana Maria, *Introdução da medicina ocidental em Macau e as receitas de segredo da Botica do Colégio de São Paulo* (Macau: Instituto Cultural de Macau, 1992).

Amaya, José Antonio, ed., *Mutis, Apóstol de Linneo. Historia de la botánica en el virreinato de Nueva Granada (1760–1783)*. 2 vols. (Bogotá: Instituto Colombiano de Antropología e Historia, 2005).

Anagnostou, Sabine, 'The International Transfer of Medicinal Drugs by the Society of Jesus (Sixteenth to Eighteenth Centuries) and Connections with the Work of Carolus Clusius,' in *Royal Netherlands Academy of Arts and Sciences*, ed. Florike Egmond, Paul Hoftijzer and Robert P. W. Visser (Amsterdam: Koniklijke Nederlandse Akademie van Wetenschappen, 2007), 293–312.

Jesuiten in Spanisch-Amerika als Übermittler von heilkundlichem Wissen, vol. 78 (Stuttgart: Wissenschaftliche Verlagsgesellschaft, 2000).

Missionspharmazie. Konzepte, Praxis, Organisation und wissenschaftliche Ausstrahlung, Sudhoffs Archiv. Zeitschrift für Wissenschaftsgeschichte. Beiheft (Stuttgart: Franz Steiner Verlag, 2011).

Anda Aguirre, Alfonso, *La Quina o Cascarilla en Loja* (Quito: Universidad Técnica Particular de Loja, 2003).

Andrés Turrión, María Luisa de, 'Quina de la Real Hacienda para el ejército español en el siglo XVIII,' in *Guerra y milicia en la España del X Conde de Aranda. Actas del IV Congreso de Historia Militar*, ed. José A. Armillas Vicente (Zaragoza: Gobierno de Aragon, Departamento de Cultura y Turismo, 1998), 415–25.

Andrés Turrión, María Luisa, and María Rosario Terreros Gómez, 'Organización administrativa del ramo de la quina para la Real Hacienda Española en el Virreinato de Nueva Granada,' in *Medicina y Quina en la España del siglo XVIII*, ed. Juan Riera Palmero (Valladolid: Universidad de Valladolid, 1997), 37–43.

Andrews, Jonathan. 'History of Medicine: Health, Medicine and Disease in the Eighteenth Century.' *Journal for Eighteenth-Century Studies* 34, no. 4 (2011), 503–15.

Andrien, Kenneth J., *Andean Worlds. Indigenous History, Culture, and Consciousness under Spanish Rule, 1532–1825* (Albuquerque: University of New Mexico Press, 2001).

The Kingdom of Quito, 1690–1830: The State and Regional Development (Cambridge: Cambridge University Press, 2002).

Anker, Peder, *Imperial Ecology: Environmental Order in the British Empire, 1895–1945* (Cambridge, Mass.: Harvard University Press, 2001).

Arens, William, *The Man-Eating Myth: Anthropology and Anthropophagy* (Oxford: Oxford University Press, 1979).

Arnold, David, *The Problem of Nature: Environment and Culture in Historical Perspective* (Oxford: Blackwell, 1996).

Attewell, Guy, 'Interweaving Substance Trajectories: *Tiryaq*, Circulation and Therapeutic Transformation in the Nineteenth Century,' in *Crossing Colonial Historiographies: Histories of Colonial and Indigenous Medicines in Transnational Perspective*, ed. Anne Digby and Waltraud Ernst (Cambridge: Cambridge Scholars Publishing, 2010), 1–20.

Austin Alchon, Suzanne, 'Tradiciones médicas nativas y resistencia en el Ecuador,' in *Saberes andinos. Ciencia y tecnología en Bolivia, Ecuador y Perú*, ed. Marcos Cueto (Lima: IEP, 1995), 15–36.

Bailyn, Bernard, 'Introduction. Reflections on Some Major Themes,' in *Soundings in Atlantic History. Latent Structures and Intellectual Currents, 1500–1830*, ed. Bernard Bailyn and Patricia L. Denault (Cambridge, Mass.: Harvard University Press, 2009), 1–43.

Baldwin, Martha, 'Expanding the Therapeutic Canon: Learned Medicine Listens to Folk Medicine,' in *Cultures of Communication from Reformation to Enlightenment*, ed. James Van Horn Melton (Ashgate: Aldershot, 2002), 239–56.

Baratas Díaz, Luis Alfredo, and Joaquín Fernández Pérez. 'Conocimiento botánico de las especies de cinchona entre 1750 y 1850: Relevancia de la obra botánica española en América.' *Estudios de historia de las tecnicas, la arqueología industrial y las ciencias* 2 (1998), 647–59.

Barbier, Jacques A., and Allan J. Kuethe, *The North American Role in the Spanish Imperial Economy: 1760–1819* (Manchester: Manchester University Press, 1984).

Barnes, Linda L., *Needles, Herbs, Gods, and Ghosts. China, Healing, and the West to 1848* (Cambridge, Mass.: Harvard University Press, 2007).

Barrera-Osorio, Antonio, *Experiencing Nature. The Spanish American Empire and the Early Scientific Revolution* (Austin: University of Texas Press, 2006).

Barrow, Mark Jr., *Nature's Ghosts. Confronting Extinction from the Age of Jefferson to the Age of Ecology* (Chicago: University of Chicago Press, 2009).

Bautista Vilar, Juan, and Ramón Lourido Díaz, *Relaciones entre España y el Magreb: siglos XVII y XVIII* (Madrid: MAPFRE, 1994).

Bayly, Christopher A., 'The First Age of Global Imperialism, c. 1760–1830.' *The Journal of Imperial and Commonwealth History* XXVI, no. 2 (1998), 28–47.

Bayr, Georg, *Hahnemanns Selbstversuch mit der Chinarinde im Jahre 1790. Die Konzipierung der Homöopathie* (Heidelberg: K.F. Haug, 1989).

Beinart, William, and Lotte Hughes, introduction to *Environment and Empire*, ed. William Beinart and Lotte Hughes, Oxford History of the British Empire Companion Series (Oxford: Oxford University Press, 2007).

Beisswanger, Gabriele, *Arzneimittelversorgung im 18. Jahrhundert. Die Stadt Braunschweig und die ländlichen Distrikte im Herzogtum Braunschweig-Wolfenbüttel* (Braunschweig: Deutscher Apotheker-Verlag, 1995).

Belich, James, John Darwin and Chris Wickham, 'Introduction. The Prospect of Global History,' in *The Prospect of Global History*, ed. James Belich, John Darwin, Margret Frenz and Chris Wickham (New York: Oxford University Press, 2016), 3–22.

Beltrão Marques, Vera Regina, *Natureza em Boiões: medicinas e boticários no Brasil setecentista* (Campinas: Editora da Unicamp / Centro de Memória–Unicamp, 1999).

Benton, Lauren, *Law and Colonial Cultures. Legal Regimes in World History, 1400–1900* (Cambridge: Cambridge University Press, 2002).

Bertucci, Paola. 'Enlightened Secrets: Silk, Intelligent Travel, and Industrial Espionage in Eighteenth-Century France.' *Technology and Culture* 54, no. 4 (2013), 820–52.

Bhattacharya, Sanjoy, 'Global and Local Histories of Medicine: Interpretative Challenges and Future Possibilities,' in *The Oxford Handbook of the History of Medicine*, ed. Mark Jackson (Oxford: Oxford University Press, 2011), 135–49.

Bjork, Katherine, 'The Link That Kept the Philippines Spanish: Mexican Merchant Interests and the Manila Trade, 1571–1815.' *Journal of World History* 9, no. 1 (1998), 25–50.

Blakemore, Richard J., 'The Changing Fortunes of Atlantic History.' *English Historical Review* CXXXI, no. 551 (2016), 851–68.

Bleichmar, Daniela, 'Atlantic Competitions. Botany in the Eighteenth-Century Spanish Empire,' in *Science and Empire in the Atlantic World*, ed. James Delbourgo and Nicholas Dew (New York: Routledge, 2008), 225–54.

'Books, Bodies, and Fields: Sixteenth-Century Transatlantic Encounters with New World *Materia Medica*,' in *Colonial Botany. Science, Commerce, and Politics in the Early Modern World*, ed. Londa Schiebinger and Claudia Swan (Philadelphia: University of Pennsylvania Press, 2005), 83–99.

Visible Empire. Botanical Expeditions and Visual Culture in the Hispanic Enlighten-ment (Chicago: University of Chicago Press, 2012).

Bolton Valencius, Conevery, *The Health of the Country. How American Settlers Understood Themselves and Their Land* (New York: Basic Books, 2002).

Boulanger, Patrick, 'Droguistes marseillais à la fin du XVIIIe siècle,' in *Herbes, Drogues et Epices en Mediterranée*, ed. George J. Aillaud (Paris: Editions du Centre National de la Recherche Scientifique 1988), 43–55.

Boumediene, Samir, *La colonisation du savoir. Une histoire des plantes médicinales du 'Nouveau Monde' (1492–1750)* (Vaulx-en-Velin: Les Éditions des Mon-des à Faire, 2016).

Brading, David, *The First America. The Spanish Monarchy, Creole Patriots, and the Liberal State 1492–1867* (Cambridge: Cambridge University Press, 1991).

Breen, Benjamin, 'Empires on Drugs: Materia Medica and the Anglo-Portuguese Alliance,' in *Entangled Empires: The Anglo-Iberian Atlantic, 1500–1830*, ed. Jorge Cañizares-Esguerra (Philadelphia: University of Pennsylvania Press, 2018), 63–81.

Brendecke, Arndt, *Imperium und Empirie. Funktionen des Wissens in der spanischen Kolonialherrschaft* (Köln: Böhlau Verlag, 2009).

Brewer, Daniel, *The Enlightenment Past. Reconstructing Eighteenth-Century French Thought* (Cambridge: Cambridge University Press, 2008).

Brilli, Catia, 'Mercaderes genoveses en el Cádiz del siglo XVIII. Crisis y reajuste de una simbiosis secular,' in *Comunidades transnacionales: colonias de merca-deres extranjeros en el mundo atlántico (1500–1830)*, ed. Ana Crespo Solana (Aranjuez: Ediciones Doce Calles, 2006), 83–102.

Brockliss, Laurence, and Colin Jones, *The Medical World of Early Modern France* (Oxford: Clarendon Press, 1997).

Brockway, Lucille H., *Science and Colonial Expansion* (New Haven: Yale Univer-sity Press, 2002 (1979)).

Bruce-Chwatt, Leonard Jan, 'Cinchona and Quinine: A Remarkable Anniver-sary.' *Interdisciplinary Science Review* 15, no. 1 (1990), 87–93.

Bruce-Chwatt, Leonard Jan, and Julian de Zulueta, *The Rise and Fall of Malaria in Europe* (Oxford: Oxford University Press, 1980).

Brug, Peter Harmen van der, 'Malaria in Batavia in the 18th Century.' *Tropical Medicine and International Health* 2, no. 9 (1997), 892–902.

  'Malaria en malaise. De VOC in Batavia in de achttiende eeuw.' PhD disserta-tion, Rijksuniversiteit te Leiden, 1994.

Bruijn, Iris, *Ship's Surgeons of the Dutch East India Company: Commerce and the Progress of Medicine in the Eighteenth Century* (Amsterdam: Amsterdam Uni-versity Press, 2009).

'BUCHAN, William, 1729–1805,' in *An Annotated Catalogue of the Edward C. Atwater Collection of American Popular Medicine and Health Reform: A–L*, ed. Christopher Hoolihan (Rochester, NY: University of Rochester Press, 2001), 130–49.

Gonzàlez Bueno, Antonio, 'Plantas y Luces: la Botánica de la Ilustración en la América Hispana,' in *La formación de la Cultura Virreinal. El siglo XVIII*, ed. Karl Kohut and Sonia V. Rose (Madrid: Iberoamericana, 2006), 197–28.

Bynum, William F., 'Cullen and the Study of Fevers in Britain, 1760–1820,' in *Theories of Fever from Antiquity to the Enlightenment*, ed. William F. Bynum and Vivian Nutton (London: Wellcome Institute for the History of Medicine, 1981), 135–47.

Bynum, William F., and Vivian Nutton, introduction to *Theories of Fever from Antiquity to the Enlightenment*, ed. William F. Bynum and Vivian Nutton (London: Wellcome Institute for the History of Medicine, 1981), vii–ix.

Cagle, Hugh, *Assembling the Tropics. Science and Medicine in Portugal's Empire, 1450–1700* (Cambridge: Cambridge University Press 2018).

'Beyond the Senegal: Inventing the Tropics in the Late Middle Ages.' *Journal of Medieval Iberian Studies* 7, no. 2 (2015), 197–217.

Cahill, David, 'Colour by Numbers: Racial and Ethnic Categories in the Viceroyalty of Peru, 1532–1824.' *Journal of Latin American Studies* 26, no. 2 (1994), 325–46.

Caillavet, Chantal, and Martin Minchom, 'Le métis imaginaire: idéaux classificatoires et stratégies socio-raciales en Amérique latine (xvie–xxe siècle).' *L'Homme* 32, no. 122–124 (1992), 115–32.

Calabozo Justel, Braulio, *El médico Coll en la corte del sultan de Marruecos (año 1800)* (Cádiz: Servicio de Publicaciones de la Universidad de Cádiz / Instituto de Cooperación con el Mundo Arabe, 1991).

Cañizares Esguerra, Jorge, *How to Write the History of the New World: Histories, Epistemologies, and Identities in the Eighteenth-Century Atlantic World* (Stanford, Calif.: Stanford University Press, 2001).

Nature, Empire, and Nation. Explorations of the History of Science in the Iberian World (Stanford, Calif.: Stanford University Press, 2006).

Canny, Nicholas, and Philip Morgan, 'Introduction. The Making and Unmaking of an Atlantic World,' in *The Oxford Handbook of the Atlantic World: 1450–1850*, ed. Nicholas Canny and Philip Morgan (Oxford: Oxford University Press, 2011), 1–17.

Carasa, Pedro, 'Welfare Provision in Castile and Madrid,' in *Health Care and Poor Relief in 18th and 19th Century Southern Europe*, ed. Ole Peter Grell, Andrew Cunningham and Bernd Roeck (Aldershot: Ashgate, 2005), 96–120.

Chaiklin, Martha, *Cultural Commerce and Dutch Commercial Culture. The Influence of European Material Culture on Japan, 1700–1850* (Leiden: CNWS, 2003).

Chakrabarti, Pratik, 'Empire and Alternatives: Swietenia febrifuga and the Cinchona Substitutes.' *Medical History* 54, no. 1 (2010), 75–94.

Medicine and Empire 1600–1960 (Basingstoke: Palgrave Macmillan, 2014).

Chaunu, Pierre, and Huguette Chaunu, *Séville et l'Atlantique (1504–1650). Première Partie: Partie Statistique. Le mouvement des navires et des marchandises entre l'espagne et l'amérique de 1504 à 1650* (Paris: S.E.V.P.E.N., 1956).

Clément, Jean-Pierre, 'El nacimiento de la higiene urbana en la America Española del siglo XVIII.' *Revista de Indias* 43, no. 171 (1983), 77–95.

Coffman, D'Maris, and Adrian Leonard, 'The Atlantic World: Definition, Theory, and Boundaries,' in *The Atlantic World: 1400–1850*, ed. D'Maris Coffman, Adrian Leonard and William O'Reilly, The Routledge Worlds (London: Routledge, 2015), 1–10.

Colley, Linda, 'Writing Constitutions and Writing World History,' in *The Prospect of Global History*, ed. James Belich, John Darwin, Margret Frenz and Chris Wickham (New York: Oxford University Press, 2016), 160–77.

Contreras, Carlos C., *El sector exportador de una economía colonial. La costa del Ecuador: 1760–1830*, Colección Tesis Historia (Quito: FLACSO, 1990).

Cook, Harold J., 'Markets and Cultures. Medical Specifics and the Reconfiguration of the Body in Early Modern Europe.' *Transactions of the Royal Historical Society* 21 (2011), 123–45.

Matters of Exchange. Commerce, Medicine and Science in the Age of Empire (Hyderabad: Orient Longman, 2008).

'Practical Medicine and the British Armed Forces after the "Glorious Revolutions."' *Medical History* 34 (1990), 1–26.

Cook, Harold J., and Timothy Walker, 'Circulation of Medicine in the Early Modern Atlantic World.' *Social History of Medicine* 26, no. 3 (2013), 337–51.

Cooper, Alix, 'The Indigenous versus the Exotic: Debating Natural Origins in Early Modern Europe.' *Landscape Research* 28, no. 1 (2003), 51–60.

Inventing the Indigenous. Local Knowledge and Natural History in Early Modern Europe (Cambridge: Cambridge University Press, 2007).

Cooper, Frederick, 'What Is the Concept of Globalization Good For? An African Historian's Perspective.' *African Affairs* 100, no. 189–213 (2001).

Cooter, Roger, 'The Turn of the Body. History and the Politics of the Corporeal.' *ARBOR. Ciencia, Pensamiento y Cultura* CLXXXVI, no. 743 (2010), 393–405.

Cortez Wissenbach, Maria Cristina, 'Gomes Ferreira e os símplices da terra: experiências sociais dos cirurgiões no Brasil-Colônia,' in *Luís Gomes Ferreira, Erário mineral*, ed. Júnia Ferreira Furtado (Rio de Janeiro: Editora FIO-CRUZ, 2002), 107–49.

Crawford, Matthew, *The Andean Wonder Drug. Cinchona Bark and Imperial Science in the Spanish Atlantic, 1630–1800* (Pittsburgh, Pa.: University of Pittsburgh Press, 2016).

'Empire's Experts: The Politics of Knowledge in Spain's Royal Monopoly of Quina (1751–1808).' Unpublished PhD dissertation, University of California, San Diego, 2009.

'An Empire's Extract: Chemical Manipulations of Cinchona Bark in the Eighteenth-Century Spanish Atlantic World.' *Osiris* 29 (2014), 215–29.

'"Para desterrar las dudas y adulteraciones": Scientific Expertise and the Attempts to Make a Better Bark for the Royal Monopoly of *Quina*.' *Journal of Spanish Cultural Studies* 8, no. 2 (2007), 193–212.

Cronon, William, 'The Trouble with Wilderness; or, Getting Back to the Wrong Nature,' in *Uncommon Ground: Rethinking the Human Place in Nature*, ed. William Cronon (New York: W. W. Norton, 1995).

Crosby, Alfred W., *The Columbian Exchange. Biological and Cultural Consequences of 1492* (Westport, Conn.: Greenwood, 1972).

Cueto, Marcos, and Steven Palmer, *Medicine and Public Health in Latin America* (Cambridge: Cambridge University Press, 2015).

Cunningham, Andrew, 'Some Closing and Opening Remarks,' in *Health Care and Poor Relief in 18th and 19th Century Southern Europe*, ed. Ole Peter Grell, Andrew Cunningham and Bernd Roeck (Aldershot: Ashgate, 2005), 1–9.

Curtin, Philip D., *Death by Migration. Europe's Encounter with the Tropical World in the Nineteenth Century* (New York: Cambridge University Press, 1989).

The Image of Africa. British Ideas and Action, 1780–1850, vol. 1 (Madison: University of Wisconsin Press, 1964).

Daston, Lorraine, 'Introduction. Speechless,' in *Things That Talk. Object Lessons from Art and Science*, ed. Lorraine Daston (New York: Zone Books, 2004), 9–24.

'Introduction. The Coming into Being of Scientific Objects,' in *Biographies of Scientific Objects*, ed. Lorraine Daston (Chicago: University of Chicago Press, 2000), 1–14.

'Science Studies and the History of Science.' *Critical Inquiry* 35, no. 4 (2009), 798–813.

Daston, Lorraine, and Fernando Vidal, 'Introduction: Doing What Comes Naturally,' in *The Moral Authority of Nature*, ed. Lorraine Daston and Fernando Vidal (Chicago: University of Chicago Press, 2004), 1–20.

De Vos, Paula. 'From Herbs to Alchemy: The Introduction of Chemical Medicine to Mexican Pharmacies in the Seventeenth and Eighteenth Centuries.' *Journal of Spanish Cultural Studies* 8, no. 2 (2007), 135–68.

'Natural History and the Pursuit of Empire in Eighteenth-Century Spain.' *Eighteenth-Century Studies* 40, no. 2 (2007), 209–39.

'The Science of Spices: Empiricism and Economic Botany in the Early Spanish Empire.' *Journal of World History* 17, no. 4 (2006), 399–427.

Delbourgo, James, and Nicholas Dew, *Science and Empire in the Atlantic World* (New York: Routledge, 2008).

Derex, Jean-Michel. 'Géographie sociale et physique du paludisme et des fièvres intermittentes en France du XVIIIe au XXe siècles.' *Histoire, économie & société* 27 (2008), 39–59.

Dobson, Mary J., *Contours of Death and Disease in Early Modern England* (Cambridge: Cambridge University Press, 1997).

'Mortality Gradients and Disease Exchanges: Comparisons from Old England and Colonial America.' *Social History of Medicine* 2, no. 3 (1989), 259–97.

Dodero, Giuseppe, 'Ippocratismo, malaria e medicina didascalica in Sardegna,' in *Pietro Antonio Leo. Di alcuni pregiudizii sulla così detta Sarda intemperie*, ed. Alessandro Riva and Giuseppe Dodero (Cagliari: CUEC, 2005), XXI–LXXVIII.

Dousset, Jean-Claude, *Histoire des médicaments des origines à nos jours* (Paris: Payot, 1985).

Drayton, Richard, *Nature's Government: Science, Imperial Britain and the 'Improvement' of the World* (New Haven: Yale University Press, 2000).

Dubois, Colette, 'Le quotidien d'une pharmacie hospitaliere: la boutique de l'Hôtel-Dieu de Carpentras,' in *Herbes, Drogues et Epices en Méditerranée*, ed. George J. Aillaud (Paris: Editions du Centre National de la Recherche Scientifique 1988), 79–114.

Duden, Barbara, *The Woman beneath the Skin. A Doctor's Patients in Eighteenth-Century Germany* (Cambridge, Mass.: Harvard University Press, 1991).

Duffy, John, 'The Impact of Malaria on the South,' in *Disease and Distinctiveness in the American South*, ed. Todd L. Savitt and James Harvey Young (Knoxville: University of Tennessee Press, 1988), 29–54.

Dunn, Frederick L., 'Malaria,' in *The Cambridge World History of Human Disease*, ed. Kenneth F. Kiple (Cambridge: Cambridge University Press, 1993), 855–62.

Eamon, William, *Science and the Secrets of Nature. Books of Secrets in Medieval and Early Modern Culture* (Princeton, N.J.: Princeton University Press, 1994).

Easterby-Smith, Sarah, *Cultivating Commerce. Cultures of Botany in Britain and France, 1760–1815* (Cambridge: Cambridge University Press 2017).

Ebrahimnejad, Hormoz, 'Medicine in Islam and Islamic Medicine,' in *The Oxford Handbook of the History of Medicine*, ed. Mark Jackson (Oxford: Oxford University Press, 2011), 169–89.

Edmond, Rod, 'Island Transactions. Encounter and Disease in the South Pacific,' in *The Global Eighteenth Century*, ed. Felicity A. Nussbaum (Baltimore: Johns Hopkins University Press, 2003), 251–62.

Eldem, Edhem, *French Trade in Istanbul in the Eighteenth Century* (Leiden: Brill, 1999).

Ellingson, Terry Jay, *The Myth of the Noble Savage* (Berkeley: University of California Press, 2001).

Elman, Benjamin A., *A Cultural History of Modern Science in China* (Cambridge, Mass.: Harvard University Press, 2006).

Elshakry, Marwa. 'When Science Became Western: Historiographical Reflections.' *Isis* 101, no. 1 (2010), 98–109.

Escobedo Mansilla, Ronald. 'El tributo de los zambaigos, negros y mulatos libres en el Virreinato peruano.' *Revista de Indias*, no. 1 (1981), 43–54.

Estes, J. Worth, 'The Reception of American Drugs in Europe, 1500–1650,' in *Searching for the Secrets of Nature. The Life and Works of Dr. Francisco Hernández*, ed. Simon Varey, Rafael Chabrán and Dora B. Weiner (Stanford, Calif.: Stanford University Press, 2000), 111–21.

Estrella, Eduardo, 'Ciencia ilustrada y saber popular en el conocimiento de la quina en el siglo XVIII,' in *Saberes Andinos. Ciencia y tecnología en Bolivia, Ecuador y Perú*, ed. Marcos Cueto (Lima: Instituto de Estudios Peruanos, 1995), 37–57.

'Introducción de la quina a la terapeutica: misión geodesica y tradición popular', *Revista de la Facultad de Ciencias Médicas – Quito* 14, no. 1–4 (1989), 52–58.

Fan, Fa-ti, *British Naturalists in Qing China. Science, Empire, and Cultural Encounter* (Cambridge, Mass.: Harvard University Press, 2004).

'The Global Turn in the History of Science.' *East Asian Science, Technology and Society: An International Journal* 6 (2012), 249–58.

Ferreira Furtado, Júnia. 'Barbeiros, cirurgiões e médicos na Minas colonial.' *Revista do Arquivo Público Mineiro* 41 (2005), 88–105.

'Tropical Empiricism. Making Medical Knowledge in Colonial Brazil,' in *Science and Empire in the Atlantic World*, ed. James Delbourgo and Nicholas Dew (New York: Routledge, 2008), 127–51.

Fett, Sharla M., *Working Cures. Healing, Health, and Power on Southern Slave Plantations* (Chapel Hill: University of North Carolina Press, 2002).

Few, Martha, *For All of Humanity. Mesoamerican and Colonial Medicine in Enlightenment Guatemala* (Tucson: University of Arizona Press, 2015).

Fields, Sherry, *Pestilence and Headcolds. Encountering Illness in Colonial Mexico* (New York: Columbia University Press, 2008).

Figueiredo, Patrick, 'A "Água de Inglaterra" em Portugal,' in *A Circulação do Conhecimento: Medicina, Redes e Impérios*, ed. Cristiana Bastos and Renilda Barreto (Lisboa: Imprensa do Instituto de Ciências Sociais, 2011), 113–29.

Findlen, Paula, 'Early Modern Things. Objects in Motion, 1500–1800,' in *Early Modern Things. Objects and Their Histories, 1500–1800*, ed. Paula Findlen (London: Routledge, 2013), 3–27.

Fisher, John, *Commercial Relations between Spain and Spanish America in the Era of Free Trade* (Liverpool: Centre for Latin American Studies, University of Liverpool, 1985).

Fissell, Mary E., 'Popular Medical Writing,' in *The Oxford History of Popular Print Culture. Cheap Print in Britain and Ireland to 1660*, ed. Joad Raymond (Oxford: Oxford University Press, 2011), 418–31.

Flynn, Dennis O., and Arturo Giráldez, 'Born Again: Globalization's Sixteenth Century Origins (Asian/Global versus European Dynamics).' *Pacific Economic Review* 13, no. 4 (2008), 359–87.

'Born with a "Silver Spoon": The Origin of World Trade in 1571.' *Journal of World History* 6, no. 2 (1995), 201–20.

Flynn, Dennis O., Arturo Giráldez, and James Sobredo, eds., *European Entry into the Pacific. Spain and the Acapulco-Manila Galleons, The Pacific World. Lands, Peoples and History of the Pacific, 1500–1900*, vol. 4 (Ashgate: Variorum, 2001).

Foster, George M., 'On the Origin of Humoral Medicine in Latin America.' *Medical Anthropology Quarterly* 1, no. 4 (2013), 355–93.

Foust, Clifford M., *Rhubarb. The Wondrous Drug* (Princeton, N.J.: Princeton University Press, 1992).

Frangakis-Syrett, Elena, *The Commerce of Smyrna in the Eighteenth Century (1700–1820)* (Athens: Centre of Asia Minor Studies, 1992).

Frevert, Ute, *Krankheit als politisches Problem 1770–1880. Soziale Unterschichten in Preußen zwischen medizinischer Polizei und staatlicher Sozialversicherung.* Kritische Studien zur Geschichtswissenschaft (Göttingen: Vandenhoeck & Ruprecht, 1984).

Frias Núñez, Marcelo, *José Celestino Mutis y la real expedición botánica del nuevo reino de Granada, 1783–1808* (Sevilla: Diputación Provincial de Sevilla, 1994).

Gänger, Stefanie, 'Circulation: Reflections on Circularity, Entity and Liquidity in the Language of Global History.' *Journal of Global History* 12, no. 3 (2017), 303–18.

'Disjunctive Circles: Modern Intellectual Culture in Cuzco and the Journeys of Incan Antiquities, c. 1877–1921.' *Modern Intellectual History* 10, no. 2 (2013), 399–414.

'In Their Own Hands. Domestic Medicine and "the Cure of all Kinds of Tertian and Quartan Fevers" in Late-Colonial Lima.' *Colonial Latin American Review* 25, no. 4 (2016), 492–511.

'World Trade in Medicinal Plants from Spanish America, 1717–1815.' *Medical History* 59, no. 1 (2015), 44–62.

Garcés, Carlos Alberto, 'Místicos, curanderos y hechiceros: Historias de afroamericanos en la sociedad del Tucumán colonial.' *Contra Relatos desde el Sur. Apuntes sobre Africa y Medio Oriente* V, no. 7 (2010), 9–26.

García-Baquero González, Antonio, *Cádiz y el Atlantico (1717–1778)*, 2 vols., vol. 1 (Cádiz: Diputación Provincial de Cádiz, 1988).

Gardeta Sabater, Pilar, *Sebastián José López Ruiz (1741–1832)* (Málaga: Universidad de Málaga, 1996).

Gentilcore, David, *Medical Charlatanism in Early Modern Italy* (Oxford: Oxford University Press, 2006).

Gerbi, Antonello, *The Dispute of the New World: The History of a Polemic, 1750–1900*, trans. Jeremy Moyle (Pittsburgh, Pa.: University of Pittsburgh Press, 1973 (1955)).

Gevitz, Norman, '"Pray Let the Medicine be Good": The New England Apothecary in the Seventeenth and Early Eighteenth Centuries,' in *Apothecaries and the Drug Trade. Essays in Celebration of the Work of David L. Cowen*, ed. Gregory J. Higby and Elaine C. Stroud (Madison, Wis.: American Institute of the History of Pharmacy, 2001), 5–27.

Geyer-Kordesch, Johanna, 'Fevers and Other Fundamentals: Dutch and German Medical Explanations c. 1680 to 1730,' in *Theories of Fever from Antiquity to the Enlightenment*, ed. William F. Bynum and Vivian Nutton (London: Wellcome Institute for the History of Medicine, 1981), 99–120.

Gil Montero, Raquel, 'Free and Unfree Labour in the Colonial Andes in the Sixteenth and Seventeenth Centuries.' *International Review for Social History* 56 (2011), 297–318.

Goffman, Daniel, 'Izmir: From Village to Colonial Port City,' in *The Ottoman City between East and West. Aleppo, Izmir, and Istanbul*, ed. Edhem Eldem, Daniel Goffman and Bruce Masters (Cambridge: Cambridge University Press, 2001), 79–135.

Gomes Ferreira, Luís, 'Erário mineral,' in *Erário mineral*, ed. Júnia Ferreira Furtado (Rio de Janeiro: Editora FIOCRUZ, 2002 [1735]), 179–767.

Gomez Maria, Rosario Terreros, and Maria Luisa Andrés Turrión, 'First Hospital Experiences with Cinchona Ordered by Spanish Court (ca. 1770).' *Revue d'histoire de la pharmacie* 84, no. 312 (1996), 363–67.

Gómez, Pablo F., *The Experiential Caribbean: Creating Knowledge and Healing in the Early Modern Atlantic* (Chapel Hill: University of North Carolina Press, 2017).

Gonzalez Gomez, Cesar, *Aspectos de la labor quinológica de los botánicos Ruiz y Pavón* (Madrid: Imprenta Góngora, 1954).

Gootenberg, Paul, 'Population and Ethnicity in Early Republican Peru: Some Revisions.' *Latin American Research Review* 26, no. 3 (1991), 109–57.

Gordon, David M., and Shepard Krech, 'Indigenous Knowledge and the Environment,' in *Indigenous Knowledge and the Environment in Africa and North America*, ed. David M. Gordon and Shepard Krech (Athens: Ohio University Press, 2012), 1–24.

Grafton, Anthony, *New Worlds, Ancient Texts. The Power of Tradition and the Shock of Discovery* (Cambridge, Mass.: Harvard University Press, 1992).

Greene, Roland, 'Petrarchism among the Discourses of Imperialism,' in *America in European Consciousness, 1493–1750* (Chapel Hill: University of North Carolina Press, 1995), 130–65.

Griffin, Clare, 'Russia and the Medical Drug Trade in the Seventeenth Century', *Social History of Medicine* 31, no. 1 (2018), 2–23.

Grove, Richard H., *Green Imperialism. Colonial Expansion, Tropical Island Edens and the Origins of Environmentalism* (Cambridge: Cambridge University Press, 1995).

'Indigenous Knowledge and the Significance of South-West India for Portuguese and Dutch Constructions of Tropical Nature.' *Modern Asian Studies* 30, no. 1 (1996), 121–43.

Guerra, Francisco, 'Medical Almanacs of the American Colonial Period.' *Journal of the History of Medicine* XVI, no. 3 (1961), 234–55.

Günergun, Feza, and Şeref Etker, 'From Quinaquina to "Quinine Law": A Bitter Chapter in the Westernization of Turkish Medicine.' *Osmanlı Bilimi Araştırmaları* XIV, no. 2 (2013), 41–68.

Hackett, L. W., *Malaria in Europe. An Ecological Study* (London: Oxford University Press, 1937 (reprinted 1944)).

Haggis, Alex, 'Fundamental Errors in the Early History of Cinchona.' *Bulletin for the History of Medicine* 10 (1941), 417–59.

Hamlin, Christopher, *Cholera. The Biography* (New York: Oxford University Press, 2009).

Hamnett, Brian R., 'Between Bourbon Reforms and Liberal Reforma: The Political Economy of a Mexican Province – Oaxaca, 1750–1850,' in *The Political Economy of Spanish America in the Age of Revolution, 1750–1850*, ed. Kenneth J. Andrien and Lyman L. Johnson (Albuquerque: University of New Mexico Press, 1994), 39–62.

Hanson, Marta, and Gianna Pomata, 'Medicinal Formulas and Experiential Knowledge in the Seventeenth-Century Epistemic Exchange between China and Europe.' *Isis* 108, no. 1 (2017), 1–25.

Haraway, Donna, 'Situated Knowledges: The Science Question in Feminism and the Privilege of Partial Perspective.' *Feminist Studies* 14, no. 3 (1988), 575–99.

Harrison, Mark, *Climates & Constitutions: Health, Race, Environment, and British Imperialism in India, 1600–1850* (New Delhi: Oxford University Press, 2002 (c1999)).

'Disease and Medicine in the Armies of British India, 1750–1830,' in *British Military and Naval Medicine*, ed. Geoffrey L. Hudson (Amsterdam: Rodopi, 2007), 87–119.

'A Global Perspective: Reframing the History of Health, Medicine, and Disease.' *Bulletin of the History of Medicine* 89, no. 4 (2015), 639–89.

'Medicine and the Management of Modern Warfare.' *History of Science* 34, no. 106 (1996), 379–410.

Medicine in an Age of Commerce and Empire. Britain and Its Tropical Colonies, 1660–1830 (Oxford: Oxford University Press, 2010).

218   Bibliography

Härter, Karl, *Policey und Strafjustiz in Kurmainz. Gesetzgebung, Normdurchsetzung und Sozialkontrolle im frühneuzeitlichen Territorialstaat* (Frankfurt a.M.: Vittorio Klostermann, 2005).

Haycock, David Boyd, and Patrick Wallis, *Quackery and Commerce in Seventeenth-Century London: The Proprietary Medicine Business of Anthony Daffy*, Medical History Supplement (London: Wellcome Trust Centre for the History of Medicine at UCL, 2005).

Hayes, Kevin J., *A Colonial Woman's Bookshelf* (Eugene, Ore.: Wipf & Stock, 1996).

Headrick, Daniel R, *The Tools of Empire: Technology and European Imperialism in the Nineteenth Century* (New York: Oxford University Press, 1981).

Heller, Dana, 'Holy Fools, Secular Saints, and Illiterate Saviors in American Literature and Popular Culture.' *CLCWeb: Comparative Literature and Culture* 5, no. 3 (2003), 1–15.

Hernández de Alba, Gonzalo, *Quinas Amargas. El sabio Mutis y la discusión naturalista del siglo XVIII* (Bogotá: Academia de Historia de Bogotá / Tercer Mundo Editores, 1991).

Hill Curth, Louise, 'Introduction: Perspectives on the Evolution of the Retailing of Pharmaceuticals,' in *From Physick to Pharmacology: Five Hundred Years of British Drug Retailing*, ed. Louise Hill Curth (Aldershot: Ashgate, 2006), 1–12.

Honigsbaum, Mark, and Merlin Willcox, 'Cinchona,' in *Traditional Medicinal Plants and Malaria*, ed. Merlin Willcox, Gerard Bodeker and Philippe Rasoanaivo (Boca Raton, Fla.: CRC Press, 2004), 21–41.

Howard, Martin R., 'Walcheren 1809: A Medical Catastrophe.' *British Medical Journal* 319, no. 7225 (1999), 1642–45.

Huguet-Termes, Teresa, 'Madrid Hospitals and Welfare in the Context of the Habsburg Empire,' in *Health and Medicine in Hapsburg Spain: Agents, Practices, Representations*, ed. Teresa Huguet-Termes, Jon Arrizabalaga and Harold J. Cook (London: The Wellcome Trust Centre for the History of Medicine at UCL, 2009), 64–85.

Huldén, Lena, 'The First Finnish Malariologist, Johan Haartman, and the Discussion about Malaria in 18th Century Turku, Finland.' *Malaria Journal* 10, no. 43 (2011), 1–7.

Huldén, Lena, Larry Huldén, and Kari Heliövaara, 'Endemic Malaria: An "Indoor" Disease in Northern Europe. Historical Data Analysed.' *Malaria Journal* 4, no. 19 (2005), 1–13.

Humphreys, Margaret, *Malaria. Poverty, Race, and Public Health in the United States* (Baltimore: Johns Hopkins University Press, 2001).

Ingold, Tim, 'Toward an Ecology of Materials.' *Annual Review of Anthropology* 41 (2012), 427–42.

Jannetta, Ann, *The Vaccinators: Smallpox, Medical Knowledge, and the 'Opening' of Japan* (Stanford, Calif.: Stanford University Press, 2007).

Jaramillo Arango, Jaime, 'Comercio y ciclos económicos regionales a fines del período colonial. Piura, 1770–1830,' in *El Perú en el siglo XVIII. La Era Borbónica*, ed. Scarlett O'Phelan Godoy (Lima: Pontificia Universidad Católica del Perú / Instituto Riva-Agüero, 1999), 37–69.

Jaramillo Baanante, Miguel, 'El comercio de la cascarilla en el norte peruano-sur ecuatoriano: evolución de impacto regional de una economía de exportación, 1750–1796,' in *El Norte en la Historia Regional. Siglos XVIII–XIX*, ed. Scarlett O'Phelan Godoy and Yves Saint-Geours (Lima: IFEA-CiPCA, 1998), 51–90.

Jaramillo-Arango, Jaime, 'A Critical Review of the Basic Facts in the History of Cinchona.' *Journal of the Linnaean Society* 53 (1949), 272–311.

Jarcho, Saul, *Quinine's Predecessor. Francesco Torti and the Early History of Cinchona* (Baltimore: Johns Hopkins University Press, 1993).

Jenner, Mark S. R., and Patrick Wallis, 'The Medical Marketplace,' in *Medicine and the Market in England and Its Colonies, c. 1450–1850*, ed. Mark S. R. Jenner and Patrick Wallis (Basingstoke: Palgrave Macmillan, 2007), 1–23.

Jorge, Arias-Schreiber Pezet, *Los médicos en la independencia del Perú* (Lima: Editorial Universitaria, 1971).

Justel Calabozo, Braulio, 'El doctor Masdevall. Protomédico del sultán Marroquí Muley Solimán.' *Al-Andalus – Magreb* II (1994), 167–202.

Jütte, Robert, 'Hanseatic Towns: Hamburg, Bremen and Lübeck,' in *Health Care and Poor Relief in 18th and 19th Century Southern Europe*, ed. Ole Peter Grell, Andrew Cunningham and Bernd Roeck (Aldershot: Ashgate, 2005), 105–25.

Kaiser, David, *Drawing Theories Apart: The Dispersion of Feynman Diagrams in Postwar Physics* (Chicago: University of Chicago Press, 2005).

Karras, Alan L., 'Transgressive Exchange. Circumventing Eighteenth-Century Atlantic Commercial Restrictions, or the Discount of Monte Christi,' in *Seascapes. Maritime Histories, Littoral Cultures, and Transoceanic Exchanges*, ed. Jerry H. Bentley, Renate Bridenthal and Kären Wigen (Honolulu: University of Hawai'i Press, 2007), 121–34.

Keeble, T. W., 'A Cure for the Ague: The Contribution of Robert Talbor (1642–81).' *Journal of the Royal Society of Medicine* 90, no. 5 (1997), 285–90.

Kellogg, Susan, *Law and the Transformation of Aztec Culture, 1500–1700* (Norman: University of Oklahoma Press, 1995).

King, Steven, 'Accessing Drugs in the Eighteenth-Century Regions,' in *From Physick to Pharmacology: Five Hundred Years of British Drug Retailing*, ed. Louise Hill Curth (Aldershot: Ashgate Publishing, 2006), 49–78.

Kiple, Kenneth F., *The Caribbean Slave. A Biological History* (Cambridge: Cambridge University Press, 1984).

Kirker, Constance L., and Mary Newman, *Edible Flowers. A Global History* (London: Reaktion Books, 2016).

Kjærgaard, Thorkild, *The Danish Revolution, 1500–1800: An Ecohistorical Interpretation* (Cambridge: Cambridge University Press, 2006).

Klauth, Carlo, *Geschichtskonstruktion bei der Eroberung Mexikos: Am Beispiel der Chronisten Bernal Diáz del Castillo, Bartolomé de las Casas und Gonzalo Fernández de Oviedo* (Hildesheim: Georg Olms Verlag, 2012).

Klein, Herbert S., 'The Portuguese Slave Trade from Angola in the Eighteenth Century.' *The Journal of Economic History* 32, no. 4 (1972), 894–918.

Klein, Ursula, and Wolfgang Lefèvre, *Materials in Eighteenth-Century Science. A Historical Ontology* (Cambridge, Mass.: MIT Press, 2007).

Klein, Wouter, and Toine Pieters, 'The Hidden History of a Famous Drug: Tracing the Medical and Public Acculturation of Peruvian Bark in Early Modern Western Europe (c. 1650–1720).' *Journal of the History of Medicine and Allied Sciences* 71, no. 4 (2016), 400–21.

Klingle, Matthew W., 'Spaces of Consumption in Environmental History.' *History and Theory. Studies in the Philosophy of History*, no. 42 (2003), 94–110.

Klooster, Wim, 'Inter-Imperial Smuggling in the Amercas, 1600–1800,' in *Soundings in Atlantic History. Latent Structures and Intellectual Currents, 1500–1830*, ed. Bernard Bailyn and Patricia L. Denault (Cambridge, Mass.: Harvard University Press, 2009), 141–80.

Knobloch, Hilda, *Der Wunderbaum im Urwald. Wie die Chinarinde zum Allgemeingut der Menschheit wurde* (Wien: Eduard Wancura Verlag, 1954).

Knotterus, Otto S., 'Malaria around the North Sea: A Survey,' in *Climate Development and History of the North Atlantic Realm*, ed. Gerold Wefer (Berlin: Springer, 2002), 339–53.

Koerner, Lisbet, 'Women and Utility in Enlightenment Science.' *Configurations* 3, no. 2 (1995), 233–55.

Kohn Goodman, Grant, *Japan: The Dutch Experience* (London: Bloomsbury, 1986).

Kopperman, Paul E., 'The British Army in North America and the West Indies, 1755–83: A Medical Perspective,' in *British Military and Naval Medicine*, ed. Geoffrey L. Hudson (Amsterdam: Editions Rodopi B.V., 2007), 51–86.

Kopytoff, Igor, 'The Cultural Biography of Things: Commoditization as Process,' in *The Social Life of Things: Commodities in Cultural Perspective*, ed. Arjun Appadurai (Cambridge: Cambridge University Press, 1986), 64–91.

Krech, Shepard, *The Ecological Indian: Myth and History* (New York: W. W. Norton, 1999).

Latour, Bruno, *Science in Action: How to Follow Scientists and Engineers Through Society* (Cambridge, Mass.: Harvard University Press, 1987).

LeCain, Timothy, 'Against the Anthropocene. A Neo-Materialist Perspective.' *International Journal for History, Culture and Modernity* 3, no. 1 (2015), 1–28.

Lee, M. R., 'Ipecacuanha: the South American Vomiting Root.' *Journal of the Royal College of Physicians of Edinburgh* 38 (2008), 355–60.

Lees, Andrew, and Lynn Hollen Lees, *Cities and the Making of Modern Europe* (Cambridge: Cambridge University Press 2007).

Lehrer, Steven, *Explorers of the Body. Dramatic Breakthroughs in Medicine from Ancient Times to Modern Science* (New York: iUniverse, 2006).

Leon Borja, Dora, 'Algunos datos acerca de la cascarilla ecuatoriana en el siglo XVIII,' in *Medicina y Quina en la España del siglo XVIII*, ed. Juan Riera Palmero (Valladolid: Universidad de Valladolid, 1997), 85–106.

Leong, Elaine, 'Making Medicines in the Early Modern Household,' *Bulletin of the History of Medicine* 82, no. 1 (2008), 145–68.

Recipes and Everyday Knowledge. Medicine, Science and the Household in Early Modern England (Chicago: University of Chicago Press).

Leong, Elaine, and Alisha Rankin, 'Introduction: Secrets and Knowledge,' in *Secrets and Knowledge in Medicine and Science, 1500–1800*, ed. Elaine Leong and Alisha Rankin (Farnham: Ashgate, 2011), 1–22.

Linden, Marcel van der, 'The "Globalization" of Labour and Working-Class History and Its Consequences,' in *Global Labour History. A State of the Art*, ed. Jan Lucassen (Bern: Peter Lang, 2006), 13–36.

Liss, Robert, 'Frontier Tales: Tokugawa Japan in Translation,' in *The Brokered World. Go-Betweens and Global Intelligence, 1770–1820*, ed. Simon Schaffer, Lissa Roberts, Kapil Raj and James Delbourgo (Sagamore Beach: Watson Publishing International, 2009), 1–48.

Livi-Bacci, Massimo, *A Concise History of World Population* (Oxford: Blackwell, 1992).

Lochbrunner, Birgit, *Der Chinarindenversuch. Schlüsselexperiment für die Homöopathie* (Essen: KVC Verlag, 2007).

López Piñero, José María, 'Los primeros estudios científicos: Nicolás Monardes y Francisco Hernández,' in *Medicina, drogas y alimentos vegetales del Nuevo Mundo. Textos e imágenes españolas que los introdujeron en Europa*, ed. José María López Piñero, José Luis Fresquet Febrer, María Luz López Terrada and José Pardo-Tomás (Madrid: Ministerio de Sanidad y Consumo, 1998), 105–46.

López Piñero, José María, and Francisco Calero, introducción to *De Pulvere Febrifugo Occidentalis Indiae (1663), de Gaspar Caldera de Heredia y la introducción de la Quina en Europa*, ed. José María López Piñero and Francisco Calero (Valencia: Instituto de Estudios Documentales e Históricos sobre la Ciencia / Universidad de Valencia, 1992), 9–11.

López Terrada, María Luz, and José Pardo Tomás, 'Las primeras noticias y descripciones de las plantas americanas (1492–1553),' in *Medicina, drogas y alimentos vegetales del Nuevo Mundo. Textos e imágenes españolas que los introdujeron en Europa*, ed. José María López Piñero, José Luis Fresquet Febrer, María Luz López Terrada and José Pardo-Tomás (Madrid: Ministerio de Sanidad y Consumo, 1998), 19–103.

Lowood, Henry, 'The New World and the European Catalog of Nature,' in *America in European Consciousness, 1493–1750*, ed. Karen Ordahl Kupperman (Chapel Hill: University of North Carolina Press, 1995), 295–323.

Macera, Pablo, *Precios del Perú XVI–XIX. Fuentes*, 3 vols., vol. 1 (Lima: Fondo Editorial / Banco Central de la Reserva, 1992).

Maehle, Andreas-Holger, *Drugs on Trial: Experimental Pharmacology and Therapeutic Innovation in the Eighteenth Century*, Clio Medica / The Wellcome Institute Series in the History of Medicine (Amsterdam: Editions Rodopi, 1999).

Majluf, Natalia. 'The Creation of the Image of the Indian in 19th-Century Peru: The Paintings of Francisco Laso (1823–1869).' Unpublished PhD dissertation, University of Texas, 1996.

Malanima, Paolo, 'Italian Cities 1300–1800. A Quantitative Approach.' *Rivista di storia Economica* XIV, no. 2 (1998), 91–126.

Manning, Susan, and Francis D. Cogliano, 'Introduction. The Enlightenment and the Atlantic,' in *The Atlantic Enlightenment*, ed. Susan Manning and Francis D. Cogliano (London: Routledge, 2008), 19–35.

Marichal, Carlos, 'Mexican Cochineal, Local Technologies and the Rise of Global Trade from the Sixteenth to the Nineteenth Centuries,' in *Global*

History and New Polycentric Approaches, ed. Manuel Perez Garcia and Lucio De Sousa, Palgrave Studies in Comparative Global History (Singapore: Palgrave Macmillan, 2018), 255–73.

Mariss, Anne, *A World of New Things. Praktiken der Naturgeschichte bei Johann Reinhold Forster* (Frankfurt: Campus Verlag, 2015).

Marland, Hilary, 'Women, Health, and Medicine,' in *The Oxford Handbook of the History of Medicine*, ed. Mark Jackson (Oxford: Oxford University Press, 2011), 484–502.

Martin, Luis, *The Intellectual Conquest of Peru. The Jesuit College of San Pablo, 1568–1767* (New York: Fordham University Press, 1968).

Martín Martín, Carmen, and José Luis Valverde, *La farmacia en la América Colonial: el arte de preparar medicamentos* (Granada: Universidad de Granada, 1995).

Martínez Cerro, Manuel, 'Don Pedro María González, navegante y erudito. Aclaratoria solicitud de licencia.' *Apuntes* 2, no. 4 (2004), 59–68.

McClellan, James E., and François Regourd, *The Colonial Machine: French Science and Overseas Expansion in the Old Regime* (Turnhout: Brepols Publishers, 2012).

McClure, Julia, *The Franciscan Invention of the New World* (Basingstoke: Palgrave Macmillan, 2016).

McNeill, John R., *Mosquito Empires. Ecology and War in the Greater Caribbean, 1620–1914* (Cambridge: Cambridge University Press, 2010).

McNeill, William H., *Plagues and Peoples* (Garden City, N.Y.: Anchor Press, 1976).

Merchant, Carolyn, *Reinventing Eden. The Fate of Nature in Western Culture* (New York: Routledge, 2013).

Mignolo, Walter D., Margaret R. Greer and Maureen Quilligan, introduction to *Rereading the Black Legend. Discourses of Religious and Racial Difference in the Renaissance Empires*, ed. Walter D. Mignolo, Margaret R. Greer and Maureen Quilligan (Chicago: University of Chicago Press, 2007), 1–27.

Miller, Joseph C., *Ways of Death. Merchant Capitalism and the Angolan Slave Trade, 1730–1830* (London: James Currey, 1988).

Milton, Cynthia E., and Ben Vinson III, 'Counting Heads: Race and Non-Native Tribute Policy in Colonial Spanish America.' *Journal of Colonialism and Colonial History* 3, no. 3 (2002), 1–18.

Minchom, Martin, 'Demographic Change in Eighteenth-Century Ecuador,' in *Equateur 1986*, ed. D. Delaunay and M. Portais (Paris: ORSTOM, 1989), 179–96.

'The Making of a White Province: Demographic Movement and Ethnic Transformation in the South of the Audiencia de Quito (1670–1830).' *Bulletin de l'Institut français d'études andines* XII, no. 3–4 (1983), 23–39.

Mintz, Sidney W., *Sweetness and Power: The Place of Sugar in Modern History* (New York: Penguin, 1985).

Monahan, Erika, 'Locating Rhubarb. Early Modernity's Relevant Obscurity,' in *Early Modern Things. Objects and Their Histories, 1500–1800*, ed. Paula Findlen (London: Routledge, 2013), 227–51.

Morse, Richard, 'The Urban Development of Colonial Spanish America,' in *The Cambridge History of Latin America*, ed. Leslie Bethell (Cambridge: Cambridge University Press, 1984), 67–104.

Moya, Luz del Alba, *Auge y Crisis de la Cascarilla en la Audiencia de Quito, Siglo XVIII* (Quito: Facultad Latinoamericana de Ciencias Sociales, Sede Ecuador, 1994).

Muñoz Garmendia, Félix, ed., *La botánica al servicio de la corona. La expedición de Ruiz, Pavón y Dombey al Virreinato del Perú (1777–1831)* (Madrid: CSIC / Real Jardín Botánico, 2003).

Murphy, Kathleen S., 'Translating the Vernacular: Indigenous and African Knowledge in the Eighteenth-Century British Atlantic.' *Atlantic Studies* 8, no. 1 (2011), 29–48.

Nappi, Carla, 'Surface Tension. Objectifying Ginseng in Chinese Early Modernity,' in *Early Modern Things. Objects and Their Histories, 1500–1800*, ed. Paula Findlen (London: Routledge, 2012), 31–52.

'Winter Worm, Summer Grass: *Cordyceps*, Colonial Chinese Medicine, and the Formation of Historical Objects,' in *Crossing Colonial Historiographies: Histories of Colonial and Indigenous Medicines in Transnational Perspective*, ed. Anne Digby, Projit B. Muhkarji and Waltraud Ernst (Newcastle upon Tyne: Cambridge Scholars, 2010), 21–35.

Naranjo, Plutarco, 'Pedro Leiva y el secreto de la quina.' *Revista Ecuatoriana de Medicina* XV, no. 6 (1979), 393–402.

Neves Abreu, Jean Luiz, 'A Colônia enferma e a saúde dos povos: a medicina das "luzes" e as informações sobre as enfermidades da América portuguesa.' *História, Ciências, Saúde-Manguinhos* 14, no. 3 (2007), 761–78.

Nos Domínios do Corpo. O saber médico luso-brasileiro no século XVIII (Rio de Janeiro: Editora Fiocruz, 2011).

Newson, Linda A., *Conquest and Pestilence in the Early Spanish Philippines* (Honolulu: University of Hawaii Press, 2009).

'Medical Practice in Early Colonial Spanish America: A Prospectus.' *Bulletin of Latin American Research* 25, no. 3 (2006), 367–91.

Newson, Linda A., and Susie Minchin, *From Capture to Sale. The Portuguese Slave Trade to Spanish South America in the Early Seventeenth Century* (Leiden: Brill, 2007).

Nichols, Robert L., 'Orthodoxy and Russia's Enlightenment, 1762–1825,' in *Russian Orthodoxy under the Old Regime*, ed. Robert L. Nichols and Theofanis George Stavrou (Minneapolis: University of Minnesota Press, 1978), 65–89.

Nieto Olarte, Mauricio, *Remedios para el imperio. Historia natural y la apropiación del Nuevo Mundo* (Bogotá: Universidad de los Andes – FLACSO-CESO, 2006).

Norton, Marcy, *Sacred Gifts, Profane Pleasures. A History of Tobacco and Chocolate in the Atlantic World* (Ithaca: Cornell University Press, 2008).

'Tasting Empire: Chocolate and the European Internalization of Mesoamerican Aesthetics.' *The American Historical Review* 111, no. 3 (2006), 660–91.

O'Hara-May, Jane, 'Foods or Medicines? A Study in the Relationship between Foodstuffs and Materia Medica from the Sixteenth to the Nineteenth Century.' *Transactions of the British Society for the History of Pharmacy* 1, no. 2 (1971), 61–97.

O'Phelan Godoy, Scarlett, *Rebellions and Revolts in Eighteenth Century Peru and Upper Peru* (Köln: Böhlau Verlag, 1985).

Orland, Barbara, and Kijan Espahangizi, 'Pseudo-Smaragde, Flussmittel und bewegte Stoffe. Überlegungen zu einer Wissensgeschichte der materiellen Welt,' in *Stoffe in Bewegung. Beiträge zu einer Wissensgeschichte der materiellen Welt*, ed. Barbara Orland and Kijan Espahangizi (Zürich diaphanes, 2014), 11–35.

Ortiz Crespo, Fernando, *La corteza del árbol sin nombre. Hacia una historia congruente del descubrimiento y difusión de la quina* (Quito: Fundación Fernando Ortiz Crespo, 2002).

Osterhammel, Jürgen, *Unfabling the East. The Enlightenment's Encounter with Asia* (Princeton, N.J.: Princeton University Press, 2019).

Die Verwandlung der Welt. Eine Geschichte des 19. Jahrhunderts (München: C.H. Beck, 2011).

'Globalizations,' in *The Oxford Handbook of World History*, ed. Jerry H. Bentley (Oxford: Oxford University Press, 2014), 89–104.

Pacheco Olivera, Reyna María, 'Análisis del intercambio de plantas entre México y Asia de los siglos XVI al XIX.' Unpublished master's thesis, Universidad Nacional Autónoma de México, 2006.

Packard, Randall M., *The Making of a Tropical Disease: A Short History of Malaria* (Baltimore: Johns Hopkins University Press, 2007).

Pádua, José Augusto, 'European Colonialism and Tropical Forest Destruction in Brazil. Environment Beyond Economic History,' in *Environmental History. As If Nature Existed*, ed. John R. McNeill, José Augusto Pádua and Mahesh Rangarajan (Oxford: Oxford University Press, 2010), 130–48.

'Tropical Forests in Brazilian Political Culture. From Economic Hindrance to Endangered Treasure,' in *Endangerment, Biodiversity and Culture*, ed. Fernando Vidal and Nélia Dias (London: Routledge, 2016), 148–74.

Pagden, Anthony, *The Fall of Natural Man. The American Indian and the Origins of Comparative Ethnology* (Cambridge: Cambridge University Press, 1982).

Palmer, Steven, *Doctors, Healers, and Public Power in Costa Rica, 1800–1940* (Durham, N.C.: Duke University Press, 2003).

Palomeque, Silvia, 'Loja en el mercado interno colonial.' *Revista Latinoamericana de Historia Económica y Social* 2 (1983), 33–47.

Park, Katherine, 'Observation in the Margins, 500–1500,' in *Histories of Scientific Observation*, ed. Lorraine Daston and Elizabeth Lunbeck (Chicago: University of Chicago Press, 2011), 15–44.

Secrets of Women. Gender, Generation, and the Origins of Human Dissection (New York: Zone Books, 2006).

Pataki, Katalin, 'Medical Provision in the Convents of Poor Clares in Late Eighteenth Century Hungary.' *Cornova* 6, no. 2 (2016), 33–58.

'Healers, Quacks, Professionals: Monastery Pharmacies in the Rural Medicinal Marketplace.' *Society and Politics* 12, no. 1 (2018), 32–49.

Perdiguero, Enrique, 'The Popularization of Medicine during the Spanish Enlightenment,' in *The Popularization of Medicine 1650–1850*, ed. Roy Porter (London: Routledge, 1992), 160–93.

Pereda-Miranda, Rogelio, Daniel Rosas-Ramírez, and Jhon Castañeda-Gómez, 'Resin Glycosides from the Morning Glory Family,' in *Progress in the Chemistry of Organic Natural Products*, ed. A. Douglas Kinghorn (Wien: Springer, 2010), 77–146.

Pérez Cantó, Pilar, 'La población de Lima en el siglo XVIII.' *Boletín Americanista* 32 (1982), 383–407.

Pérez, Carlos, 'Quinine and Caudillos: Manuel Isidoro Belzu and the Cinchona Bark Trade in Bolivia, 1848–1855.' Unpublished PhD dissertation, University of California, 1998.

Perez, Stanis, 'Louis XIV et le quinquina.' *Vesalius* IX, no. 2 (2003), 25–30.

'Les médicins du roi et le quinquina aux XVIIe et XVIIIe siècles,' in *La santé des populations civiles et militaires. Nouvelles approches et nouvelles sources hospitalières, XVIIe–XVIIIe siècles,* ed. Élisabeth Belmas and Serenella Nonnis-Vigilante (Villeneuve-d'Ascq: Presses Universitaires du Septentrion, 2010), 179–89.

Petitjean, Martine, and Yves Saint-Geours, 'La ecomomía de la cascarilla en el Corregimiento de Loja (Segunda mitad del siglo XVIII – Principios del siglo XIX).' *Revista Cultural del Banco Central del Ecuador* 5, no. 15 (1983), 15–49.

Pfister, Ulrich, and Christine Fertig, 'Coffee, Mind and Body: Global Material Culture and the Eighteenth-Century Hamburg Import Trade,' in *The Global Lives of Things. The Material Culture of Connections in the Early Modern World,* ed. Anne Gerritsen and Giorgio Riello (London: Routledge, 2016), 221–40.

Philip, Kavita, 'Imperial Science Rescues a Tree: Global Botanic Networks, Local Knowledge and the Transcontinental Transplantation of Cinchona.' *Environment and History* 1, no. 2 (1995), 173–200.

Pickering, Michael, 'Experience as Horizon: Koselleck, Expectation and Historical Time.' *Cultural Studies* 18, no. 2–3 (2004), 271–89.

Pocock, John G. A., *Barbarism and Religion,* vol. 4, *Barbarians, Savages and Empires* (Cambridge: Cambridge University Press, 2008).

Podgorny, Irina, 'The Elk, the Ass, the Tapir, Their Hooves, and the Falling Sickness: A Story of Substitution and Animal Medical Substances.' *Journal of Global History* 13, no. 1 (2018), 46–68.

'From Lake Titicaca to Guatemala: The Travels of Joseph Charles Manó and His Wife of Unknown Name,' in *Nature and Antiquities. The Making of Archaeology in the Americas,* ed. Philip L. Kohl, Irina Podgorny and Stefanie Gänger (Tucson: University of Arizona Press, 2014), 125–44.

Pohl, Hans, *Die Beziehungen Hamburgs zu Spanien und dem Spanischen Amerika in der Zeit von 1740 bis 1806,* Vierteljahrschrift für Sozial- und Wirtschaftsgeschichte (Wiesbaden: Franz Steiner Verlag GmbH, 1963).

Poloni-Simard, Jacques, *El Mosaico Indígena. Movilidad, estratificación social y mestizaje en el corregimiento de Cuenca (Ecuador) del siglo XVI al XVIII* (Quito: ABYA-YALA, 2006).

Pomeranz, Kenneth, *The Great Divergence: China, Europe, and the Making of the Modern World* (Princeton, N.J.: Princeton University Press, 2000).

Porter, Philip W., 'Note on Cotton and Climate: A Colonial Conundrum,' in *Cotton, Colonialism, and Social History in Sub-Saharan Africa,* ed. Allen F. Isaacman and Richard L. Roberts (London / Portsmouth N.H.: James Currey / Heinemann, 1995), 43–49.

Porter, Roy, 'The Eighteenth Century,' in *The Western Medical Tradition,* ed. Lawrence I. Conrad, Michael Neve, Vivian Nutton, Roy Porter and Andrew Wear (Cambridge: Cambridge University Press, 1995), 371–476.

Porter, Roy, and Dorothy Porter, *Patient's Progress. Doctors and Doctoring in Eighteenth-Century England* (Cambridge: Polity Press, 1989).

Purcell, Nicholas, 'Unnecessary Dependences. Illustrating Circulation in Premodern Large-Scale History,' in *The Prospect of Global History*, ed. James Belich, John Darwin, Margret Frenz and Chris Wickham (New York: Oxford University Press, 2016), 65–79.

Raj, Kapil, 'Beyond Postcolonialism … and Postpositivism. Circulation and the Global History of Science.' *Isis* 104 (2013), 337–47.

Ramos, Gabriela, 'Indian Hospitals and Government in the Colonial Andes.' *Medical History* 57, no. 2 (2013), 186–205.

Ramos Cárdenas, Gabriela, and Yanna Yannakakis, introduction to *Indigenous Intellectuals: Knowledge, Power, and Colonial Culture in New Spain and the Andes*, ed. Gabriela Ramos Cárdenas and Yanna Yannakakis (Durham: Duke University Press, 2014), 1–20.

Ramos Gómez, Luis Javier, *El viaje a América (1735–1745), de los tenientes de navío Jorge Juan y Antonio de Ulloa, y sus consecuencias literarias* (Madrid: Consejo Superior de Investigaciones Científicas, Instituto Gonzalo Fernández de Oviedo, 1985).

Rasoanaivo, Philippe, Colin W. Wright, Merlin L. Willcox, and Ben Gilbert, 'Whole Plant Extracts versus Single Compounds for the Treatment of Malaria: Synergy and Positive Interactions.' *Malaria Journal* 10, no. 1 (2011), 1–12.

Redondo Gómez, Maruja, *Cartagena de Indias: cinco siglos de evolución urbanística* (Bogotá: Fundación Universidad de Bogotá Jorge Tadeo Lozano, 2004).

Reinarz, Jonathan, and Rebecca Wynter, 'The Spirit of Medicine: The Use of Alcohol in Nineteenth-Century Medical Practice,' in *Drink in the Eighteenth and Nineteenth Centuries*, ed. Barbara Schmidt-Haberkamp and Susanne Schmid (Routledge: Pickering & Chatto, 2014), 127–39.

Reis, Jaime, and Conceição Andra de Martins, 'Prices, Wages and Rents in Portugal 1300–1910' (Instituto de Ciências Sociais da Universidade de Lisboa, 2009), http://pwr-portugal.ics.ul.pt/?page_id=56.

Riera Palmero, Juan, 'Epidemias y comercio americano de la quina en la España del s. XVIII', in *Capítulos de la medicina española ilustrada. Libros, cirujanos, epidemias y comercio de quina* (Valladolid: Universidad de Valladolid, 1992), 81–112.

  Fiebres y paludismo en la España Ilustrada. Félix Ibáñez y la epidemia de La Alcarria, 1784–1792 (Valladolid: Universidad de Valladolid, Secretariado de Publicaciones, 1984).

  José Masdevall y la medicina española ilustrada. Enseñanza, epidemias y guerra a finales del siglo XVIII (Valladolid: Secretariado de Publicaciones, 1980).

  'La Medicina en la España del siglo XVIII,' in *Medicina y Quina en la España del siglo XVIII*, ed. Juan Riera Palmero (Valladolid: Universidad de Valladolid, 1997), 9–34.

  'Quina y malaria en la España del siglo XVIII.' *Medicina & Historia. Revista de estudios históricos de las ciencias médicas* 52 (1994), 6–28.

Riley, James C., *The Eighteenth-Century Campaign to Avoid Disease* (Basingstoke & London: Macmillan, 1989).

Risse, Guenter B., *New Medical Challenges during the Scottish Enlightenment*, vol. 78, *Clio medica: The Wellcome Institute Series in the History of Medicine* (Amsterdam: Rodopi, 2005).

Roberts, Lissa, and Simon Schaffer, preface to *The Mindful Hand. Inquiry and Invention from the Late Renaissance to Early Industrialisation*, ed. Lissa Roberts, Simon Schaffer and Peter Dear (Amsterdam: Koninklijke Nederlandse Akademie van Wetenschappen, 2007), xiii–xxvii.

Robertson, Roland, 'Glocalization: Time-Space and Homogeneity-Heterogeneity,' in *Global Modernities*, ed. Mike Featherstone, Scott Lash and Roland Robertson (London: Sage, 1995), 30.

Robinson, Howard, 'Substance,' in *Stanford Encyclopedia of Philosophy*, edited by Edward N. Zalta, 2014.

Rockefeller, Stuart Alexander, 'Flow.' *Current Anthropology* 52, no. 4 (2011), 557–78.

Rosenberg, Charles E., *Explaining Epidemics and Other Studies in the History of Medicine* (Cambridge: Cambridge University Press, 1992).

Rosenthal, Jean-Laurent, and R. Bin Wong, *Before and Beyond Divergence: The Politics of Economic Change in China and Europe* (Cambridge, Mass.: Harvard University Press, 2011).

Rosner, Erhard, 'Gewöhnung an die Malaria in chinesischen Quellen des 18. Jahrhunderts.' *Sudhoffs Archiv* 68, no. 1 (1984), 43–60.

Ross, Corey, *Ecology and Power in the Age of Empire: Europe and the Transformation of the Tropical World* (New York: Oxford University Press, 2017).

Ross Griffel, Margaret, 'ZUMA, or, the Tree of Health,' in *Operas in English: A Dictionary*, ed. Margaret Ross Griffel (Plymouth: Scarecrow Press, 2013).

Rousseau, G. S., and Roy Porter, 'Introduction: Approaching Enlightenment Exoticism,' in *Exoticism in the Enlightenment*, ed. G. S. Rousseau and Roy Porter (Manchester: Manchester University Press, 1990), 1–22.

Rudwick, Martin J. S., *The Meaning of Fossils. Episodes in the History of Palaeontology* (Chicago: University of Chicago Press, 1976).

Rui Pita, João, and Ana Leonor Pereira, 'A arte farmacêutica no século XVIII, a farmácia conventual e o inventário da Botica do Convento de Nossa Senhora do Carmo (Aveiro).' *Ágora. Estudos Clássicos em Debate* 14, no. 1 (2012), 227–68.

Ruiz Vega, Paloma, 'La quina en la expedición geodésica al Virreinato de Perú (1734–1743),' in *Las Cortes de Cádiz, la Constitución de 1812 y las independencias nacionales en América*, ed. Antonio Colomer Viadel (Valencia: Universidad Politécnica de Valencia, 2011), 673–683.

Rutten, A. M. G., *Dutch Transatlantic Medicine Trade in the Eighteenth Century under the Cover of the West India Company* (Rotterdam: Erasmus Publishing, 2000).

Safier, Neil, *Measuring the New World. Enlightenment Science and South America* (Chicago: University of Chicago Press, 2008).

Sallares, Robert, *Malaria and Rome: A History of Malaria in Ancient Italy* (Oxford: Oxford University Press, 2002).

Salvador Vázquez, Manuel, 'Las quinas del norte de Nueva Granada,' in *Enfermedad y muerte en América y Andalucía (siglos XVI–XX)*, ed. José Jesús Hernández Palomo (Sevilla: CSIC, 2004), 403–425.

'Mutis y las quinas del norte de Nueva Granada,' in *Medicina y Quina en la España del siglo XVIII*, ed. Juan Riera Palmero (Valladolid: Universidad de Valladolid, 1997), 47–55.

Samanta, Arabinda, *Malarial Fever in Colonial Bengal 1820–1939. Social History of an Epidemic* (Kolkata: Firma KLM Private Limited, 2002).

Sanches de Almeida, Danielle, 'Entre lojas e boticas: O comércio de remédios entre o Rio de Janeiro e Minas Gerais (1750–1808).' Universidade de São Paulo, 2008.

Sánchez, Susy, 'Clima, hambre y enfermedad en Lima durante la guerra independentista (1817–1826),' in *La independencia del Perú. De los Borbones a Bolívar*, ed. Scarlett O'Phelan Godoy (Lima: PUCP Instituto Riva-Agüero, 2001), 237–63.

Sanchez-Albornoz, Nicolás, 'El trabajo indigena en los Andes: teorías del siglo XVI.' *Revista ecuatoriana de historia económica* 2 (1987), 153–81.

'The Population of Colonial Spanish America,' in *The Cambridge History of Latin America*, ed. Leslie Bethell (Cambridge: Cambridge University Press, 1984), 3–36.

Schaffer, Simon, Lissa Roberts, Kapil Raj, and James Delbourgo, introduction to *The Brokered World. Go-Betweens and Global Intelligence, 1770–1820*, ed. Simon Schaffer, Lissa Roberts, Kapil Raj and James Delbourgo (Sagamore Beach: Watson Publishing International, 2009), ix–xxxviii.

Schatzki, Theodore, 'Nature and Technology in History.' *History and Theory*, no.42 (2003), 82–93.

Schiebinger, Londa, *Plants and Empire* (Cambridge, Mass.: Harvard University Press, 2004).

'Prospecting for Drugs. European Naturalists in the West Indies,' in *The Postcolonial Science and Technology Studies Reader*, ed. Sandra Harding (Durham, N.C.: Duke University Press, 2011), 110–26.

Schupbach, William, 'The Fame and Notoriety of Dr. John Huxham.' *Medical History* 25, no. 25 (1981), 415–21.

Secord, James A., 'Knowledge in Transit.' *Isis* 95, no. 4 (2004), 654–72.

Sempat Assadourian, Carlos, 'La despoblación indígena en Perú y Nueva España durante el siglo XVI y la formación de la economía colonial.' *Historia Mexicana* 38, no. 3 (1989), 419–53.

Sepkoski, David, 'Extinction, Diversity, and Endangerment,' in *Endangerment, Biodiversity and Culture*, ed. Fernando Vidal and Nélia Dias, Routledge Environmental Humanities Series (New York: Routledge, 2015), 62–86.

Shammas, Carole, 'Changes in English and Anglo-American Consumption from 1550 to 1800,' in *Consumption and the World of Goods*, ed. Roy Porter and John Brewer (London: Routledge, 1993), 177–205.

Shapin, Steven, *A Social History of Truth. Civility and Science in Seventeenth-Century England* (Chicago: The University of Chicago Press, 1994).

Shefer-Mossensohn, Miri, *Ottoman Medicine. Healing and Medical Institutions 1500–1700* (Albany: State University of New York Press, 2009).

Sheridan, Richard B., *Doctors and Slaves. A Medical and Demographic History of Slavery in the British West Indies, 1680–1834* (Cambridge: Cambridge University Press, 1985).

Smith, Frederick H., *Caribbean Rum: A Social and Economic History* (Gainesville: University Press of Florida, 2005).

Sneader, Walter, *Drug Discovery: The Evolution of Modern Medicines* (Chichester: John Wiley & Sons, 1985).

Snyder, Jon R., *Dissimulation and the Culture of Secrecy in Early Modern Europe* (Berkeley: University of California Press, 2009).

Sousa Dias, José Pedro, *A água de Inglaterra: paludismo e terapêutica em Portugal no século XVIII* (Lisboa: Caleidoscópio, 2012).

Sowell, David, Review of *The Andean Wonder Drug: Cinchona Bark and Imperial Science in the Spanish Atlantic, 1630–1800*, by Matthew James Crawford. *American Historical Review* 122, no. 3 (2017), 899–900.

Spary, Emma, 'Health and Medicine in the Enlightenment,' in *The Oxford Handbook of the History of Medicine*, ed. Mark Jackson (Oxford: Oxford University Press, 2011), 82–99.

Spinney, Erin, 'British Military Medicine during the Long Eighteenth Century: A Relationship between Preventive and Reactionary Medicine, Supply, and Empir.' Master's thesis, University of New Brunswick, 2011.

Stange, Marion, 'Governing the Swamp: Health and Environment in Eighteenth-Century Nouvelle-Orleans.' *French Colonial History* 11 (2010), 1–21.

Steele, Arthur Robert, *Flowers for the King: The Expedition of Ruiz and Pavón and the Flora of Peru* (Durham, N.C.: Duke University Press, 1964).

Stolberg, Michael, *Homo patiens. Krankheits- und Körpererfahrung in der Frühen Neuzeit* (Köln: Böhlau, 2003).

'Learning from the Common Folks. Academic Physicians and Medical Lay Culture in the Sixteenth Century.' *Social History of Medicine* 27, no. 4 (2014), 649–67.

'Medical Popularization and the Patient in the Eighteenth Century,' in *Cultural Approaches to the History of Medicine. Mediating Medicine in Early Modern and Modern Europe*, ed. Willem de Blécourt and Cornelie Usborne (Basingstoke: Palgrave Macmillan, 2004), 89–107.

Tabakoğlu, Hüseyin Serdar, 'The Impact of the French Revolution on the Ottoman-Spanish Relations.' *Turkish Studies* 3, no. 1 (2008), 335–54.

Tandeter, Enrique, *Coercion and Market: Silver Mining in Colonial Potosí, 1692–1826* (Albuquerque: University of New Mexico Press, 1993).

Taylor, Norman, *Cinchona in Java: The Story of Quinine* (New York: Greenberg, 1945).

Totelin, Laurence M. V., 'The World in a Pill. Local Specialties and Global Remedies in the Graeco-Roman World,' in *The Routledge Handbook of Identity and the Environment in the Classical and Medieval Worlds*, ed. Rebecca Futo Kennedy and Molly Jones-Lewis (London: Routledge, 2016), 151–70.

Unschuld, Paul, *Medicine in China. A History of Pharmaceutics* (Berkeley: University of California Press, 1986).

Valera Candel, Manuel, *Proyección internacional de la ciencia ilustrada española. Catálogo de la producción científica española publicada en el extranjero 1751–1830* (Murcia: Universidad de Murcia, Servicio de Publicaciones, 2006).

Varey, Simon, Rafael Chabrán and Dora B. Weiner, eds., *Searching for the Secrets of Nature. The Life and Works of Dr. Francisco Hernández* (Stanford, Calif.: Stanford University Press, 2000).

Vega Ugalde, Silvia, 'Cuenca en los movimientos independentistas.' *Revista del Archivo Nacional de Historia, Sección del Azuay* 6 (1986), 9–48.

Vermeir, Koen, and Dániel Margócsy, 'States of Secrecy: An Introduction.' *British Journal for the History of Science* 45, no. 2 (2012), 153–64.

Villamarin, Juan A., and Judith E. Villamarin, *Indian Labor in Mainland Colonial Spanish America* (Newark: University of Delaware Press, 1975).

Vogel, Morris J., introduction to *The Therapeutic Revolution. Essays in the Social History of Medicine*, ed. Morris J. Vogel and Charles Rosenberg (Philadelphia: University of Pennsylvania Press, 1979), vii–xiii.

The Invention of the Modern Hospital. Boston 1870–1930 (Chicago: University of Chicago Press, 1980).

Vries, Jan de, 'The Limits of Globalization in the Early Modern World.' *Economic History Review* 63, no. 3 (2010), 710–33.

Wachtel, Nathan, *Sociedad y ideología: ensayos de historia y antropología andinas* (Lima: Instituto de Estudios Peruanos, 1973).

Wain, Harry, *A History of Preventive Medicine* (Springfield: Charles C. Thomas, 1970).

Walker, Timothy, 'The Early Modern Globalization of Indian Medicine: Portuguese Dissemination of Drugs and Healing Techniques from South Asia on Four Continents, 1670–1830.' *Portuguese Literary & Cultural Studies* no. 17/18 (2010), 77–97.

'The Medicines Trade in the Portuguese Atlantic World: Acquisition and Dissemination of Healing Knowledge from Brazil (c. 1580–1800).' *Social History of Medicine* 26 (2013), 403–31.

Wallis, Patrick, 'Exotic Drugs and English Medicine: England's Drug Trade, c. 1550 – c. 1800.' *Social History of Medicine* 25, no. 1 (2011), 20–46.

Ward, Candace, *Desire and Disorder. Fevers, Fictions, and Feeling in English Georgian Culture* (Lewisburg, Pa.: Bucknell University Press, 2007).

Warren, Adam, *Medicine and Politics in Colonial Peru: Population Growth and the Bourbon Reforms* (Pittsburgh, Pa.: University of Pittsburgh Press, 2010).

Webb, James L. A., *Humanity's Burden. A Global History of Malaria* (Cambridge: Cambridge University Press, 2009).

The Long Struggle against Malaria in Tropical Africa (New York: Cambridge University Press, 2014).

Weidmann, Almuth, *Die Arzneiversorgung der Armen zu Beginn der Industrialisierung im deutschen Sprachgebiet, besonders in Hamburg*, Braunschweiger Veröffentlichungen zur Geschichte der Pharmazie und der Naturwissenschaften (Stuttgart: Deutscher Apotheker-Verlag, 1982).

Williams, Donovan, 'Clements Robert Markham and the Introduction of the Cinchona Tree into British India, 1861.' *The Geographical Journal* 128, no. 4 (1962), 431–42.

Withers, Charles W. J., *Placing the Enlightenment. Thinking Geographically about the Age of Reason* (Chicago: University of Chicago Press, 2007).

Wood, Peter, *Black Majority* (New York: Knopf, 1974).

Wrigley, Edward Anthony, and Roger S. Schofield, *The Population History of England: 1541–1871: A Reconstruction* (Cambridge: Cambridge University Press, 1989).

Wrigley, Richard, *Roman Fever. Influence, Infection and the Image of Rome, 1700–1870* (New Haven: Yale University Press, 2013).

Young, James Harvey, 'Patent Medicines and the Self-Help Syndrome,' in *Medicine without Doctors. Home Health Care in American History*, ed. Guenter B. Risse, Ronald L. Numbers and Judith Walzer Leavitt (New York: Science History Publications, 1977), 95–116.

The Toadstool Millionaires: A Social History of Patent Medicines in America before Federal Regulation (Princeton, N.J.: Princeton University Press, 1972 (1961)).

Zwingelberg, Tanja, 'Medizinische Topographien und stadthygienische Entwicklungen von 1750–1850, dargestellt an den Städten Berlin und Hamburg,' in *Natur und Gesellschaft. Perspektiven der interdisziplinären Umweltgeschichte*, ed. Manfred Jakubowski-Tiessen and Jana Sprenger (Universitätsverlag Göttingen: Universitätsverlag Göttingen, 2014), 115–40.

Index